Atomic Layer Processing

Atomic Layer Processing

Semiconductor Dry Etching Technology

Thorsten Lill

WILEY-VCH

Author

Dr. Thorsten Lill
Lam Research Corporation
VP Emerging Etch Technologies & Systems
4400 Cushing Parkway
94538 Freemont, CA
United States

Library of Congress Card No.:
applied for

British Library Cataloguing-in-Publication Data
A catalogue record for this book is available from the British Library.

Bibliographic information published by the Deutsche Nationalbibliothek
The Deutsche Nationalbibliothek lists this publication in the Deutsche Nationalbibliografie; detailed bibliographic data are available on the Internet at <http://dnb.d-nb.de>.

Print ISBN: 978-3-527-34668-4
ePDF ISBN: 978-3-527-82418-2
ePub ISBN: 978-3-527-82420-5
oBook ISBN: 978-3-527-82419-9

Printed and bound by CPI Group (UK) Ltd, Croydon, CR0 4YY

C9783527346684_250925

The manufacturer's authorized representative according to the EU General Product Safety Regulation is Wiley-VCH GmbH, Boschstr. 12, 69469 Weinheim, Germany, e-mail: Product_Safety@wiley.com.

Contents

List of Abbreviations

Symbols

A	surface area and parameter in DG model
AR	aspect ratio
B	parameter in DG model
c	concentration
C	capacitance
CD	critical dimension
D	diffusion coefficient
d	depth, thickness
DC	direct current
dc	duty cycle
E	energy and Young's modulus
$\mathcal{E}$	electric field
EPC	etch per cycle
EPE	edge placement error
ER	etching rate
ERNU	etching rate non-uniformity
$G°$	standard Gibbs free energy
GPC	growth per cycle
$\mathcal{H}$	magnetic field
$H°$	standard enthalpy
h	height
h_G	gas phase transport coefficient
I	current
J	particle flux
K	transmission probability
k	constant, coefficient: for instance reaction rate or sputtering coefficient
M	atomic mass
N	number. For instance: number of molecules, number of adsorbed surface sites, etc.

n	density (for gases)
r	distance between atoms or radius
R	reaction rate
RIE	reactive ion etching
R_p,ΔR_p	projected range and straggle
S	etching synergy
$S°$	standard entropy
s	sticking coefficient
SR	sputtering rate
T	temperature
t	time
V	voltage or electric potential
v	velocity
V_{LJ}	Lennard–Jones potential energy
w	width
X	reactance
Z	atomic number

Greek Symbols

θ	angle with respect to surface normal
σ	cross section
ω	circular frequency
τ	characteristic time
κ	dielectric constant
Δ	difference
ε	energy difference, for instance depth of potential energy well
α, β	etch amount in step A and B of an ALE process
Σ	film stress
$\Delta\Phi_{\mathrm{Mott}}$	Mott potential (eV)
Γ	sputtering yield
Θ	surface coverage
ν	volume
λ	wavelength

Subscripts

0	denotes initial value
a	activation
A	adsorption, adsorbate
b	bottom
B	bias
c	capacitive

ca	cathode
col	collision
D	desorption
DC	direct current
dense	dense features
diff	diffusion
diss	dissociation
e	electron
G	gas
i	ion
iso	isolated features
im	impact
in	incoming
iz	ionization
kin	kinetic
M	maximum
m	minimum
n	neutrals
ox	oxidation
p	plasma
RF	radiofrequency
S	surface
sol	solution
sh	sheath
sp	sputtering
sw	sidewall
t	top
th	threshold
out	outgoing
ox	oxide, oxidation
w	wall

Acronyms

AC	alternating current
AFM	atomic force microscopy
ALE	atomic layer etching
ALD	atomic layer deposition
AR	aspect ratio
ARDE	aspect ratio dependent etching
BARC	bottom antireflective coating
BCA	binary collision approximation
BEOL	back end of line

BPS	bounded plasma system
BPSG	boron phosphorous silicon glass
BST	barium strontium titanate: $Ba_{1-x}Sr_xTiO_3$
CAIBE	chemically assisted ion beam etching
CBRAM	conductive bridge random access memory
CCP	capacitively coupled plasma
CD	critical dimension
CDE	chemical downstream etching
CFSTR	continuous flow stirred tank reactor
CM	Cabrera–Mott oxidation model
CMOS	complementary metal–oxide–semiconductor MOSFET fabrication process
CMP	chemical mechanical polishing
CVD	chemical vapor deposition
DARC	dielectric antireflective coating
DC	direct current
DFT	density functional theory
DG	Deal–Grove oxidation model
DMAC	dimethyl aluminum chloride
DRAM	dynamic random access memory
ECP	electro copper plating
ECR	electron cyclotron resonance
ESC	electrostatic chuck
FEOL	front end of line
FeRAM	ferroelectric random access memory
FET	field effect transistor
FG	floating gate flash device
FinFET	fin field effect transistor
FTIR	Fourier transform infrared spectroscopy
GAA	gate-all-around (transistors)
GST	phase change material comprised of germanium, antimonium, and tellurium
HPEM	hybrid plasma equipment model
IAD	ion angular distribution
IBE	ion beam etching
ICP	inductively coupled plasma
IED	ion energy distribution
IIP	ion-ion plasma
ILD	inter-layer dielectric
LEIS	low energy ion spectroscopy
LELE	Litho–Etch–Litho–Etch multipatterning
LER	line edge roughness
LWR	line width roughness
LSS	Lindhard, Scharff, and Schiott theory
MD	molecular dynamics

MEMS	micro-electromechanical systems
MEOL	mid end of line
MMP	mixed Mode Pulsing
MRAM	magnetic random access memory
MOSFET	metal oxide semiconductor field effect transistor
NAND	logic gate with "false" output if all inputs are "true." This type of logic gates is used in flash memory devices. 3D NAND is an implementation of flash memory devices where the gates are stacked in the third dimension inside tall vertical channels
ONON	oxide–nitride–oxide–nitride 3D NAND
OPOP	oxide–polysilicon–oxide–polysilicon 3D NAND
OxRAM	metal oxide resistive random access memory
PIC	particle-in-cell plasma model
PVD	physical vapor deposition
PCM	phase change memory
PSD	power spectral density
PZT	lead zirconate titanate: $Pb(Zr_xTi_{1-x})O_3$
QCM	quartz crystal microbalance
ReRAM	resistive random access memory
RF	radio frequency
RG	replacement gate flash
RIBE	reactive ion beam etching
RIE	reactive ion etching
SADP	self-aligned double patterning
SAQP	self-aligned quadruple patterning
SCM	storage class memory
SE	spectroscopic ellipsometry
SEM	scanning electron microscopy
SIMS	secondary ion mass spectrometry
SIT	sidewall image transfer
SRIM	"stopping and range of ions in matter" program
SOS	spacer-on-spacer implementation of self-aligned quadruple patterning
STI	shallow trench isolation
TCP	transformer coupled plasma
TEM	transmission electron microscopy
TMA	trimethylaluminum
TPD	temperature programmed desorption
TRIM	"transport of ions in matter" program
TSV	though silicon via
TWB	tailored waveform bias
UHV	ultra-high vacuum
VUV	vacuum ultraviolet light
ZBL	Ziegler, Biersack, and Littmark model

Constants

e	elementary charge: $1.602\,176\,62 \times 10^{-19}$ C
ε_0	vacuum permittivity: $8.854\,187\,812\,8(13) \times 10^{-12}$ F/m
ε	dielectric constant
k_B	Boltzmann constant
R	universal gas constant
N_A	Avogadro constant

Units

Å	angstrom (length)
C	coulomb (charge)
°C	centigrade (temperature)
deg, °	degree (angle)
eV	electron volt (energy)
F	Farad (standard unit of capacitance in the SI system)
h	hour (time)
K	Kelvin (temperature)
L	Langmuir (surface coverage)
m	meter (length)
min	minute (time)
Pa	pascal (pressure)
s	second (time)
Torr	pressure

1 Introduction

People have been scratching, engraving, and carving stone, wood, bones, and other materials since the dawn of time to record information and to create art. These early forms of material removal can possibly be viewed as the origins of etching technology.

The importance of etching throughout history can be illustrated with a few remarkable examples. Hammurabi's code of law was inscribed into a stone stele at around 1754 BCE and is one of the earliest and influential legal tests. Carved woodblocks were applied to print paper money during the Tang dynasty in China during the second half of the first millennium CE. Michelangelo's statue of David is an embodiment of the European renaissance. All these etching techniques use *physical* energy to remove material.

Chemical etching techniques using acids evolved in medieval Europe to decorate armor with greater detail. Selected areas of a surface were covered by soft "maskants," which could be easily removed with sharp objects and the exposed areas were removed by "etchants." One of the greatest etchers of all times was Rembrandt who created around 290 prints. Many of his etching plates still survive.

John Senebier discovered in 1782 that certain resins lost their solubility to turpentine after exposure to light. This allowed to create early forms of photomasks and ultimately led to the development of photographic methods. Paul Eisler invented the printed and etched circuit board in 1936. Etching was also instrumental for the realization of the first integrated circuits by Jack Kilby and Robert Noyce in 1958. The words "etch" and "etching" figure 11 times in Kilby's seminal US patent 3 138 743 "Miniaturized electronic circuits" (Kilby 1959).

Originally, integrated circuits were etched with wet chemical methods using photoresists as a mask. While these methods can be directional for some single crystal materials and selected etchants, removal of amorphous materials etch proceeds in all directions roughly with the same rate. This kind of etching is also called isotropic. It works only for features where the lateral dimension is much larger than the thickness of the material to be etched. This property is obviously an obstacle for device shrinking. Another drawback of wet etching is the creation of large amounts of toxic waste.

Atomic Layer Processing: Semiconductor Dry Etching Technology, First Edition. Thorsten Lill.

To overcome these challenges, dry plasma etching methods were introduced into the manufacturing of integrated semiconductor devices in the 1980s. When a plasma is in contact with a solid surface a phenomenon called sputtering occurs, which causes material removal. Sputtering was discovered by W.R. Grove in 1852. Physical sputtering with noble gas plasmas was used in the 1960s in the electronics industry. When the wafer is placed on a radio frequency (RF) powered electrode, ions are accelerated, and the sputter rate can be increased to make the method more productive (Coburn and Kay 1972). However, physical sputtering is still too slow to be useful in the manufacturing of semiconductor devices. It also critically lacks the selectivity to mask and stop materials.

Chemistry provided the necessary performance boost. The development of chemical plasma etching started with stripping of photoresists in oxygen RF plasmas (Irving et al. 1971). Soon, fluorine and chlorine plasma were tested to etch a wide range of materials. An increase of the silicon etching rate by a factor of 10–20 was observed when replacing argon with fluoro-chloro-hydrocarbon gases (Hosokawa et al. 1974). The term "reactive ion etching" (RIE) was coined in the mid 1970s for etching technologies involving chemically reactive plasmas where the wafer is placed on an RF-powered electrode. Initially, the mechanism of the etch rate enhancement was not understood even though the benefits were clearly demonstrated in experiments (Bondur 1976). Coburn and Winters found that "the magnitude of the etch rates which are observed are such that the enhancement caused by ion bombardment cannot be easily explained by simply superimposing a physical sputtering process onto the chemical etching process" (Coburn and Winters 1979). Their seminal experiments demonstrated the existence of synergy between the ion and neutral fluxes. Synergy is also a key concept in atomic layer etching (ALE) with atomic layer fidelity. We will use this concept throughout this book.

Production-worthy etching reactors took hold in the semiconductor industry with the introduction of batch RIE reactors based on developments at Bell Labs. An overview of the evolution plasma etching equipment can be found in a review article by Donnelly and Kornblit (2013). The 1990s saw the introduction of single wafer etching reactors, which improved wafer to wafer repeatability and overall process control. This decade was also the time of search for the best source technology for the large number of rapidly emerging applications. The first single wafer etching reactors were simple parallel plate reactors with RF power applied to the wafer pedestal. Some embodiments featured etch rate enhancing magnetic fields.

High-density plasmas powered by transformer-coupled plasma (TCP) or inductively coupled plasma (ICP) established themselves as the tools of choice for silicon and metal etching. Medium-density capacitively coupled plasma (CCP) sources proved superior for etching of silicon oxide and other dielectric materials. CCP reactors found widespread application with the introduction of damascene metallization in the end of the 1990s, which created a large market for etching of materials with low relative dielectric constant, the so-called low-k materials.

The 2000s were the decade of continuous improvement of uniformity across the wafer by means of radial uniformity tuning knobs for ion flux, neutral flux, and temperature. This was driven by the transition from 200 to 300 mm wafers and escalating uniformity requirements to satisfy Moore's Law. The last decade was characterized by a strong focus on within die and feature scale performance. This is caused by the transition from traditional Moore's Law scaling to vertical device scaling, which drives devices with increasingly high aspect ratios such as 3D NAND flash and fin field effect transistors (FinFET's).

One of the solutions to within die performance challenges is "time domain processing," for instance plasma pulsing, and mixed mode pulsing (MMP) where RF power and gas flows are pulsed. Time domain processing necessitates that all subsystems operate repeatably on second timescales and faster. This is an enormous engineering challenge considering all the parameters that need to be controlled with the large number of process parameters including the radial tuning knobs. Model-based process controllers and machine learning process development algorithms are being introduced.

As semiconductor devices are shrinking to sub-10 nm dimension, etching technologies with atomic-scale fidelity are required. Here fidelity refers to the degree of matching to the intent of design engineers in shape and composition (Kanarik et al. 2015). ALE , which has been studied in laboratories for 30 years, promises to deliver this level of performance. The first report on ALE was published in Yoder's US patent 4 756 794 entitled "Atomic layer etching" (Yoder 1988). After a first wave of research during the 1990s, a second wave of interest and development started in the mid-2000s driven by the need for etching technologies with infinite selectivity and the ability to remove controlled amounts of material down to sub-monolayer resolution.

A variety of etching technologies were discussed under the umbrella of "ALE" including very slow RIE processes, radical and vapor etching. This lack of common understanding and terminology in the etching community slowed the development of true ALE. A definition of ALE as an etching process comprising of at least two self-limited steps was adopted during a Sematech workshop on ALE in April 2014. This definition is in analogy to its counterpart of atomic layer deposition (ALD). Many of the established concepts in ALD were adopted in ALE. The separation of the etching process into self-limiting steps breaks the trade-offs caused in RIE by simultaneous ion and neutral fluxes. The result is improved uniformity across the wafer, across features with different critical dimension called aspect ratio dependent etching (ARDE), and surface smoothness (Kanarik et al. 2015). It also greatly simplifies the process and makes ALE accessible to a rigorous fundamental understanding.

This book covers the latest research and developments of directional and isotropic ALE and puts them into the context of established dry etching technologies for semiconductor devices. In this book, we will introduce etching technologies in the order of increasing complexity. We will begin with critical elementary surface processes, followed by single species etching technologies (thermal etching and radical etching), sequential multi-species etching (ALE), and multi-species continuous processing (RIE). Finally, we will review plasmas and other methods to produce the species we discussed in the first half of the book.

This structure does not consider the chronological order of discovery or the size of the market of the various etching technologies. Novel ALE will be studied before classical RIE. Directional ALE is introduced as a simplified embodiment of RIE, which is amiable to a rigorous treatment. Salient RIE properties will be presented as the result of a lack of self-limitation of continuous processing where all species fluxes are on all the time. The goal is to understand RIE on an atomic level as rigorously as possible to illuminate the "black box" that RIE still is today (Winters et al. 1977; Gottscho et al. 1999).

Specific etching applications such as gate etching, contact etching, or 3D NAND channel hole etching will be introduced as examples for the mechanisms discussed without attempting to give a comprehensive description of the process challenges and solutions. The emergence and evolution of semiconductor devices and the corresponding etching applications is simply too fast paced, and such an attempt would be outdated within a few years. Rather, the intend of this book is to provide an atomic level understanding of all dry etching technologies, which will hopefully help to develop specific solutions for existing and emerging semiconductor devices.

Plasmas are the method of choice to generate ions and radicals used in dry etching. In this book, the plasma and source technologies are covered to a level of detail sufficient enough to understand how they impact the species fluxes to the etching surface. For deeper understanding, we refer to the seminal monography on plasma technology and materials processing is Liebermann's monography (Lieberman and Lichtenberg 2005).

References

Bondur, J.A. (1976). Dry process technology (reactive ion etching). *J. Vac. Sci. Technol.* 13: 1023–1029.

Coburn, J.W. and Kay, E. (1972). Positive-ion bombardment of substrates in rf diode glow discharge sputtering. *J. Appl. Phys.* 43: 4965–4971.

Coburn, J.W. and Winters, H.F. (1979). Ion- and electron-assisted gas-surface chemistry – an important effect in plasma etching. *J. Appl. Phys.* 50: 3189–3196.

Donnelly, V.M. and Kornblit, A. (2013). Plasma etching: yesterday, today, and tomorrow. *J. Vac. Sci. Technol., A* 31: 050825 1–48.

Gottscho, R.A., Cooperberg, D., and Vahedi, V. (1999). The black box illuminated. *Workshop on Frontiers in Low Temperature Plasma Diagnostics III (LTPD)*, Saillon, Switzerland.

Hosokawa, N., Matsuzaki, R., and Asamaki, T. (1974). RF sputter-etching by fluoro-chloro-hydrocarbon gases. *Jpn. J. Appl. Phys. Suppl.* 2: 435–438.

Irving, S.M., Lemons, K.E., and Bobos, G.E. (1971). Gas plasma vapor etching process. US Patent 3,615,956.

Kanarik, K.J., Lill, T., Hudson, E.A. et al. (2015). Overview of atomic layer etching in the semiconductor industry. *J. Vac. Sci. Technol., A* 33: 020802 1–14.

Kilby, J.S. (1959). Miniaturized electronic circuits. US Patent 3,138,743.

Lieberman, M.A. and Lichtenberg, A.J. (2005). *Principles of Plasma Discharges and Materials Processing*, 2e. Wiley.

Winters, H.F., Coburn, J.W., and Kay, E. (1977). Plasma etching – a "pseudo-black-box" approach. *J. Appl. Phys.* 48: 4973–4983.

Yoder, M.N. (1988). Atomic layer etching. US Patent 4,756,794.

2

Fundamentals

2.1 Important Performance Metrics of Etching Processes

Etching is a process of material removal and requires breaking of bonds and removing of atoms from the surface of a solid material. Except for physical sputtering, dry etching processes for semiconductor devices deploy chemical reactions. To understand how the technologies work, it is important to understand the mechanisms and elementary steps. In dry etching, chemically active species are delivered to the surface through the gas phase and adsorb on the surface. It is also possible to bombard the surface with chemically active ions (see Section 7.1.2). The role of the adsorbents is to weaken the bonds of the surface atoms to the bulk material. Energy is provided simultaneously or sequentially to break the bonds and to remove the reaction products. This can be achieved by thermal energy or kinetic energy of ions and atoms as well as by other means such as photons or electrons. Sufficiently high thermal or kinetic energy also leads to pure physical removal such as evaporation and sputtering, respectively. Evaporation and physical sputtering are, however, of limited usefulness in the etching of semiconductor devices.

Figure 2.1 illustrates the effect of adsorption of chemically active species with a schematic representation of a solid surface with adsorbed atoms. The bond energy between adsorbate and surface atoms is denoted with E_A, the bond energy between surface and bulk atoms with E_S, and between bulk atoms with E_O. Adsorbed atoms form bonds with the surface atoms with an energy E_A. This weakens the bonds of the surface atoms to the bulk characterized by E_S relative to the bonds in the bulk, E_O. When sufficient energy is provided, the weakest bonds will break. If these are the bonds characterized by E_S, the result is etching. Breaking of the other bonds will result in desorption (in case E_A is weakest) or sputtering/evaporation of the bulk material (if E_O is the weakest bond).

This is a greatly simplified, phenomenological representation, which ignores among others the actual structure of the surface and the kinetics of the reactions and activation barriers. The surface may also be amorphous or damaged and intermixed with chemically active species. Reactive layers can extend beyond the first atomic layer of the etching material. In that case, the entire reactive layer is characterized by weak chemical bonds.

Atomic Layer Processing: Semiconductor Dry Etching Technology, First Edition. Thorsten Lill.

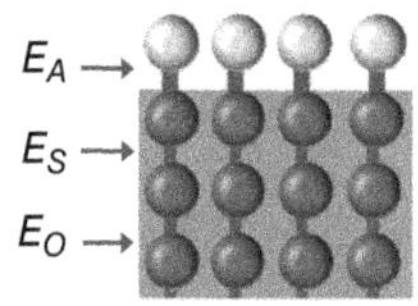

Figure 2.1 Schematic representation of a surface with adsorbed atoms with the bond energy between adsorbate and surface atoms E_A, between surface and bulk atoms E_S, and between bulk atoms E_O. Condition for etching: $E_S < E_O$ and E_A. Source: Lill et al. (1994).

The energy that is applied to break bonds can be thermal energy, kinetic energy of ions, chemical energy of atoms, molecules, and radicals, as well as chemical or thermal energy of photons, and electrons. Even the mechanical energy of the tip of an atomic force microscope has been reported to induce the etching of a chemically modified surface (Chen et al. 2018).

In this framework for etching, one or more fluxes of particles and energy sources interact with the surface simultaneously or in sequence to modify the surface and to break bonds of the material to be removed. Depending on what particles and energy sources are used and whether they interact with the surface simultaneously or in sequence, dry etching technologies can be classified into thermal etching, thermal isotropic atomic layer etching (ALE), radical etching, directional or ion-assisted ALE, reactive ion etching (RIE), and ion beam etching (IBE). We will introduce a classification of etching technologies in Section 2.9.

Etching processes of semiconductor devices form 3D structures that can consist of several different materials. Extensive, application-specific tables of requirements must be met when etching advanced semiconductor devices. Here is a list of common important etching requirements.

2.1.1 Etching Rate (ER)

This is the most important etching performance parameter. It is expressed as the change in thickness of the etching film as a function of time and is usually measured in nm/min. For a sputtering process, this rate can also be expressed as a function of the sputtering yield Γ, which is the ratio of ejected atoms and ions to impinging ions. For chemically enhanced processes such as RIE, the dependence of the etching rate (ER) on the incoming fluxes is much more complex (see Section 7.1).

For ALE, the term etching rate is replaced by the term "etching per cycle" (EPC) expressed in nm or Å because the process is cyclic and removes a well-defined amount of material every cycle. This approach is analogous to the nomenclature in thin film deposition where continuous technologies such as chemical vapor deposition (CVD) are characterized by a deposition rate while atomic layer deposition (ALD) features "growth per cycle" (GPC).

2.1.2 Etching Rate Nonuniformity (ERNU)

Advanced semiconductor devices are manufactured on wafers with 300 mm diameter. Etching rate uniformity requirements are very stringent. This parameter is expressed as the ratio of the etching rate difference across the wafer divided by the average etching rate as percentage or as the standard deviation. Typically, etching rate nonuniformity (ERNU) is expressed in % of 1 sigma of the standard deviation. ERNU can be introduced by nonuniformities of any of the incoming species fluxes across the wafer or at the extreme edge of the wafer.

2.1.3 Selectivity

Very few etching applications are limited to etching one material only. Masks made of slower etching materials are needed to etch 3D structures such as trenches or holes. Selectivity of material 1 to material 2 is expressed as the ratio of their etching rates ER_1/ER_2 or their etching amount per cycle EPC_1/EPC_2. Etching technologies are characterized by an "intrinsic" selectivity. This is the selectivity without deposition sub-reactions. The intrinsic selectivity is a function of "excess" removal energy. For instance, thermal processes deploy neutral molecules at thermal energies with just about enough energy to break the bonds of the etching material. Because chemical pathways are very different for different materials, etching processes with infinite selectivity can be designed. In contrast, RIE uses ions with energies of several hundred electron-volt (eV). This is much more than the energy needed to break the critical bonds of the etching material. Bonds of materials that the process should be selective to will also be broken. The result is a finite selectivity based on the etching rate difference. Therefore, depositing species are added to RIE processes to deposit a protective layer where etching is not desired while etching in locations where it is desired.

2.1.4 Profile

The use of ions provides the etching process with directionality or anisotropy. In combination with a selective etching mask, this allows to create vertical or nearly vertical features. The resulting shape of the vertical cross section is called an etching profile. Ideally, this cross section is a square, but various artifacts can lead to profile deformations.

One of the reasons for nonideal profiles is the effect of neutral species that can etch in vertical and horizontal directions. The latter is called isotropic etching and can be suppressed among others by means of reducing the temperature, choosing less reactive neutral species, and by adding inhibiting species that suppress isotropic etching (Coburn 1994). Figure 2.2 shows common conventions for how to characterize an etching profile. These terms are not rigorously defined in the etching community and can vary from organization to organization.

2.1.5 Critical Dimension (CD)

The term critical dimension (CD) is applied to the space of a trench, the width of a line, or the diameter of a hole. Depending on where the measurement is taken, top and bottom CDs can be distinguished. If the profile is bowed, the largest extent of this profile deformation is called bow CD. Many more specific CD-related terms can be introduced in a similar fashion. The term CD can also be applied to the

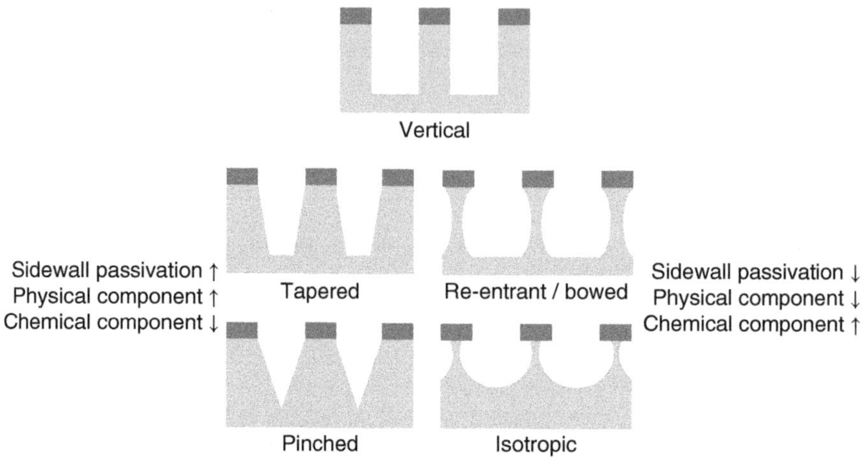

Figure 2.2 Classification of the most common etching profiles.

incoming resist mask. The difference between the lithography CD and etch CD is called CD bias or ΔCD. The uniformity of the CD bias across the wafer is directly related to ERNU. The repeatability of the CD is a function of the wafer-to-wafer and chamber-to-chamber repeatability.

2.1.6 Line Width and Edge Roughness (LWR and LER)

Line edge roughness (LER) is the roughness of only one edge and considers the contributions from line bending. Line width roughness (LWR) is defined as the variance of the width of a line. This parameter considers the roughness of both edges of the line. Because line bending impacts both edges of a line, LWR typically does not include it.

2.1.7 Edge Placement Error (EPE)

Edge placement error (EPE) is defined as the relative displacement of the edges of two features from their intended target position. CD bias, LER, and LWR are components of EPE, which also considers lithography overlay and other contributions.

2.1.8 Aspect Ratio-Dependent Etching (ARDE)

This parameter measures the change in etching rate as the feature evolves and the aspect ratio increases. The aspect ratio (AR) for a feature with vertical sidewall is defined as the ratio of the feature width and depth ($AR = w/d$). For non-vertical profiles, the average or minimum width is also used. The root cause of aspect ratio-dependent etching (ARDE) is the transport of etching species to the etching surface inside the feature. For etching processes with one type of etching species, for instance for low-pressure radical etching, the etching rate slows down as the aspect ratio increases because transport becomes the limiting step. RIE uses the combined effect of ions and neutrals for etching. Ions and neutral species have different angular distributions and hence the fluxes to the etch front are attenuated to different degrees as the aspect ratio of the feature evolves (Gottscho et al. 1992). We will investigate the root causes for ARDE for each etching technology in Sections 3.1, 4.2, 5.2, 6.2, 7.2, and 8.1.

2.2 Physisorption and Chemisorption

A chemically modified surface layer is a precondition for chemically enhanced etching (see Figure 2.1). In dry etching, this modified layer is formed by adsorption, specifically chemisorption, diffusion, and ion implantation.

Molecules, atoms, or radicals approach the surface with thermal energy where they bounce off (scatter) or adhere to form bonds and thus convert kinetic into potential energy. When incoming particles form weak bonds, for instance van der Waals and hydrogen bonds, with the surface the process is called physisorption. The adsorption process is classified as chemisorption when covalent bonds are formed. Adsorption processes can be illustrated by drawing the potential energy as a function of the distance between the surface atoms (adsorbent) and the incoming atom or molecule (adsorbate). The potential energy is the sum of attractive and repulsive forces between the adsorbate and adsorbent and can be mathematically expressed, for instance, by the Lennard-Jones or 12–6 potential:

$$V_{\mathrm{LJ}} = \varepsilon \left[\left(\frac{r_m}{r} \right)^{12} - 2 \left(\frac{r_m}{r} \right)^{6} \right] \tag{2.1}$$

Here, ε is the depth of the potential well, r is the distance between the atoms, and r_m the distance at which the potential reaches its minimum. Figure 2.3 shows the potential energy as a function of the distance r between the adsorbate and adsorbent. As

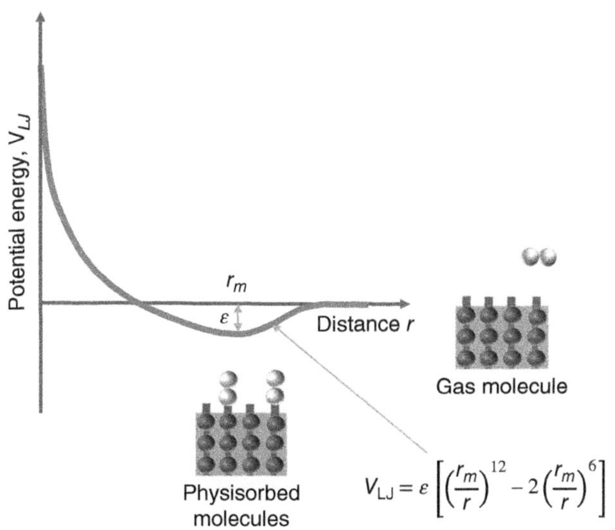

Figure 2.3 Lennard-Jones potential for adsorbent–adsorbate system.

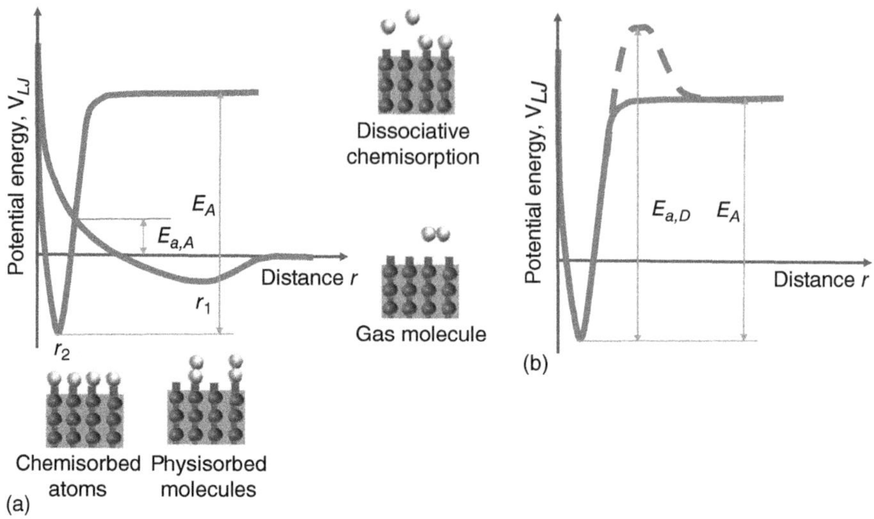

Figure 2.4 Lennard-Jones diagram for dissociative adsorption.

the adsorbate molecule is brought closer to the surface, the interaction is first attractive until repulsive forces prevail. The equilibrium distance is reached when the net potential energy is at a minimum. This representation is valid for the formation of chemical bonds in general.

This concept can be applied to adsorption of chemical species in dry etching: molecules and radicals. The diagram in Figure 2.4 is called Lennard-Jones diagram for dissociative adsorption of molecules (see Figure 5.1 for adsorption of radicals). Because chemically stable molecules have saturated bonds, they first interact with the surface via long-ranging van der Waals forces with an equilibrium distance r_1. When the molecule is moved closer to the surface, the atoms that form the molecule start to form individual bonds with the surface and as a result, the molecule dissociates. A new equilibrium distance r_2 with a lower potential energy characterizes these chemisorbed atoms.

An energy barrier $E_{a,A}$ is formed by the overlapping L–J curves for the molecule and the atom. This energy barrier is the activation energy for adsorption. It separates the equilibrium states for physisorption and chemisorption and is the activation barrier for dissociation. With sufficient kinetic energy, which means at high enough temperature, this barrier can be overcome. When the temperature of the surface is high enough, some species can desorb from the surface. Because the molecule is dissociated into atoms, the species that leave the surface leave as atoms.

Frequently used gases in dry etching are, for instance, Cl_2, H_2, and O_2. These gases dissociate into radicals, which are atoms or molecule fragments with unpaired electrons that make them highly reactive. The existence of an activation barrier for dissociative adsorption means that the process is temperature dependent and can be described by an Arrhenius equation:

$$R_A = k_0 e^{-E_{a,A}/RT}(1 - \theta_A) \tag{2.2}$$

Here, R_A is the adsorption rate, k_0 is a rate constant, $E_{a,A}$ the activation energy for adsorption, R the universal gas constant, and θ_A the adsorption surface coverage. The universal gas constant equals the Boltzmann constant multiplied by the Avogadro constant: $R = k_B N_A$.

2.3 Desorption

Desorption is the process of a molecule or atom leaving a surface. Depending on the nature of the adsorbate–adsorbent bond and the form of energy deployed to break it, a multitude of desorption mechanisms exist. These desorption mechanisms are very important for etching.

In the case of thermal desorption, the desorption rate is a function of the surface temperature. The temperature of a solid state material is a measure of the vibrational energy of the atoms in the solid. The energies of the atoms are not all equal, they follow a distribution. The atomic vibrations are related to sound waves and are called phonons in their quantum mechanical particle representation. Energies of non-interacting quantum mechanical phonons, for instance, are distributed according to the Bose–Einstein statistics. The implication of the existence of a distribution is that for a given temperature, there is a probability that the energy stored in the vibration of a bond will lead to its breakage. The lower the bond energy, the higher the probability for breaking the bond. Atoms that are bonded to the surface by weak van der Waals or hydrogen bonds (bond energy typically in the 10–100 meV rage) will desorb at much lower temperatures than chemisorbed atoms or molecules, which form stronger covalent bonds to the surface (bond energy typically around 1 eV). This is the reason why physisorption occurs at relatively low temperatures. When the vibrational energy of the atoms is increased by increasing the surface temperature, statistically, the weakest bonds will break first. In homogenous materials all bonds are similar and therefore the bulk material will sublimate at high enough a temperature.

Chemisorption weakens the surface-to-bulk bonds. This is because the strength of a covalent bond between atoms is proportional to the probability of the shared electrons to be located between these atoms. Hence, adsorbates that bond to a surface atom weaken its bond to the surface by shifting the electron density. This can lead to the condition shown in Figure 2.1. Increasing the temperature will break the bonds between the surface and bulk atoms at a lower temperature than for the homogeneous bulk material. As a result, only the top layer of the solid material is removed. Thus, chemisorption enables chemically assisted etching via thermal desorption as illustrated by the Lennard-Jones diagram in Figure 2.5. A new L–J potential curve is added as a dashed line to represent the potential of a newly formed molecule containing at least one surface atom and another surface atom.

The desorption rate can be described by Arrhenius equation:

$$R_D = k_0 e^{-E_{a,D}/RT} \theta_A \tag{2.3}$$

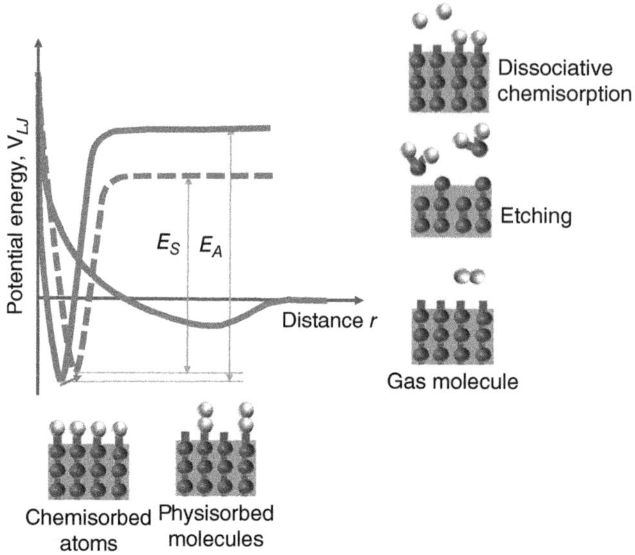

Figure 2.5 Lennard-Jones diagram for thermal etching.

Here, R_D is the desorption or etching rate, and $E_{a,D}$ the activation energy for desorption or etching. The activation energy $E_{a,D}$ is related to the adsorption energy E_A and surface energy E_S for desorption and thermal etching, respectively. $E_{a,D}$ considers the kinetics of the reaction and the corresponding energy barriers, while E_A and E_S represent only the enthalpies of formation without considering activation barriers (see Figure 2.4). This is, of course, a very simplified representation.

2.4 Surface Reactions

Dissociative adsorption is a surface-mediated reaction, but it must not necessarily lead to removal of surface atoms or etching. For etching to occur, the atoms on the surface must rearrange into molecules that contain at least one surface atom. These newly formed molecules must desorb at lower energies than the adsorbed molecules or atoms themselves. For instance, F_2 can physisorb on a silicon surface, and dissociate into chemisorbed F atoms that can form SiF_x. SiF_x has a weaker bond to the silicon than F. Therefore, in this system, heating of the fluorine-covered silicon sample leads to the removal of SiF_2 and SiF_4 (Engstrom et al. 1988).

In Figure 2.5, the transition of the reaction coordinate from the system chemisorbed atom/surface to the reaction product/surface is indicated by an arrow. In the simplest case, this is just a change in what bond is represented. In reality, the transition may involve chemical reactions to rearrange bonds and to form new molecules. In this case, the transition between desorption coordinates involves chemical reactions with their own reaction coordinates and energy barriers. If the energy barrier of the reaction is higher than the barrier for desorption, this reaction will be the rate-limiting step.

Removal of fluorine from a previously fluorine-etched silicon surface without losing any additional silicon atoms is a challenge. In this case, yet another surface reaction, for instance, with water vapor, must be used to just remove the fluorine in the form of HF. Multiple surface reactions involving two or more molecules or atoms are also the mechanism of thermal ALE and will be covered in Section 4.1.

The potential for surface reactions to proceed can be predicted by calculating the Gibbs free energy, ΔG°:

$$\Delta G^\circ = \Delta H^\circ - T \cdot \Delta S^\circ \tag{2.4}$$

Here, ΔH° is the change in enthalpy and represents the heat of reaction. ΔS° is the change in entropy for the reaction and reflects the change in order of the system. The superscript denotes standard temperature and pressure, which are 273.15 K and 10^5 Pa. The Gibbs free energy of a chemical reaction can be calculated using commercially available software programs.

When ΔG° in Eq. (2.4) is negative, the reaction is thermodynamically favorable and is said to be spontaneous. Spontaneous reactions are needed to realize thermal etching and thermal ALE. Thermodynamic calculations are therefore a first test of the feasibility of a thermal etching reaction. The individual values for ΔH° and ΔS° provide guidance for the temperature behavior of an etching reaction. A negative ΔH° means that the reaction is exothermic. The equilibrium will be shifted in favor of etching at lower temperatures. If ΔS° is negative, the reaction is also more favorable at lower temperatures. Even when an etching reaction is thermodynamically favorable, the reaction kinetics may prevent meaningful etching rates. The reaction pathway may include activation energy barriers that have to be overcome. These barriers determine the reaction rate. The calculation of reaction kinetics is much more complicated.

Surface reactions that involve more than one adsorbate can be classified by the underlying kinetics as shown in Figure 2.6. In the Langmuir–Hinshelwood mechanism, two molecules adsorb on neighboring sites and the adsorbed molecules undergo a reaction. For Elay–Rideal reactions, only one of the molecules adsorbs and the other one reacts with it directly from the gas phase, for instance, by direct impact.

2.5 Sputtering

Sputtering is the process of ejecting surface atoms by means of bombardment by energetic (several 10 eV) particles such as ions or fast atoms. Besides etching, sputtering is used in deposition (physical vapor deposition, PVD) and in analytical techniques, for instance, secondary mass spectrometry (SIMS). Sputtering was discovered in 1852 but the underlying mechanisms were not understood until about 100 years later. It was not until the discovery of the so-called Wehner spots that sputtering was attributed to a sequence of atomic collision processes rather than the result of local evaporation (Wehner 1955).

Sputtering is commonly described by collision cascade theory, which was developed originally as a tool to determine the amount of radiation damage generated by

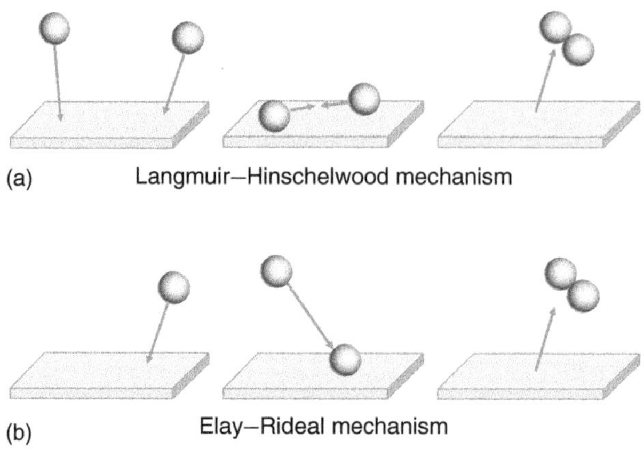

Figure 2.6 Illustration of Langmuir–Hinshelwood (a) and Eley–Rideal (b) surface reaction mechanisms.

fast neutrons (Sigmund 2012). This kind of a cascade can be theoretically treated using the binary collision approximation (BCA) simulation approach, which calculates the energy transfer as binary elastic collisions between two atoms:

$$\frac{E_2}{E_1} = \frac{4M_1M_2}{(M_1 + M_2)^2}\cos^2\theta \tag{2.5}$$

The scattering angle θ is a function of the alignment between the two atoms, called impact parameter, and the interatomic potentials. The implication of Eq. (2.5) is that energy transfer is the most efficient for atoms with similar mass. Very light atoms such as hydrogen do not transfer energy efficiently. They can penetrate deep into the material and create damage deep inside the material. This is important, for instance, when choosing plasmas containing hydrogen, such as HBr for etching. An example is damage below the gate oxide for HBr containing RIE overetches (see Section 7.3.2).

In a collision cascade, if momentum is imparted to a surface atom with an energy more than the surface binding energy, that atom can be ejected from the surface depending on the direction of its momentum vector. The energy of the impinging ion is shared among several recoil atoms. The minimum energy for which atoms are sputtered from the surface is larger than the bond energy, typically about 10 times larger. For instance, the bond energy for silicon is 4.7 eV (Yamamura and Tawara 1996) while the lowest energy at which silicon sputtering is observed for Ar ion impact at normal incidence is between 30 and 40 eV (Oostra et al. 1987). The latter value is called the sputtering threshold energy E_{th}.

There are two important implications of sputtering via collision cascades. Firstly, sputtering is always accompanied by some damage to the sputtered material. Vacancies and interstitials are generated in crystalline materials. Depending on the ion energy and dose, the surface region of the crystalline material can become amorphous. Molecular dynamics (MD) simulations of argon ions with energies between 10 and 200 eV reveal a dynamic balance between ion-induced damage

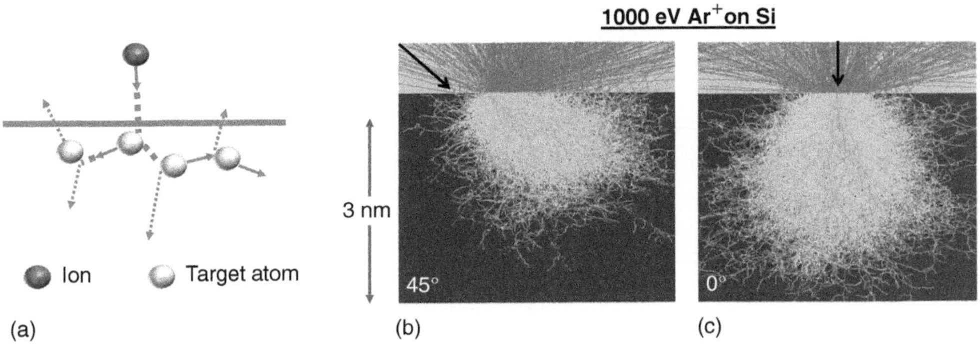

Figure 2.7 Left: Series of collision processes leading to sputtering of atoms 1 and 4 (a). Source: Adapted from Sigmund (2012). The case of impinging ion recoiling from the first collision and sputtering a surface atom in a second collision is also called "single knock-on mechanism" and occurs for ion energies in the 10–30 eV range, i.e. around the sputtering threshold. Right: Collision cascades in silicon created by 1000 eV Ar^+ ions impinging the surface of silicon at an angle of 0° (b) and 45° (c). Source: Berry et al. (2020).

and recrystallization of the surface (Graves and Humbird 2002). A surface that was amorphized at 200 eV can be recrystallized at 10 eV but the recrystallized region is not free of defects. The fact that sputtering used energies exceeding bond energy levels greatly also reduced the selectivity of etching technologies involving ions. This has decremental consequences for etching selectivity (see Section 7.2).

Secondly, due to the nature of the collision cascade, impinging ions penetrate the solid surface not only in the vertical direction but also in the lateral direction. This effect is called straggle. The atoms that are set in motion by the primary ion also exhibit straggle. Figure 2.7 shows the trajectories of displaced atoms that represent the collision cascade for 1000 eV Ar^+ ions in normal direction (0° to the surface) and 45° to the surface (Berry et al. 2020). The lateral motion of the collision cascade is clearly visible. Straggle can cause damage to the sidewall of a feature etched with RIE (Eriguchi et al. 2014).

The metric for sputtering efficiency is called sputtering yield. It equals the number of sputtered atoms per incoming ion. The sputtering rate can be calculated knowing the ion flux. The sputtering yield Γ is proportional to the difference of the square roots of the ion energy E_i and sputtering threshold energy E_{th} (Steinbruechel 1989):

$$\Gamma(E) = k(E_i^{1/2} - E_{\text{th}}^{1/2}) \tag{2.6}$$

Here, k is a constant depending on the combination of impinging ion and target material. According to Eq. (2.6), the sputtering yield increase slows down at higher energies. More complex relationships have been proposed (Eckstein and Preuss 2003). Sputtering yield curves for large numbers of systems can be found in the literature (Yamamura and Tawara 1996; Eckstein 2007).

Sputtering threshold energies are difficult to measure precisely because the yield goes to zero near the threshold (Hotston 1975). Theoretical or semiempirical models are useful to extrapolate data measured at energies with appreciable sputtering

Table 2.1 Sputtering thresholds for Ar^+ ion bombardment of materials important for etching of semiconductor devices.

Material	Sputtering threshold (eV)	Source
Silicon	27	Yamamura and Bohdansky (1985)
	~20	Oehrlein et al. (2015)
SiO_2	>50	Oostra et al. (1987)
	30	Todorov and Fossum (1988)
	45	Oehrlein et al. (2015)
	65	Kaler et al. (2017)
Carbon	34	Nishi et al. (1979)
Tungsten	20 (simulation)	Nakamura et al. (2016)
	28	Nishi et al. (1979)
Titanium	19	Nishi et al. (1979)
Tantalum	32	Nishi et al. (1979)
Molybdenum	35	Nishi et al. (1979)

yields. One of the equations is given below (Mantenieks 1999):

$$\frac{E_{\text{th}}}{E_O} = 4.4 - 1.3 \log\left(\frac{M_2}{M_1}\right) \tag{2.7}$$

Here, E_O is the sublimation or bond energy of the target material, M_1 the mass of the impinging ion, and M_2 the atomic mass of the target material. The sputtering threshold is higher for materials with higher sublimation or bond energy and for heavier ions. Sputtering thresholds for argon bombardment of materials used in semiconductor manufacturing are listed in Table 2.1.

The influence of the bond energy on sputtering threshold implies that ion sputtering can be used to remove a surface layer that was weakened by adsorption. This is the mechanism that enables directional or ion-assisted ALE and will be covered in detail in Section 6.1.

The sputtering yield is inversely proportional to the surface binding energy of the target ions. If the ion resides on the very surface of the material, this binding energy is E_S. If the sputtered atom originates from below the surface, the binding energy is E_O. Frequently, the heat of sublimation is used in calculations of the sputtering yield of various materials using the following equation (Sigmund 1969):

$$\Gamma = \frac{3f(M_1, M_2, \theta_i)4M_1M_2E_i}{4\pi^2(M_1 + M_2)^2E_O} \tag{2.8}$$

Here, f is a function of M_1 and M_2 as well as of the incident angle θ_i and E_O is the binding energy of the bulk atoms or the heat of sublimation.

While the collision cascade reaches deep into the solid, sputtered atoms and molecules originate from the surface. The depth of origin of the sputtered atoms is

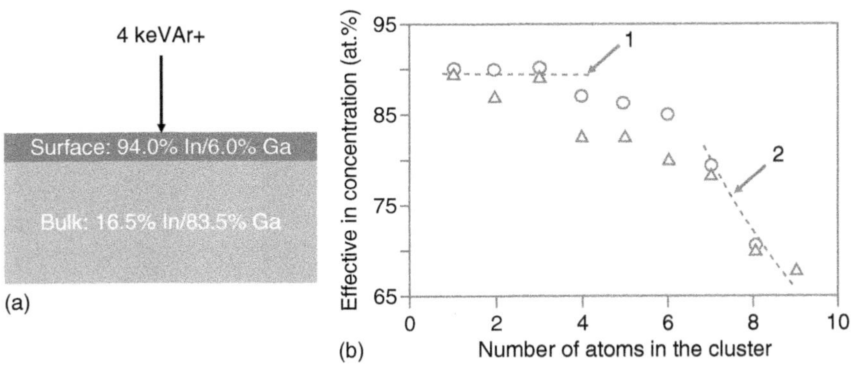

Figure 2.8 Effective indium concentration of sputtered neutral (triangles) and ionic (dots) clusters as a function of cluster size for 4 keV Ar^+ ions on a 16.5% indium in gallium liquid eutectic alloy. Curve 1 is calculated by a model based on the assumption that the depth of origin is constant. Curve 2 represents a fit for emission of the clusters from a cylindrical region with a cross section of 5.7 atoms. Source: Lill et al. (1994).

insensitive to beam properties but dependent on the material. It is quite small, of the order of an interatomic distance (Sigmund 2012).

The depth of origin for sputtered atoms and clusters was measured directly for 4 keV impact of Ar^+ of a liquid eutectic InGa alloy where the surface layer contains 94% indium while the bulk contains 16.5% indium (Lill et al. 1994). The fact that the alloy is liquid allowed the surface to restore equilibrium composition during ion bombardment. Because of this particular property, the composition of the mixed In and Ga clusters informs about the depth of their origin. The results are shown in Figure 2.8. While atoms, dimers, and trimers originate from the very top surface, clusters containing six and more atoms are removed from a cylindrical region with a cross section of roughly 5.7 atoms. Because the abundance of clusters diminishes very quickly as a function of the number of atoms, almost all sputtered species originate from the very top surface. This is despite the fact that collision cascades for keV ion bombardment reach several nanometers deep into the surface as shown in Figure 2.7b. The implication for etching with ions is that while predominantly atoms from the very surface are removed, collision cascades can damage the bulk material several nanometers deep.

Sputtering yields and rates are functions of ion impact angle. The reason is visualized in Figure 2.7 where the collision cascades for 1000 eV Ar^+ impact at 0° and 45° onto silicon are depicted. The collision cascade for the 45° impact angle is closer to the surface and hence creates more sputtering events.

Sputtering rates for 500 eV Ar^+ ion bombardment of materials used in the manufacturing of magnetic random access memory (MRAM) devices as a function of ion beam incidence angle are shown in Figure 2.9 (Ip et al. 2017). The general shape of these curves is characterized by a plateau of relative insensitivity around normal incidence, a peak around 45°–60°, and a rapid decay in sputtering yield for oblique incidence beyond 70°. For crystalline materials, local peaks caused by the orientation

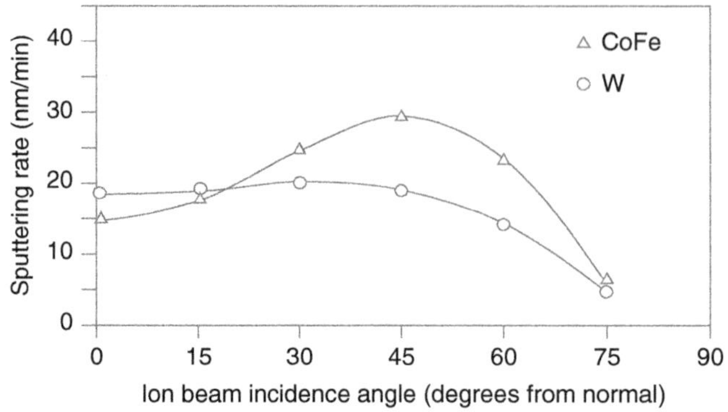

Figure 2.9 Sputtering rates for 500 eV Ar^+ ion bombardment of CoFe and *W* as a function of ion incidence angle. Source: Based on Ip et al. (2017).

of the crystal lattice can be observed (Sigmund 2012). Rough surfaces will introduce distortions to the angular dependence of sputtering rates.

This behavior of the angular dependence has two very important implications for etching. First, it is impossible to etch vertical profiles by means of physical ion bombardment under normal ion incidence. This would be the case for ions generated in a plasma that is in direct contact with the sputtering surface (see Section 9.1). Profiles etched by physical sputtering exhibit tapered sidewalls because the sputtering yield is 0° for 90°. To achieve a vertical profile, chemical effects must be utilized. This approach will be discussed in Section 7.2. In the absence of chemistry, ion beam tools with variable ion incidence impact angle are used (see Section 8.1).

Secondly, grazing incidence ions that impact angles close to 90° do not result in sputtering but scattering of the ions. This effect is important for transporting ions and fast atoms to the bottom of high aspect ratio features and will be discussed in Section 7.3.3.

Many materials used in semiconductor manufacturing are compounds, for instance, SiO_2, SiN, HfO_2, and others. Sputtering of multicomponent materials is more complex. For instance, ion bombardment of an alloy yields compositional changes over a depth range that may considerably exceed the depth of origin of sputtered atoms (Sigmund 2012). Sputtering of compound materials is preferential, which is to say nonstoichiometric. The ratio of sputtering yields of a binary material can be expressed by the following equation (Sigmund 2012):

$$\frac{\Gamma_j}{\Gamma_k} \sim \frac{c_i}{c_j}\left(\frac{M_k}{M_j}\right)^{2n}\left(\frac{E_{O,k}}{E_{O,j}}\right)^{1-2n} \tag{2.9}$$

Here, j and k denote the two components of the material, Γ is the sputtering yield, M the atomic mass, E_O the binding energy, and n is a power exponent characterizing the energy loss cross section at low energies. According to Eq. (2.9), the element with the lower mass and the lower binding energy will sputter preferentially. This has implications for directional or ion-assisted ALE and will be discussed in detail in Section 6.1.2. More in-depth information about sputtering yields of compound materials can be found in a review paper by Seah and Nunney (2010).

2.6 Implantation

Ions or fast atoms impinging a surface with energies of several 100 eV or higher create collision cascades and displacement of atoms as described in section 2.5. High-energy particles will also modify the composition of the material by stopping and accumulating around a characteristic depth. In RIE, adsorption and implantation of reactive species occur simultaneously. Ion implantation can create device damage during etching with RIE. The effects of implantation on RIE results will be discussed in Section 7.3.2.

Implantation can also be deployed in the modification step in thermal ALE of metals (Chen et al. 2017b). Implanted atoms can form chemical bonds with the surrounding matrix and form a new compound material. With appropriate choice of implanted elements, the bonds in the implantation region can be weakened and etching can be enabled. This effect will be covered in Section 6.1.1.

To understand the effects of implantation on etching, it is important to know the depth profile of implanted ions as a function of their energy and mass as well as the atomic mass of the solid. Questions like "How deep will hydrogen travel into the silicon below a gate oxide during gate etching?" or "What ion energy should I use to oxidize just 1 nm iron in an ALE modification step?" require knowledge about ion implantation fundamentals.

The distribution of the implanted ions can, in most cases, be approximated by a Gaussian distribution. The depth where the implant concentration peaks is called projected range R_p and the width of the distribution, longitudinal straggle ΔR_p. The projected range is a function of how fast a particle slows down while traveling in the solid. There are two stopping mechanisms: nuclear and electronic stopping. Heavier and slower atoms collide with the substrate atoms via binary elastic collisions. This effect is called nuclear stopping. It leads to atoms displacement and crystalline damage. Lighter and faster atoms with energies above 2 keV interact with the electronic shells of the substrate atoms causing inelastic energy losses. The kinetic energy lost via electron stopping is converted into heat.

The first attempt to develop a stopping and range theory was made by Lindhard, Scharff, and Schiott. Their approach is commonly called the Lindhard, Scharff, and Schiott (LSS) theory (Lindhard et al. 1963). Several semiempirical stopping power formulas including nuclear and electronic stopping have been developed since then, among others the stopping model by Ziegler, Biersack, and Littmark (ZBL model) (Ziegler et al. 1985). The widely used "transport of ions in matter" (TRIM) and "stopping and range of ions in matter" (SRIM) programs leverage the ZBL model. Table 2.2 shows projected ranges and straggle of ion/solid combinations for energies relevant to etching.

Table 2.2 shows projected ranges and straggle for implantation of H^+ and Br^+ ions into amorphous carbon, silicon, and tungsten. These values were calculated using TRIM (Berry 2020). Several trends are visible. Hydrogen ions penetrate much deeper than bromine ions. The projected ranges increase in the order Si > C > W. Typically ions penetrate deeper into materials with smaller atomic mass. However, the projected range in carbon is smaller than silicon because the atomic density for

Table 2.2 Projected ranges and straggle for systems and energies relevant for etching of semiconductor devices.

	200 eV		500 eV		1000 eV		5000 eV	
	R_p (Å)	ΔR_p (Å)	R_p (Å)	ΔR_p (Å)	R_p (Å)	ΔR_p (Å)	R_p (Å)	ΔR_p (Å)
Amorphous silicon								
H^+	68	36	128	68	217	107	760	290
Br^+	18	4	25	7	35	10	80	27
Amorphous carbon								
H^+	50	27	105	51	190	86	690	215
Br^+	15	2	22	3	30	5	70	15
Tungsten								
H^+	37	20	62	34	100	53	510	160
Br^+	6	3	8	5	12	7	26	16

Source: Berry et al. (2020).

carbon is much larger than for silicon. Values of 1.3×10^{23} atoms/cm^3 for carbon and 5×10^{22} atoms/cm^3 for silicon were used in the calculations.

To modify a thickness of about 1 nm, ion energies of 100 eV are required. This is well above the sputtering threshold of most materials (see Table 2.1). This means that implantation with meaningful ranges is accompanied by sputtering. If implantation is used to modify a surface to enable etching, the material loss must be taken into consideration (Chen et al. 2017a). Sputtering also limits the maximum concentration of the implanted species. A model for the maximum concentration for energies in the keV range can be found in Liau and Mayer (1978). Liau finds that, as a rule of thumb, one can estimate that the maximum concentration will be proportional to the preferential sputter factor divided by the total sputtering yield. Since lower mass elements tend to be preferentially sputtered, one can achieve a higher concentration of heavy elements than of lighter elements in the substrate.

2.7 Diffusion

Diffusion is the movement of molecules from a region of higher concentration to a region of lower concentration. Etching creates high concentration of reactive species at the surface via adsorption (see Section 2.2) and implantation (see Section 2.6). Concentration differences between surface and bulk of an etching feature will drive diffusion processes, which may be intended or unintended depending on the specific etching technology.

Until recently, surface oxidation caused by diffusion was not a very important mechanism in etching because devices were rather large and RIE and other etching technologies proceed at near room temperature where diffusion is relatively slow.

This is changing for advanced devices, which can be only a few nanometers across. Species continue to diffuse from the surface into the bulk during the time after etching and before cleaning. Another reason for the increased importance of diffusion as a mechanism relevant for etching is the emergence of metal-based devices such as MRAM and phase change memories (PCM). These devices are sensitive to ambient air exposure due to diffusion of oxygen and the resulting material oxidation (see Sections 7.3.4 and 8.2).

Diffusion of etching and ambient gases causes chemical oxidation of device materials. Two oxidation theories can be applied to semiconductor devices: Deal–Grove (DG)-based oxidation models for silicon (Deal and Grove 1965) and the Cabrera–Mott (CM) model for oxidation of metals (Cabrera and Mott 1948). Both models assume different driving forces for diffusion, chemical potentials in the case of DG and electrostatic forces for CM. The nature of these driving forces results in different rate dependences on time and gas pressure with profound implication for etching applications. In the following paragraphs, we will briefly describe both models.

The success of silicon-based integrated circuits can be attributed, in no small part, to the fact that high-quality silicon oxide films, the so-called thermal oxides, can be produced by direct oxidation of silicon. Thermal oxidation was instrumental in the implementation of Jean Hoerni's planar process, which positioned Fairchild Semiconductor as the leading company of a nascent semiconductor industry in the late 1950s (Hoerni 1962). Silicon doping, which is necessary to create semiconductor junctions in specific places, was achieved by first thermally growing a layer of silicon dioxide on the surface of a silicon wafer, and then wet etching windows in the oxide where the junctions were needed. The first (metal oxide semiconductor field effect transistor (MOSFET) was reported by Kahng and Atalla at Bell Labs in 1960 (Kahng and Atalla 1960). In the MOS transistor, signal amplification is achieved by controlling the conduction of charge carriers at the interface between silicon and thermally grown silicon oxide. Local oxidation of silicon (LOCOS) was introduced in the early 1970s (Appels et al. 1970).

These examples illustrate the tremendous importance of oxygen diffusion in silicon for the semiconductor industry. The DG model allowed to predict the thickness of thermally grown silicon oxide as a function of process parameters such as temperature, time, and pressure. The DG model considers fluxes of oxidizing species from the gas phase to the surface, from the surface to the oxidizing interface, and the flux of oxidants consumed by the oxidation reaction. These three fluxes must be equal to fulfill the requirement of conservation of fluxes. The following expression can be derived (Deal and Grove 1965):

$$d_{ox}^2 + Ad_{ox} = B(t - t_0) \tag{2.10}$$

Here, t_0 is a shift in the time coordinate corresponding to an initial oxide layer and A and B are parameters. For short oxidation times, the oxidation time is a linear function of time:

$$d_{ox} \approx \frac{B}{A}t \tag{2.11}$$

The ratio B/A is called the linear DG parameter:

$$\frac{B}{A} = \frac{k_{ox}h_G}{k_{ox} + h_G}\left(\frac{c_G}{N_{ox}}\right) \tag{2.12}$$

Here, h_G is the gas phase transport coefficient, k_{ox} is the oxidation rate constant, c_G is the concentration of the oxidant in the gas phase, and N_1 is the number of oxidant molecules incorporated into a unit volume of the oxide layer.

Under conditions present in etching, gas phase transport is not the rate-limiting step for oxidation. The unit of pressure multiplied by time is called Langmuir. An exposure of a surface at a pressure 10^{-6} Torr for one second corresponds to one Langmuir, 1 L. An exposure dose of 1 L creates about one monolayer of adsorbates if the sticking coefficient is 1. Typical transfer and etching pressures are 4 orders of magnitude higher in the order of 10^{-2} Torr. For very large h_G, the linear rate constant is proportional to the oxidation rate constant k_{ox}.

For long oxidation times, Eq. (2.10) is reduced to a parabolic dependence on time:

$$d_{ox} \approx \sqrt{Bt} \tag{2.13}$$

The parabolic DG parameter B is a function of pressure of the oxidant in the gas phase:

$$B = \frac{2D_{ox}c_G}{N_{ox}} \tag{2.14}$$

Here, D_{ox} is the diffusion coefficient of the oxidant. D_{ox} is an Arrhenius function of temperature and activation energy $E_{a,\,ox}$.

$$D_{ox} = D_{0,ox}e^{-E_{a,ox}/RT} \tag{2.15}$$

The main implication of the DG model for etching is that this type of oxidation and hence device damage can be effectively suppressed by operating in vacuum at low temperatures. However, some etching processes such as resist stripping use temperatures of several hundred degrees centigrade. When wafers are not cooled sufficiently before leaving the vacuum side of the tool, the oxygen and water vapor in the atmosphere can cause DG oxidation (Figure 2.10).

The DG model is not accurate in predicting the initial stages of oxidation with thicknesses below about 10 nm. This regime is obviously important for advanced

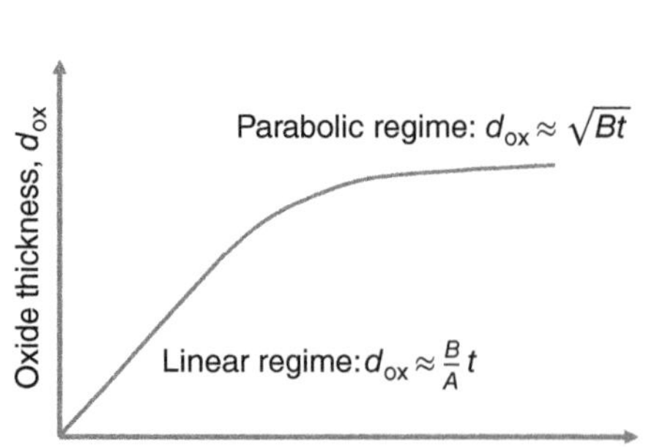

Figure 2.10 Schematic illustration of oxidation depth vs time according to Deal–Grove model.

devices with dimension of 10 nm or less. Various improvements to the DG model have been developed for low thin-film oxidation.

Oxidation of metals can be described by the CM model (Cabrera and Mott 1948; Fehlner 1984). This model is based on the following assumptions: (i) oxidation is sustained by diffusion of ions rather than neutrals. (ii) Oxygen ions (O^-) are generated by electrons that tunnel from the metal through the metal oxide to the oxide surface to ionize adsorbed oxygen atoms. (iii) This establishes a uniform field within the oxide, which is proportional to the potential difference between the oxide and metal layers, the so-called Mott potential $\Delta\Phi_{Mott}/e$. (iv) The Mott potential drives transport of the slower metal ions to the metal oxide/metal interface where they react and contribute to the metal oxide film growth. The overall reaction can be written as $1/2O_2$ (gas) + 2e (metal) → O_2^- (surface). (v) The film is so thin that the effect of any space charge set up by the dissolved ions is negligible.

Because $\Delta\Phi_{Mott}$ is created by electron tunneling, the CM model is restricted to thin films, typically below 10 nm. It can be extended to thicker films when electron transport via thermionic emission or via semiconducting oxides are assumed. The rate-limiting step is the injection of a metal ion into an interstitial position in the metal oxide or the creation of a metal vacancy at the metal oxide surface. Adsorption is very fast for the same reasons mentioned above for the DG model. Oxygen ion diffusion is also very fast because it is driven by $\Delta\Phi_{Mott}$. However, the electrical field is inversely proportional to the film thickness. Therefore, CM oxide growth is characterized by the following properties: (i) CM oxidation is initially extremely rapid, but after a few minutes or hours drops to very low or negligible values after forming a 2–10 nm thin film. CM oxidation appears to be a quasi self-limiting process. (ii) CM oxidation occurs at low temperatures. Aluminum exhibits this behavior at room temperature; copper, iron, barium, and other metals do the same at liquid air temperatures (Cabrera and Mott 1948). (iii) CM oxidation is pressure independent for typical vacuum conditions used in semiconductor manufacturing (several tens of mTorr in the processing and vacuum transfer chamber; atmospheric pressure during transport from tool to tool). (iv) Because $\Delta\Phi_{Mott}$ is in the order of 1–2 V (Cabrera and Mott 1948), the influence of temperature is only weak. However, the CM model predicts the existence of a critical temperature above which oxide growth is not self-limiting. Above this temperature, oxide growth slows down but does not completely stop. This critical temperature is approximately equal to $(H_{sol} + E_{a,\,diff}) \times 39\kappa$, where H_{sol} is the enthalpy of solution for positive ions, $E_{a,\,diff}$ the activation energy for diffusion, and κ the dielectric constant.

For the CM model, the oxidation time is an inverse logarithmic function of time:

$$d_{ox}^{-1} = k_1 - k_2 \ln(t) \tag{2.16}$$

where k_1 and k_2 are parameters. Figure 2.11 shows experimental results for CM oxidation of aluminum at 10 °C (Cabrera and Mott 1948).

CM oxidation is very important and potentially decremental for etching of advanced features for three reasons: (i) the self-limiting oxide thickness is in the range of the final dimensions. (ii) It happens within minutes. (iii) It cannot be suppressed by lowering the pressure or temperature.

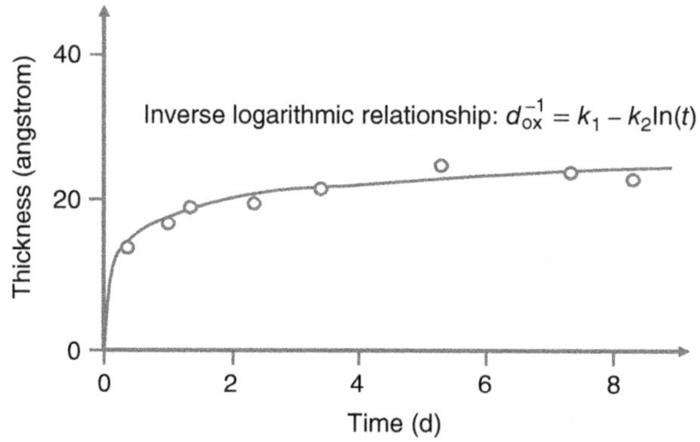

Figure 2.11 Thickness of oxidized aluminum as a function of time showing inverse logarithmic behavior according to the Cabrera–Mott model. Source: Based on Cabrera and Mott (1948).

Not all metals oxidize via CM mechanism. Figure 2.12 shows experimental results obtained with X-ray photoelectron spectroscopy (XPS) after oxidizing cobalt, copper, iron, palladium, and platinum in a 2.5 mTorr oxygen plasma (Chen et al. 2017a). The plasma was inductively coupled and excited with an RF power of 500 W, which resulted in an ion energy of 5 eV, which is too low to cause sputtering. The data analysis suggests that the reactive metals cobalt, copper, and iron follow the inverse logarithmic rate law (CM) while oxidation of the noble metals palladium and platinum can be described best by a parabolic relationship (DG).

2.8 Transport Phenomena in 3D Features

Advanced semiconductor devices contain structures with aspect ratios of 50 and higher. Figure 2.13 shows a hole structure with 80 : 1 aspect ratio etched into single crystal silicon to form capacitors for the so-called Deep Trench Dynamic Random

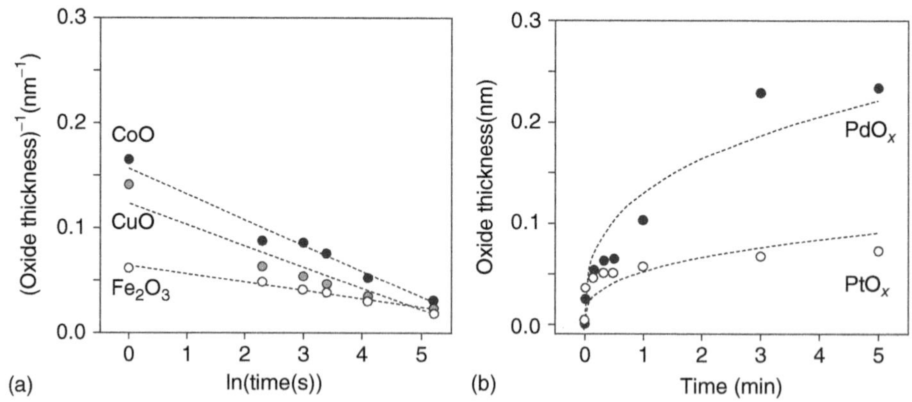

Figure 2.12 Experimental and modeling results for oxidation of various metals. Source: Chen et al. (2017a). © 2017, American Vacuum Society.

Access Memory (DRAM) devices (Lill and Joubert 2008). Clearly, transport of ions and reactive neutrals to the etch front of reaction products out of such high aspect ratio structures is a challenge for this kind of structure.

2.8.1 Neutral Transport

Under RIE and ALE process conditions, mean free paths for collisions in the gas phase are much longer than the characteristic dimensions of the microstructure. Excluding fast neutrals created by charge-exchange reactions with ions in the bulk plasma, the neutral angular distribution is nearly isotropic and the energy distribution is nearly Maxwellian. If only neutrals with a trajectory within the solid angle created by the point of impact on the bottom of the feature and the opening on the

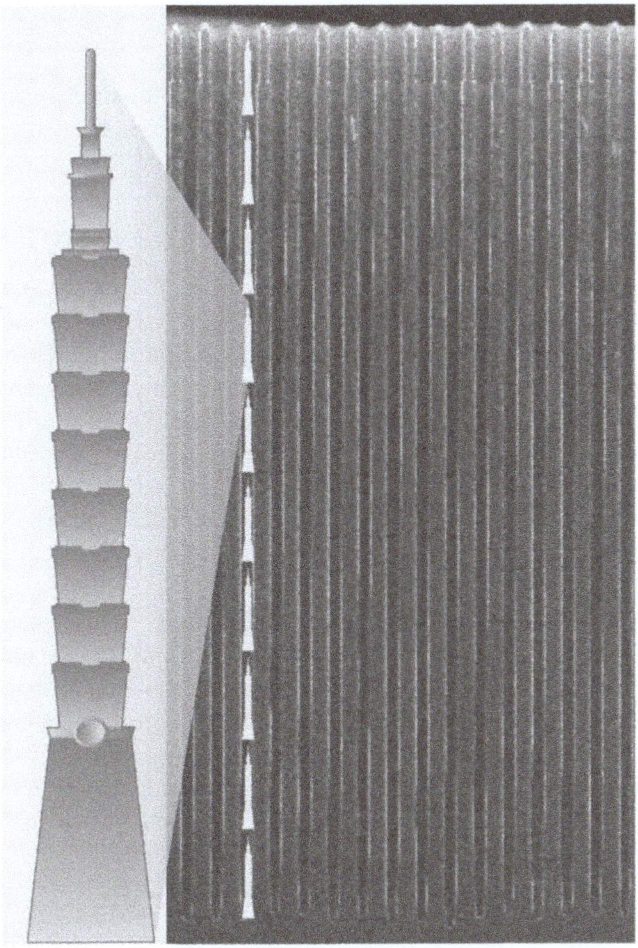

Figure 2.13 Hole and trench microstructures in advanced DRAM and NAND devices reach very high aspect ratios. This image shows holes etched in silicon with aspect ratios of 80 : 1. The Taipei 101 tower, scaled down to the trench width, would fit 10 times into this trench. Source: Lill and Joubert (2008).

top from were to reach the etch front, etching rates would attenuate extremely fast. Therefore, this effect of neutral shading is not the only effect governing the transport of neutrals inside the etching feature (Gottscho et al. 1992). Additional neutrals reach the bottom via a mechanism called Knudsen transport. Knudsen transport considers the interaction of neutrals with the feature sidewalls.

The ratio of the neutral flux on the bottom and the top of a feature can be expressed as a function of the sticking coefficient s and the transmission probability K that a randomly directed molecule incident at the top of a vertical-walled round hole or trench will reach the other end (Coburn and Winters 1989):

$$J_{n,t} - (1 - K)J_{n,t} - K(1 - s)J_{n,b} = sJ_{n,b} \tag{2.17}$$

The first term in this expression is the neutral flux on the top of the feature, the second term represents the fraction of the incident flux that is reflected back out of the feature without reaching the bottom surface, the third term represents the species that reach the bottom surface but do not react and eventually escape through the open end, and the term on the right side of the equation represents the species consumed by etching the bottom surface (Coburn and Winters 1989). The attenuation of the neutral flux can be written as

$$\frac{J_{n,b}}{J_{n,t}} = \frac{K}{K + s - Ks} \tag{2.18}$$

Tables of K for various aspect ratio tubes and slots can be found in standard texts on vacuum technology. Equation (2.18) describes ARDE for neutral or radical etching. Neutral ARDE is very sensitive to the sticking coefficient s. Figure 2.14 depicts the etching rate of polysilicon as a function of aspect ratio for an inductively coupled fluorine plasma inside high aspect ratio structures lined with silicon oxide for two different process conditions. The instantaneous etching rates were measured using in situ interferometry. Effective sticking coefficients of fluorine atoms with the silicon oxide surface between 0.03 and 0.11 can be derived when fitting the experimental data with Eq. (2.18). The graph also shows the worst ARDE case for a sticking coefficient of 1 with a solid line.

Knudsen transport assumes that the species are reflected "diffusely," which means with roughly the same probability in all directions. Obviously, the fluxes to the bottom of the feature would be larger if neutrals can maintain as much forward momentum as possible upon interaction with the feature sidewall. How neutrals interact with solid surfaces has been studied extensively in molecular beam-surface scattering experiments. When a gaseous particle collides with a solid surface, it can interact either elastically or inelastically. In the case of elastic scattering, the reflected angle will be close to the impact angle, which is called specular reflection. This is the desired scenario for transporting neutrals into a high aspect ratio feature. However, because most neutrals used in etching are reactive, they will interact with the surface and the scattering events will most likely be inelastic.

For inelastic scattering, three scenarios can be distinguished. In the first case, the molecules can lose enough of their energy to become trapped or adsorbed on the surface (see Section 2.2). If the adsorbed molecule is subsequently desorbed, it will

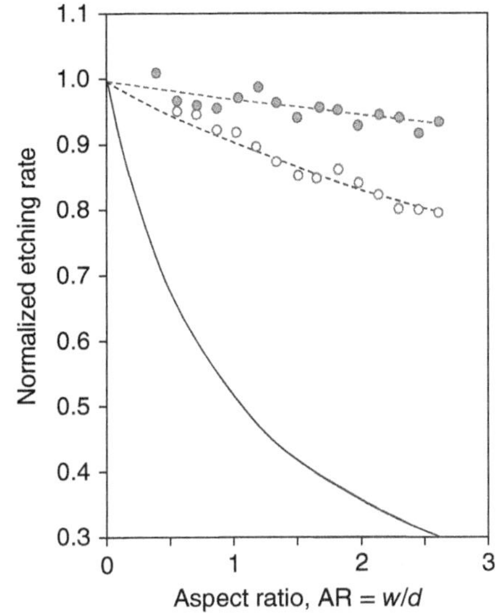

Figure 2.14 Normalized etch rate as a function of aspect ratio for a fluorine plasma etching polysilicon out of trenches lined with silicon oxide. Source: Lill et al. (2001).

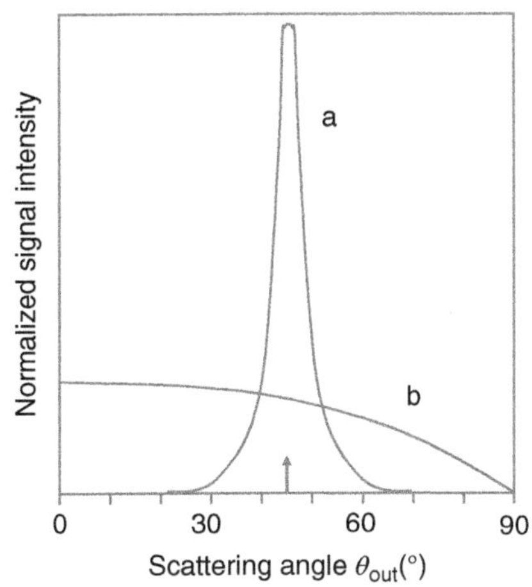

Figure 2.15 Angular distributions of neutral beams. Curve a: Angular distribution of specularly scattered beam with $\theta_{out} = \theta_{in}$. Arrow indicates angle of incidence θ_{in}. Curve b: Scattered beam with cosine angular distribution.

continue to contribute to the neutral flux inside the etching feature. Since these molecules have had a chance to equilibrate with the surface, they are likely to desorb with a cosine spatial distribution and with Maxwellian velocities characteristic of the surface temperature (Somorjai and Brumbach 1973). This case is depicted by curve b in Figure 2.15. This means they will be less directional than the original neutrals. This is the scenario considered in Knudsen transport.

Secondly, the molecules can lose some of their energy but still be scattered directly back into the gas phase. The scattering angle depends on the details of the interaction but in general, these neutrals scatter close to specular as illustrated by curve a in Figure 2.15. These scattered molecules continue their movement toward the bottom of the feature. In a third type of interaction, molecules may lose insufficient energy to adsorb but also not scatter immediately. They become molecules "hopping" or diffusing along the surface.

Conceptually, the best etching results are obtained when neutrals scatter elastically or specular, which means without chemical interaction, but stick with high probability to the etching front. These seem to be two contradictory requirements. The fact that the scattering is grazing on the sidewall and normal to the etching front is helpful. For directional etching, the composition of the sidewall is different from the etching front because a passivation layer protects the sidewall. This can also help to reduce the sticking coefficient of the neutrals to the sidewall while maintaining reactivity at the etching surface. While we understand that these mechanisms exist, practical realization is a challenge. Much is left to empirical testing, for instance the exploration of very high (>150 °C) and very low (<50 °C) temperatures in the case of RIE. Scattering of molecules will become an even more important mechanism for thermal or isotropic ALE because large and fragile molecules can be formed (see Section 4.2).

Transport of neutrals via surface diffusion is important for etching. Adsorbed atoms can hop from one adsorption site to the next empty site when they are thermally activated. The corresponding surface diffusion coefficient can be described by an Arrhenius equation similar to the bulk diffusion coefficient in Eq. (2.14):

$$D_S = D_{0,S}e^{-E_{a,S}/RT} \tag{2.19}$$

The activation energy for surface diffusion, $E_{a,S}$, is the energy barrier height for an adsorbed atom to move from one adsorption site to the next. This energy is 5–20% of the heat of desorption, E_A. Surface diffusion can be classified depending on coverage Θ. For low Θ, tracer diffusion occurs, while for medium and high Θ, the process is called chemical diffusion. The latter mechanism is applicable to etching because etching processes are designed to create large fluxes of reactive species with meaningful sticking coefficients. Attractive and repulsive interactions between adsorbed atoms influence chemical surface diffusion rates.

The existence of surface diffusion could play a role in the etching of features that have such high aspect ratios that only a negligible flux of neutrals reaches the etching front at the bottom directly without collision with the feature sidewall. Neutrals can potentially adsorb on the sidewall at a temperature low enough to prevent desorption but high enough to allow chemical surface diffusion. Once they reach the etching bottom of the feature, they accumulate there and contribute to etching. This is a mechanism that could explain some of the high etching rates of high aspect ratio features in silicon oxide and silicon nitride at cryogenic temperatures.

2.8.2 Ion Transport

Because of their charge, ions can be accelerated toward the etching surface and they have several orders of magnitude higher kinetic energy than neutrals. It appears to be a much simpler task to create a large ion flux at the bottom of a high aspect ratio feature. However, ion densities in low-temperature plasmas are orders of magnitude lower than neutral densities (see Section 9.1). Hence, narrowing the angular distribution is an ongoing effort in etching technology development. This will be discussed in Section 7.3.3.

Based on how ions are generated in a plasma and accelerated in the plasma sheath or via additional extraction grids, ions have a certain finite angular distribution. This means that some of the ions will impinge the feature sidewall where they get implanted, scatter, or sputter surface atoms. A large percentage of scattering vs. sputtering events on the sidewall is of course preferred for etching of vertical profiles with low ARDE.

As was shown in Section 2.5, sputtering yield goes to zero for ion incidence angles close to 90° from the surface normal. At these impact angles, ions scatter off the surface without transferring enough energy to the surface to cause meaningful sputtering. This is shown in Figure 2.16 where modeling results are shown for the angular dependence of the sputtering rate (a) and the ion reflection coefficient (b) for 600 eV Xe^+ ions colliding with silicon (Teichmann et al. 2014). The graphs illustrate the transition from implantation and sputtering to scattering for incidence angles above 60° from the surface normal. For incidence angles of over 60°, virtually all ions are scattered.

While this behavior makes sense intuitively based on experience with macroscopic projectiles, the underlying physics on an atomic level is intricate, and a closed theory to handle extremely grazing incidence scattering of ions is not available. Various sophisticated theories shed light on various aspects of surface scattering (Winter 2002).

Some insights can be gained from studies of ion scattering from single crystal surfaces, which are perfectly ordered and periodic. When an incident ion collides with a target atom on such a surface, it gets repelled by the interatomic potential. Some of the space behind the target ion is not accessible for the scattered incident ion. This effect is called shadowing or blocking and gives rise to a so-called scattering or blocking cone. An example of such a cone is shown in a two-dimensional representation in Figure 2.17a. Here, the impact parameter of a single ion–atom collision is varied continuously. The forbidden space behind the target atom has the shape of a paraboloid, the radius r of which can be expressed by Stensgaard et al. (1978)

$$r = 2\sqrt{\frac{Z_1 Z_2 e^2 d}{E_i}} \tag{2.20}$$

Here, E_i is the kinetic energy of the incoming ion, Z_1 and Z_2 are the atomic numbers of the incoming ion and the target or surface atom, e is the elementary charge, and d is the distance between neighboring atoms. The ion flux is increased near the edge of the shadow cone resulting in a focusing effect. For a certain incident angle, called the critical angle, the edges of the cones intersect their neighboring atoms as shown in Figure 2.17b. The result is that the focused ion flux impacts the next surface atom and is scattered again. This effect is called surface channeling and is related to channeling of implanted ions that can be subjected to the same effect at internal planes

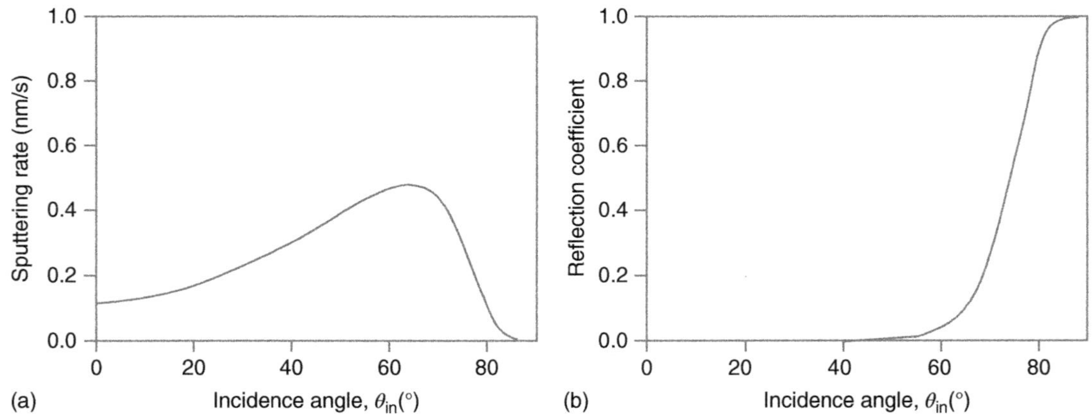

Figure 2.16 Sputtering rate (a) and reflection coefficient (b) for silicon as a function of incidence angle for different material systems with 600 eV Xe^+ based on TRIM calculations. The sputtering rate was calculated using an ion current density of 300 μA/cm^2. Source: Teichmann et al. (2014).

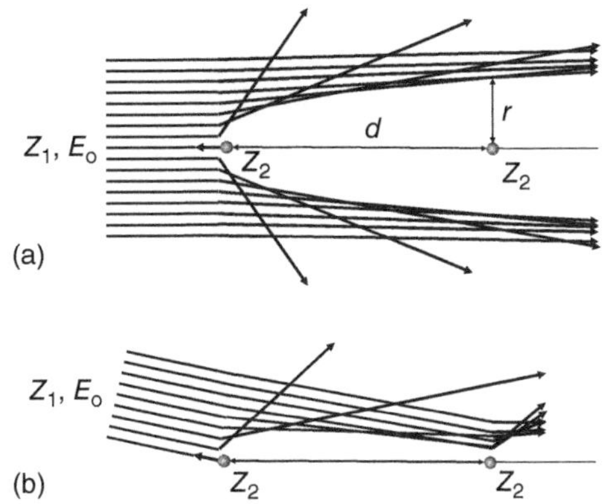

Figure 2.17 (a) Schematic illustration of a blocking or scattering cone. (b) Schematic illustration of ion surface channeling.

of single crystal materials. Surface channeling leads to nearly complete scattering of all incoming ions and focusing of the scattered ions.

Sidewall scattering is an important mechanism for RIE etching. Changes in the sidewall scattering treatment in feature scale simulation while holding bulk plasma conditions constant can account for the dramatic differences in feature shape (Vyvoda et al. 2000). Bowing of the structure can occur where scattered ion fluxes impinge the sidewall (see Section 7.3.3).

However, the effect of surface channeling does not directly apply to scattering on feature sidewalls that are amorphous or rough. Nevertheless, some insights can be gained from this ideal case. Equation (2.20) predicts channeling at larger incidence angles with respect to the sidewall surface normal for higher ion energies. This is one of the reasons why high aspect ratio etching requires high ion energies. The dependence of scattered ion flux and energy for amorphous surfaces can be modeled using Monte Carlo simulations (Berry et al. (2020). This will be discussed in more detail in Section 7.3.3.

Ion beam scattering is also being used to probe surface imperfections such as adsorbates, impurities, and steps as well as the electronic stopping power. One such analytics technique is, for instance, low-energy ion spectroscopy (LEIS).

Scattering ions interact with the electronic structure of the surface atoms and can exchange charges with the scattering surface. This contributes to inelastic energy losses. The probability of charge exchange depends on the effective interaction time of the projectile with the target electrons, which is longer under grazing surface channeling conditions and for lower ion energies. Charge transfer is also enhanced for materials with lower work functions such as metals. Ions with low velocities are mostly neutralized during scattering.

Experimental results have shown that for ion energies in the keV range only 1–5% are charged after scattering off metal surfaces (Robin 2003). This effect is used, for instance, in the generation of neutral beams by means of surface neutralization in grids with high AR and slightly tilted holes (Park et al. 2005; Samukawa 2006). During etching of high aspect ratio features, surface neutralization discharges ions in the first interaction with the feature sidewall. They continue their path toward the etching front as neutrals immune to local charges (Huang et al. 2019). Ions will see charges predominantly at the top of the feature before the first scattering event and hence it is important to neutralize surface charges caused by flux imbalances between electrons and positively charged ions. This can be accomplished using plasma pulsing (see Section 9.5).

2.8.3 Transport of Reaction Products

Reaction products of chemically assisted etching technologies are frequently referred to as "volatile." From an etching technology point of view, reaction products must meet two conditions to be considered "volatile": they must desorb from the surface and they must not redeposit on the way out of the feature to be removed from the wafer surface. Generally, if the removal process occurs in a thermodynamic equilibrium and the composition of the etching front and the sidewall are the same, both conditions are met easily. This is the case for most thermal and radical etching technologies as well as thermal ALE. For these techniques a distinct sidewall does not exist because the etching is isotropic, i.e. proceeds in all directions.

The angular distribution of desorbed molecules from an amorphous surface is nearly a cosine function and reaction products will collide to a large degree with the feature sidewalls. Depending on the complexity and specific structure of the etching product, the probability for partial dissociation can be meaningful, which could lead to sticking in a subsequent collision. This will be discussed in more detail for thermal ALE with complex molecules as reaction products (see Section 4.1).

The situation is very different for ion-assisted desorption, which is not an equilibrium process. The energy of most sputtered species is in the order of magnitude of the surface binding energy E_s, which is below 10 eV. Some collision cascades can result in sputtered species with energies of several tens of eV. Experimental studies have shown that the angular distribution of sputtered species is a function of the ion energy. For instance, the angular distribution for polycrystalline rhodium under 5 keV Ar^+ ion bombardment is a cosine function for 2 eV sputtered atoms and a cosine squared function for 12 eV sputtered atoms (Baxter et al. 1985). Etching surfaces are exposed to fluxes of ions and are mostly amorphous. Hence, we can assume that the angular distribution is between cosine and cosine squared and hence much broader than the incoming ion flux, which is within 1° for both RIE and IBE. Removal of sputtered species out of high aspect ratio features is therefore a challenge.

Etching by sputtering without chemical assistance will lead to severe redeposition. This is one of the reasons why IBE is not the technology of choice for etching of semiconductor devices. In RIE and directional ALE, reactive species are introduced, which modify the surface and form halogenated or otherwise fully or partially oxidized molecules. For instance, sputtering of chlorinated silicon surfaces will result in SiCl, $SiCl_2$, $SiCl_3$, and $SiCl_4$ species (Gou et al. 2010). These species have an ion energy of a few electron volts and a near cosine angular distribution. When they collide with sidewall surfaces, they will stick with a probability that is inversely correlated to the number of chlorine atoms in the sputtered molecules (Kiehlbauch and Graves 2003). Therefore, sufficient chlorination of the etching front is important to control sidewall redeposition. Another approach is to use halogen radicals, for instance Cl^* or F^*, to etch redeposited materials isotropically. Adding radicals to remove redeposited material is however not an option for directional ALE because it would introduce a process step that is not self-limited (see Section 6.2).

2.9 Classification of Etching Technologies

The naming and classification of an increasing number of etching technologies that are being proposed and already used for patterning of semiconductor devices is sometimes confusing. This situation is becoming exaggerated by the emergence of various ALE implementations. A logical approach to classify all etching technologies is therefore needed.

All etching technologies for semiconductor devices, with the exception of nonreactive IBE, have a chemical component to weaken the top surface and require energy to remove surface atoms. Based on these common characteristics, it makes sense to use three criteria for classification: (i) the method of surface modification to reduce the surface binding energy from E_O to E_S (see Figure 2.1), (ii) the method or kind of energy for removal, and (iii) whether modification and removal processes proceed simultaneously or in spatially or temporally separated, self-limited, sequential steps. The third criterion delineates ALE methods from continuous etching technologies.

Table 2.3 lists the most relevant continuous etching technologies. RIE is by far the most widely used and important etching method. RIE utilizes reactive plasmas that are in contact with the wafer. The wafer surface is exposed to simultaneous fluxes of reactive and nonreactive ions, neutrals, radicals, electrons, and photons. The surface is modified by reactive ion implantation, as well as adsorption and diffusion of radicals and neutrals. Simultaneously, removal is accomplished by sputtering and desorption. Desorption can be purely thermal, but it can also be enhanced by photons and electrons that are generated by the plasma. To create RIE processes that meet all requirements is a highly complex, and still mostly empirical exercise of balancing all particle and energy fluxes to the wafer surface. We will study RIE in detail in Chapter 7.

Radical etching does not use ions in the removal process. The surface is modified by radicals, which are atoms or molecules with unpaired electrons. This makes them highly reactive. The weakened top surface layer is removed via thermal desorption. This technology has historically also been called chemical downstream etching (CDE) because the physical component of ion sputtering is absent or greatly suppressed. In many tool embodiments, the so-called downstream sources are used to generate radicals. Radical etching will be covered in Chapter 5.

Thermal etching uses molecules to modify the surface via adsorption. Some adsorbed atoms diffuse into subsurface layers and modify more than one layer. As in the case of radical etching, removal is accomplished via thermal desorption. This

Table 2.3 Classification of continuous etching technologies.

Removal	Modification		
	Implantation	**Radical adsorption/diffusion**	**Neutral adsorption/diffusion**
Sputtering	RIBE, RIE	RIE	CAIBE, RIE
Desorption	RIE	Radical etching, RIE	Thermal etching, RIE

technology is sometimes also referred to as "vapor etching." Chapter 3 is dedicated to Thermal Etching.

Table 2.3 also lists two chemically enhanced IBE technologies, Reactive Ion Beam Etching (RIBE) and Chemically Assisted Ion Beam Etching (CAIBE). RIBE uses reactive ions while IBE uses noble gas ions, which are chemically not active. In both cases, only ions are involved in the process. The surface is modified by implanting reactive ions into the surface and the removal is accomplished by sputtering by either reactive or nonreactive ions if a gas mixture is used. The benefit of RIBE over RIE is that, because ions are accelerated via grids and via a plasma in direct contact with the wafer surface, the ion incidence angle can be varied. To make RIBE production worthy, the challenge of grid erosion has to be solved, among others.

CAIBE uses reactive gases to modify the surface and noble gas ions to sputter the weakened surface layer. The pressure of the reactive background gas must be low enough to prevent gas-phase collisions between the accelerated ions and the background gas molecules. IBE and its derivate technologies of RIBE and CAIBE will be described in Chapter 8.

Next, we will classify etching technologies where surface modification and removal processes are separated in time or space. ALE technologies belong to this category. ALE promises to meet the requirement of atomic-scale fidelity for manufacturing of integrated devices of the 10 nm node and below. In analogy to ALD, ALE is characterized by self-limitation of separate processing steps (George 2010; Kanarik et al. 2015; Faraz et al. 2015). Self-limitation gives ALE the property of synergy (Kanarik et al. 2017). For ALE processes with a synergy of $S = 100\%$, none of the steps etches the surface when applied individually. Etching is only observed when the steps are executed in sequence.

Self-limitation makes ALE processes flux independent if the dose of species is high enough to reach saturation. Flux nonuniformities are compensated across the wafer when saturation is reached in all locations of the wafer. Differences in the transport mechanisms between ions and neutrals, which result in ARDE for RIE, are less important. As a result, ALE processes have in principle excellent uniformity and no ARDE. If all steps are nondirectional, ALE is conceptually also perfectly isotropic, which means etching proceeds with the same rate in all directions inside a feature. Deviations from this ideal behavior will be discussed in Chapters 4 and 6.

Table 2.4 shows the classification of ALE technologies. The classification delineates between the various surface modification and removal methods. The

Table 2.4 Classification of atomic layer etching technologies.

Removal	Modification		
	Implantation	Radical adsorption/diffusion	Neutral adsorption/diffusion
Sputtering		Directional ALE	
Desorption 1 (direct)		Radical etch	Thermal etch
Desorption 2 (reaction assisted)	Directional ALE	Plasma-assisted isotropic ALE	Thermal isotropic ALE

overarching characteristic of all of these ALE processes is that modification and removal proceed in spatially or temporally separated, self-limited, sequential steps.

It is insightful to subdivide desorption removal into "Desorption 1 (Direct)" and "Desorption 2 (Reaction-Assisted)" for ALE processes. While these two types of desorption most likely also exist for continuous etching methods such as RIE, they are not implemented deliberately and hence have been grouped together for simplicity in Table 2.3.

The fundamentals of "direct" desorption are described in Section 2.3. For direct desorption of a modified layer in an ALE regime, the temperature of the wafer must be cycled. The modification step must proceed at a lower temperature than the removal step or else surface modification and removal proceed simultaneously, and the result is one of the continuous etching technologies in Table 2.3, for instance, thermal or radical etching. The challenge for ALE with temperature cycling is that the cycling between steps must be fast. Because the EPC is typically only a few nanometers, a full cycle should not take more than a few seconds. Depending on the temperature range to be covered in each temperature cycle, the engineering challenges can be daunting.

As an alternative to thermal ALE with temperature cycling, desorption can also be induced by a second surface reaction (see Section 2.4). In this approach, new adsorbates form at the surface, which can desorb at the same temperature as the surface modification step. This desorption mechanism is labeled "Reaction-Assisted" in Table 2.3. Reaction-assisted desorption is the underlying mechanism for several thermal isotropic ALE reactions such as Chelation/Condensation ALE (see Section 4.1.1), Ligand Exchange ALE (see Section 4.1.2), Conversion ALE (see Section 4.1.3), and Oxidation/Fluorination ALE (see Section 4.1.4).

An example Oxidation/Fluorination ALE is the etching of TiN with alternating exposures of O_3 and HF at temperatures of 250 °C and higher (Lee and George 2017). In the first step of this process, O_3 adsorbs at the surface and dissociates forming oxygen atoms. They substitute nitrogen at the top surface and convert TiN into TiO_2. TiO_2 has a boiling point of over 2900 °C and will not desorb. In a second step, HF reacts with TiO_2 to form TiF_4, which desorbs above a temperature of 250 °C in vacuum (Lee and George 2017). Because HF is not reactive enough to form TiF_4 in a direct reaction with TiN, the removal step is self-limited by the thickness of the TiO_2 layer, which is limited by the depth of oxidation. Experiments show that the TiO_2 layer forms a diffusion barrier that impedes further oxidation and leads to a self-limiting or decelerating oxidation reaction either by DG or CM oxidation mechanisms as discussed in Section 2.7 (Lee and George 2017).

When removal by direct or reaction-assisted desorption is combined with neutral adsorption, the resulting ALE process is called Thermal Isotropic ALE. The term "thermal" denotes that the removal process involves thermal processes, and "isotropic" reflects the property of this process to proceed in all directions with similar EPC. Instead of neutrals, radicals can be used for surface modification. This class of ALE processes is labeled "Plasma-Assisted Thermal Isotropic ALE" because the main approach to create radicals is by plasma activation. Plasma-Assisted Thermal Isotropic ALE will be discussed in Section 4.3.

Table 2.4 also shows three classes of "Directional ALE." These ALE processes use ions either in the removal step applying sputtering (see Section 2.5) or in the modification step using ion implantation (see Section 2.6). The term "directional" reflects the fact that ions are being accelerated toward the surface and hence travel in a preferred direction. In the ALE implementation where they are used to sputter the modified surface, the sputtering rate will be higher at the bottom of the feature. The sidewall will mostly see glancing ion incidence and scattering (see Section 2.8.2). When ions are implanted to modify the surface, this also happens predominantly at the bottom of the feature because ions are scattered off the sidewall. Chapter 6 is dedicated to Directional ALE. The combinations of modification via implantation and removal by sputtering or direct desorption are not filled in Table 2.4. While etching approaches based on the combination of these mechanisms are potentially possible, they have not been demonstrated or considered as viable options for etching of semiconductor devices.

Table 2.3 covers continuous etching processes and Table 2.4 covers sequential etching with separated, self-limited steps. Of course, etching process steps can be separated in time or space without the conditions of self-limitation and in fact such processes are widely used in manufacturing of semiconductor devices. Most of these processes are the so-called pulsed RIE process. Pulsing can be realized via repeating cycles of alternating plasma properties called plasma pulsing or a combination of plasma and gas flow changes called mixed mode pulsing (MMP). These techniques will be discussed in Section 7.1.1. Pulsed RIE processes can exhibit a certain degree of self-limitation. Depending on the synergy S of the pulsed RIE process, these processes can also be viewed as quasi Directional ALE (see Chapter 6).

Problems

P2.1 Analyze equation (2.2) in the context of a Thermal Etching process which is limited by the adsorption step.

P2.2 Calculate E_{th} for sputtering of silicon with argon using Eq. (2.7) and binding energy for silicon E_0 of 4.7 eV. Compare with the experimental results in Table 2.1.

P2.3 Using Eq. (2.18), calculate the relative flux of neutrals to the bottom of a round tube with an aspect ratio of 50 for several reaction probabilities between 0 and 1. Assume a transmission probability of 0.025 29. This value can be found on page 36 of the third edition of "A User's Guide to Vacuum Technology" (O'Hanlon 2003).

P2.4 Using Eq. (2.3), calculate the change in desorption rate R_D when the activation energy $E_{a,D}$ is reduced by a factor of x (x<1). Interpret the result in the context of a Thermal Etching process which is limited by desorption of etching products.

P2.5 A spacecraft enters the argon-rich atmosphere of a hypothetical extrasolar planet. The heat shield is made of tungsten. Is it possible that fast neutral impact initiates sputtering? Use Table 2.1 in your calculations.

References

Appels, J., Kooi, E., Paffen, M.M. et al. (1970). Local oxidation of silicon and its application in semiconductor-device technology. *Philips Res. Rep.* 25: 118–132.

Baxter, J.P., Schick, G.A., Singh, J. et al. (1985). Angular distribution of sputtered particles. *J. Vac. Sci. Technol., A* 4: 1218–1221.

Berry, I.L., Park, J.C., Kim, J.K. et al. (2020). Patterning of embedded STT-MRAM devices: challenges and solutions. *ECS and SMEQ Joint International Meeting 2018*, Cancun, Mexico

Berry, I.L., Kim, Y., and Lill, T. (2020). High aspect ratio etch challenges and co-optimized etch and deposition solutions. *Seminar at KAIST*, June 2020.

Cabrera, N. and Mott, N.F. (1948). Theory of the oxidation of metals. *Rep. Prog. Phys.* 12: 163–184.

Chen, J.K.C., Altieri, N.D., Kim, T. et al. (2017a). Directional etch of magnetic and noble metals. I. Role of surface oxidation states. *J. Vac. Sci. Technol., A* 35: 05C304 1–6.

Chen, J.K.C., Altieri, N.D., Kim, T. et al. (2017b). Directional etch of magnetic and noble metals. II. Organic chemical vapor etch. *J. Vac. Sci. Technol., A* 35: 05C305 1–8.

Chen, L., Wen, J., Zhang, P. et al. (2018). Nanomanufacturing of silicon surface with a single atomic layer precision via mechanochemical reactions. *Nat. Commun.* 9: 1542 1–7.

Coburn, J.W. (1994). Surface-science aspects of plasma-assisted etching. *Appl. Phys. A* 59: 451–458.

Coburn, J.W. and Winters, H.F. (1989). Conductance considerations in the reactive ion etching of high aspect ratio features. *Appl. Phys. Lett.* 55: 2730–2732.

Deal, B.E. and Grove, A.S. (1965). General relationship for the thermal oxidation of silicon. *J. Appl. Phys.* 36: 3770–3778.

Eckstein, W. (2007). Sputtering yields. In: *Sputtering by Ion Bombardment*, Topics in Applied Physics, vol. 110 (eds. R. Behrisch and W. Eckstein), 33–187. Berlin, Heidelberg: Springer-Verlag.

Eckstein, W. and Preuss, R. (2003). New fit formulae for the sputtering yield. *J. Nucl. Mater.* 320: 209–213.

Engstrom, J.R., Nelson, M.M., and Engel, T. (1988). Thermal decomposition of a silicon-fluoride adlayer: evidence for spatially inhomogeneous removal of a single monolayer of the silicon substrate. *Phys. Rev. B* 37: 6563–6566.

Eriguchi, K., Matsuda, A., Takao, Y., and Ono, K. (2014). Effects of straggling of incident ions on plasma-induced damage creation in "fin"-type field-effect transistors. *Jpn. J. Appl. Phys.* 52: 03DE02 1–6.

Faraz, T., Roozeboom, F., Knoops, H.C.M., and Kessels, W.M.M. (2015). Atomic layer etching: what can we learn from atomic layer deposition? *ECS J. Solid State Sci. Technol.* 4: N5023–N5032.

Fehlner, F.P. (1984). Low temperature oxidation of metals and semiconductors. *J. Electrochem. Soc. Solid State Sci. Technol.* 131: 1645–1652.

George, S.M. (2010). Atomic layer deposition: an overview. *Chem. Rev.* 110: 111–131.

Gottscho, R.A., Jurgensen, C.W., and Vitkavage, D.J. (1992). Microscopic uniformity in plasma etching. *J. Vac. Sci. Technol., B* 10: 2133–2147.

Gou, F., Neyts, E., Eckert, M. et al. (2010). Molecular dynamics simulations of Cl^+ etching on a Si(100) surface. *J. Appl. Phys.* 107: 113305 1–6.

Graves, D.B. and Humbird, D. (2002). Surface chemistry associated with plasma etching processes. *Appl. Surf. Sci.* 192: 72–87.

Hoerni, J.A. (1962). Method of manufacturing semiconductor devices. US Patent 3,025,589.

Hotston, E. (1975). Threshold energies for sputtering. *Nucl. Fusion* 15: 544–547.

Huang, S., Huard, C., Shim, S. et al. (2019). Plasma etching of high aspect ratio features in SiO_2 using $Ar/C_4F_8/O_2$ mixtures: a computational investigation. *J. Vac. Sci. Technol., A* 37: 031304 1–26.

Ip, V., Huang, S., Carnevale, S.D. et al. (2017). Ion beam patterning of high density STT-RAM devices. *IEEE Trans. Magn.* 53: 2400104 1–5.

Kahng, D. and Atalla, M.M. (1960). Silicon-silicon dioxide field induced surface devices. *IRE-AIEE Solid-State Device Research Conference.*

Kaler, S.S., Lou, Q., Donnelly, V.M., and Economou, D.J. (2017). Atomic layer etching of silicon dioxide using alternating C_4F_8 and energetic Ar^+ plasma beams. *J. Phys. D: Appl. Phys.* 50: 234001 1–11.

Kanarik, K.J., Lill, T., Hudson, E.A. et al. (2015). Overview of atomic layer etching in the semiconductor industry. *J. Vac. Sci. Technol., A* 33: 020802 1–14.

Kanarik, K.J., Tan, S., Yang, W. et al. (2017). Predicting synergy in atomic layer etching. *J. Vac. Sci. Technol., A* 35: 05C302 1–7.

Kiehlbauch, M.W. and Graves, D.B. (2003). Effect of neutral transport on the etch product lifecycle during plasma etching of silicon in chlorine gas. *J. Vac. Sci. Technol., A* 21: 116–126.

Lee, Y. and George, S.M. (2017). Thermal atomic layer etching of titanium nitride using sequential, self-limiting reactions: oxidation to TiO_2 and fluorination to volatile TiF_4. *Chem. Mater.* 29: 8202–8210.

Liau, Z.L. and Mayer, J.W. (1978). Limits of composition achievable by ion implantation. *J. Vac. Sci. Technol.* 15: 1629–1635.

Lill, T. and Joubert, O. (2008). The cutting edge of plasma etching. *Science* 319: 1050–1051.

Lill, T., Callaway, W.F., Pellin, M.J., and Gruen, D.M. (1994). Abundance and depth of origin of neutral and ionic clusters sputtered from a liquid gallium-indium eutectic alloy. *Phys. Rev. Lett.* 73: 1719–1722.

Lill, T., Grimbergen, M., and Mui, D. (2001). In situ measurement of aspect ratio dependent etch rates of polysilicon in an inductively coupled fluorine plasma. *J. Vac. Sci. Technol., B* 19: 2123–2128.

Lindhard, J., Scharff, M., and Schiott, H.E. (1963). Range concepts and heavy ion ranges (Notes on atomic collisions). *Mat. Fys. Medd. Dan. Vid. Selsk.* 33: 1–42.

Mantenieks, M.A. (1999). Sputtering Threshold Energies of Heavy Ions. *NASA report NASA/TM-1999-209273.*

Nakamura, H., Saito, S., and Ito, A.M. (2016). Sputtering yield of noble gas irradiation onto tungsten surface. *J. Adv. Simul. Sci. Eng.* 3: 165–172.

Nishi, M., Yamada, M., Suckewer, S., and Rosengaus, E. (1979). *Measurements of Sputtering Yields for Low-Energy Plasma Ions*. Princeton University Plasma Lab, Publication, PPPL-1521.

Oehrlein, G.S., Metzler, D., and Li, C. (2015). Atomic layer etching at the tipping point: an overview. *ESC J. Solid State Sci. Technol.* 4: N5041–N5053.

O'Hanlon, J.F. (2003). *A User's Guide to Vacuum Technology*, 3e. Wiley.

Oostra, D.J., van Ingen, R.P., Haring, A. et al. (1987). Near threshold sputtering of Si and SiO_2 in a Cl_2 environment. *Appl. Phys. Lett.* 50: 1506.

Park, S.D., Lee, D.H., and Yeom, G.Y. (2005). Atomic layer etching of Si(100) and Si(111) using Cl_2 and Ar neutral beam. *Electrochem. Solid-State Lett.* 8: C106–C109.

Robin, A. (2003). Trajectory and channeling effects in the scattering of ions off a metal surface: probing the electronic density corrugation at a surface by grazing axial ion channeling. PhD thesis. University Osnabrueck.

Samukawa, S. (2006). Ultimate top-down etching processes for future nanoscale devices: advanced neutral-beam etching. *Jpn. J. Appl. Phys.* 45: 2395–2407.

Seah, M.P. and Nunney, T.S. (2010). Sputtering yields of compounds using argon ions. *J. Phys. D: Appl. Phys.* 43: 253001 1–13.

Sigmund, P. (1969). Theory of sputtering. I. Sputtering yield of amorphous and polycrystalline targets. *Phys. Rev.* 184: 384–416.

Sigmund, P. (2012). Recollection of fifty years with sputtering. *Thin Solid Films* 520: 6031–6049.

Somorjai, G.A. and Brumbach, S.B. (1973). The interaction of molecular beams with solid surfaces. *Proceedings of the International Summer Institute in Surface Science*, Milwaukee, WI, USA.

Steinbrüchel, C. (1989). Universal energy dependence of physical and ion-enhanced chemical etch yields at low ion energy. *Appl. Phys. Lett.* 55: 1960–1962.

Stensgaard, I., Feldman, L.C., and Silverman, P.J. (1978). Calculation of the backscattering-channeling surface peak. *Surf. Sci.* 77: 513–522.

Teichmann, M., Lorbeer, J., Frost, F., and Rauschenbach, B. (2014). Ripple coarsening on ion beam-eroded surfaces. *Nanoscale Res. Lett.* 9 (439): 1–8.

Todorov, S.S. and Fossum, E.R. (1988). Sputtering of silicon dioxide near threshold. *Appl. Phys. Lett.* 52: 365–367.

Vyvoda, M.A., Li, M., Graves, D.B. et al. (2000). Role of sidewall scattering in feature profile evolution during and HBr plasma etching of silicon. *J. Vac. Sci. Technol., B* 18: 820–833.

Wehner, G.K. (1955). Sputtering of metal single crystals by ion bombardment. *J. Appl. Phys.* 26: 1056–1057.

Winter, H. (2002). Collisions of atoms and ions with surfaces under grazing incidence. *Phys. Rep.* 367: 387–582.

Yamamura, Y. and Bohdansky, J. (1985). Few collisions approach for threshold sputtering. *Vacuum* 35: 561–571.

Yamamura, Y. and Tawara, H. (1996). Energy dependence of ion-induced sputtering yields from monatomic solids at normal incidence. *At. Data Nucl. Data Tables* 62: 149–345.

Ziegler, J.F., Biersack, J.P., and Littmark, U. (1985). *The Stopping and Range of Ions in Solids*. New York: Pergamon.

3

Thermal Etching

3.1 Mechanism and Performance Metrics of Thermal Etching

This technology is sometimes also referred to as "vapor etching" because some reactants used in the process are liquid under atmosphere pressure and are "vaporized" in the vacuum chamber. The mechanism of thermal etching is schematically illustrated in Figure 2.5 with its elementary steps of physisorption, chemisorption, surface reaction, and desorption.

For dissociative chemisorption of a molecule to occur, it must first physisorb onto the surface and then overcome the activation barrier. This is called a precursor mediated adsorption model. The activation energy for physisorption is small and the potential well ε is shallow (see Figure 2.3). At higher temperatures, physisorbed molecules desorb as fast as they adsorb and hence may not take the next step to chemisorb. At the same time, the presence of and activation barrier to the chemisorbed state, $E_{a,A}$, means that the process is activated by temperature. Therefore, the chemisorption rate for a surface drops at higher temperatures for many systems. It may have a maximum at an optimum temperature or may have a more complicated shape. However, the important behavior is the drop in chemisorption rate with temperature according to the precursor mediated adsorption mode. This is shown schematically in Figure 3.1.

Figure 3.1 also shows schematically the curve for desorption as an exponentially raising curve according to Eq. (2.3). The thermal etching process works in the temperature range where the chemisorption curve of the reactants and desorption curves of the reaction products overlap. This region represents the process window for a thermal etching process.

3.1.1 Etching Rate and ERNU

The rate for thermal etching is limited by the elementary step with the lowest rate. If the process is transport or adsorption limited, the pressure can be increased to increase the etching rate. Flow uniformity across the wafer is extremely critical to achieve good etching rate nonuniformity (ERNU). If the process is reaction or desorption limited, the temperature can be raised to increase the etching rate. Wafer

Atomic Layer Processing: Semiconductor Dry Etching Technology, First Edition. Thorsten Lill.

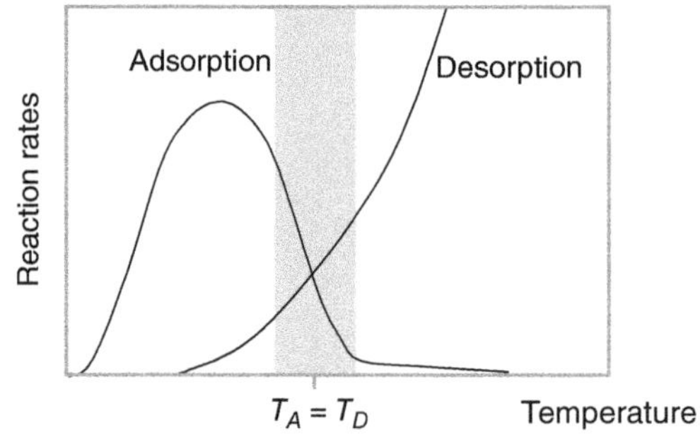

Figure 3.1 Schematic illustration of the temperature process window for thermal etching.

temperature uniformity is then the main parameter to improve ERNU in this case. Some thermal etching processes, for instance etching of SiO_2 with HF, have incubation times before the etching starts. The time needed to start the etching can depend on the surface state and the history of the wafer. This can negatively impact the repeatability of the etching rate or amount.

3.1.2 Selectivity

The intrinsic selectivity of thermal etching processes is typically very high. Together with thermal atomic layer etching (ALE), thermal etching is the class of etching processes with the highest intrinsic selectivity because the "excess energy" is close to zero. The downside of such high selectivity is that thermal etching can be prone to surface defects. For instance, surface contaminations or small amounts of residues can block thermal etching processes and form large defects of un-etched material. Thermal etching of silicon with various fluorine-containing gases has been reported to result in a rough surface due to local etch blocking (Ibbotson et al. 1984).

3.1.3 Profile and CD Control

Thermal etching is an isotropic etching process because reactants and reaction products move with thermal energies and in random direction. When a thermal etching process is applied to a surface with a masking structure, nearly spherical or cylindrical shapes are formed for hole and trench masks, respectively (see Figure 2.2). Isotropic etching processes are useful to remove materials either from a blanket surface or from structures that are confined by materials for which the given thermal etching process has high selectivity.

The critical dimension (CD) for thermal etching process is typically the depth of recess or the thickness of remaining material for partially etched films. As mentioned above, thermal etching processes are not well suited for removal of precise thicknesses if they have surface condition-sensitive incubation times. For complete removal processes where etch stop layers exist in all directions, a critical dimension can be, for instance, the thickness loss of a stopping layer if the selectivity is less than infinite.

3.1.4 ARDE

Aspect ratio dependent etching (ARDE) can be effectively reduced by increasing pressure. The goal is to make the surface reaction and desorption steps the rate-limiting steps.

For thermal etching processes with more than one gas, for instance, etching of SiO_2 with HF and H_2O, the partial pressure of both gases must be high enough to avoid a transport-limited process regime. Because ARDE is small at high enough pressures, even very small holes in a stopping layer can lead to devastating isotropic etching beyond that layer.

3.2 Applications Examples

Archetypical applications of thermal etching are etching of silicon with fluorine and etching of SiO_2 with HF. Both applications are used in the manufacturing of micro-electromechanical systems (MEMS) devices where high etch rates are needed for undercutting etching processes.

Let us first review etching of silicon by fluorine. Using the framework in Figure 2.1, a fluorinated silicon surface is characterized by values of $E_A = 5.6$ eV, $E_O = 5.7$ eV (Dean 1999), and $E_S = 0.9$ eV (Engstrom et al. 1988). Among the three energies, E_s is the smallest and hence desorption will lead to etching according to our simplified etching framework.

Engstrom's data were extracted from temperature programmed desorption (TPD) experiments under ultrahigh vacuum conditions and the value of E_s is, strictly speaking, an activation energy (see Figure 2.4). Engstrom's TPD experiments show that a temperature of about 800 K is required to desorb SiF_2 and SiF_4 molecules for F_2 doses of up to 7 and 3 L, correspondingly. For doses above 3 L, SiF_4 is observed at room temperatures. These results can be explained by dissociation and diffusion of fluorine atoms into the silicon subsurface layers and the formation of a fluorosilyl layer, which is the source of SiF_4. The last monolayer desorbs at 800 K as SiF_2 and SiF_4.

Some important implications of this behavior are that silicon, which is etched with fluorine gas or fluorine radicals, is always fluorine terminated, that this fluorine can only be directly desorbed at temperatures higher than usually used in semiconductor manufacturing, and that the removal of the fluorine will lead to loss of more silicon. Post etch treatment steps are needed to lower the fluorine removal temperature.

For safety reasons, XeF_2 is frequently used in thermal etching of silicon instead of F_2. Both XeF_2 and F_2 molecules undergo dissociative chemisorption to produce fluorine atoms that react with the fluorinated surface to form SiF_4. The reaction paths and the corresponding activation energies must be very different for the two reactants because an etching difference of 4 orders of magnitude has been reported (Winters and Coburn 1979). The etching rate of silicon with XeF_2 is several micrometers per minute, which is very fast compared to reactive ion etching (RIE). Much more information about this process can be found in a review article (Donnelly

Table 3.1 Silicon etching rates and reaction probabilities of interhalogens. n-Type (100) silicon wafers were used in the experiments.

Reactant	Pressure (Torr)	Etching rate (Å/min)	Reaction probability
ClF_3	4.7	5500	4.5×10^{-5}
BrF_3	1.0	50 000	2.4×10^{-3}
BrF_5	8.1	11 800	7.8×10^{-5}
IF_5	4.6	9900	1.3×10^{-4}
ClF	5.0	<10	$<6 \times 10^{-8}$
XeF_2	0.2	45 300	1.2×10^{-2}
F_2	10	3	9×10^{-9}
F	0.2	4600	4.1×10^{-4}

Source: Ibbotson et al. (1984).

2017). Besides F_2 and XeF_2, other fluorine-containing interhalogen compounds can be used to etch silicon. A list is provided in Table 3.1 (Ibbotson et al. 1984).

Thermal etching of silicon with XeF_2 is sensitive to the doping level of silicon. Heavily n-doped silicon etches faster than p-doped, lightly n-doped, and undoped silicon. This has been explained by applying the Cabrera–Mott (CM) diffusion model to the fluorosilyl layer, which is important for this etching process (Winters and Haarer 1987; Winters et al. 2007). Etching occurs at the surface of the fluorosilyl layer and silicon must diffuse through this layer. Doping of silicon with n-type dopants increases the value of the Mott potential $\Delta\Phi_{Mott}$, which is the driving force for diffusion (see Section 2.7). This effect is even stronger for etching of silicon with chlorine. Even under ion bombardment, the chlorine etches n-doped silicon faster than undoped silicon. For fluorine, the effect is greatly diminished when ions are involved. We will discuss simultaneous Reactive Ion Etching (RIE) of dual-doped silicon gates in Section 7.3.2.

SiO_2 does not spontaneously etch in F_2 or XeF_2. This makes it a great stopping or masking material for silicon etching with these gases. To etch SiO_2 thermally, HF gas can be used. Early studies reported that both HF and H_2O were needed to achieve meaningful etch rates even though H_2O is the reaction product (Holmes and Snell 1966). The reaction can be described using the following chemical equation:

$$SiO_2 + 4HF \rightarrow SiF_4 + H_2O \tag{3.1}$$

One of the mechanisms that was evoked to explain the need for H_2O is that it forms a multilayer physisorbed layer on the surface, which acts as a solvent medium for HF (Helms and Deal 1992). This layer is formed from water vapor in the gas phase and initiates the reaction. As the etching reaction proceeds, it is replenished by H_2O, which is a reaction product. Under these conditions, etching rates of 1000–2000 Å/min have been measured (Holmes and Snell 1966).

The challenge of this process is precipitation of fluorosilicic acid, H_2SiF_6, on the surface, which slows down the etching rate over time and needs to be rinsed after the process is finished. When the temperature is increased to values where multilayer H_2O adsorption is suppressed, etching is also strongly suppressed. Overall, this process is difficult to control because the formation of the initial multilayer H_2O layer is not repeatable and because reaction products interfere with the reaction.

It is possible to etch SiO_2 with anhydrous HF if chemisorbed H_2O is present at the surface at the start of the process or if the surface is hydrogen terminated. Once the reaction started, the reaction is sustained by H_2O formed in the reaction. Alcohols such as CH_3OH can be used to hydrate the SiO_2 surface at elevated temperatures where H_2O does not accumulate. Etching rates of 200 Å/min at 25 °C and 10 Å/min at 95 °C have been reported for a pressure of 100 Torr for thermal oxide (Ruzyllo et al. 1993). The etching rate repeatability was greatly improved at higher temperatures and lower pressures in the absence of a physisorbed layer at the surface. Under the same conditions, heavily boron- and phosphorus-doped SiO_2, the so-called boron phosphorus silicon glass (BPSG), etched with a rate of 6000 Å/min. This means that etching selectivities of several hundred to one for different types of SiO_2 can be achieved. Thermal etching processes can be extremely selective.

Another important dry etching approach for SiO_2 involves the formation of ammonium hexafluorosilicate, $(NH_4)_2SiF_6$. This compound sublimates in vacuum at temperatures above 100 °C. It can be formed when a mixture of NH_3 and NF_3 is heated in an applicator tube to temperatures above 600 °C (Ogawa et al. 2002). At these temperatures, NF_3 decomposes and forms free fluorine, which combines with NH_3 to form NH_4F in the gas phase. This gas etches SiO_2. There is no plasma involved in this process; it is a thermal process where the reactive species are formed by heating of the feed gas mixture. The process can be operated in continuous mode above 100 °C wafer temperature. It can also be operated sequentially, where $(NH_4)_2SiF_6$ salt is formed during the reaction of NH_4F with the silicon surface at a temperature below 100 °C and removed in a second step where the wafer is rapidly heated above the sublimation temperature. In this mode of operation, this process can be considered a thermal ALE process, albeit an imperfect one, because the growth rate of the salt follows a parabolic law (Ogawa et al. 2002) and therefore is not perfectly self-limited. The underlaying surface reaction can also be used in radical etching processes as will be shown in Chapter 5.

The examples for thermal etching of silicon and $SiO_{2,}$ which have been discussed so far, leverage the properties of halogens for etching. Halogens such as fluorine, chlorine, and bromine are missing one electron to complete their outer electron shell. Their electron affinity, which is the change in energy when an electron is added to a neutral atom to form a negative ion, makes them the most electronegative among all elements. This means they release more energy than any other element when accepting a free electron. If this electron is captured from another element in a chemical reaction, the result is a very strong bond. When the element that donates the electron is a surface atom, the overall electron

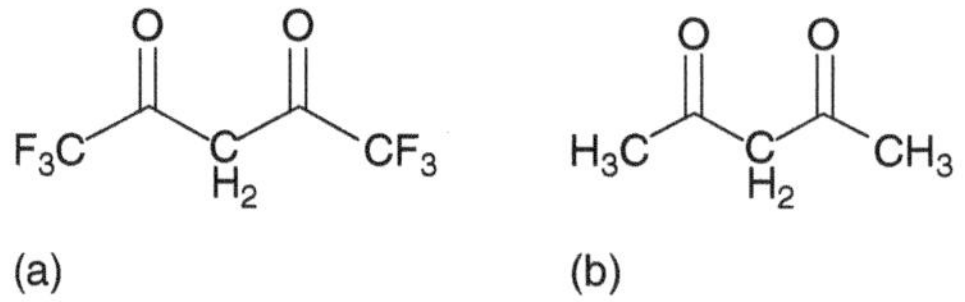

Figure 3.2 Structural formulas of the keto tautomers of Hhfac (a) and Hacac (b).

density around the surface atom is shifted toward the halogen adsorbate atom and away from the bonds that tie the surface atom to the bulk. This weakens the bonds between the surface atom and the bulk and enables etching as illustrated in Figure 2.1. This is the reason why the majority of all etching processes for metals and semiconductors use halogen gases. This includes thermal, radical, and RIE. Oxygen needs two electrons to complete the outer shell and is also a useful etching gas, especially for etching of carbon and carbon-based polymers such as photoresists. Because of the strong electron affinity of halogens and oxygen, their surface reactions are frequently themodynamically favorable and lead to volatile (which is to say "can be desorbed with thermal energies") reaction products.

When the surface halogens and oxides do not desorb at thermal energies, secondary reactions must be deployed to complete the etching process. These reactions have much more complicated chemical pathways. An archetypical example is the etching of Fe_2O_3 with hexafluoroacetylacetone (Hhfac) via a chelation reaction at temperatures between 140 and 400 °C (George et al. 1996):

$$6Hhfac_{(G)} + Fe_2O_{3(S)} \rightarrow 2Fe(hfac)_{3(G)} + 3H_2O_{(G)} \tag{3.2}$$

Chelation is the formation of chemical bonds between a single central metal ion and organic molecules, each of which forms multiple bonds with the central atom to become ligands. In this specific reaction, the ligands are called chelants. Figure 3.2 shows the structural formulas for the keto tautomer of Hhfac. It also shows the related molecule acetylactone (Hacac), which can be used to thermally etch ZnO (Mameli et al. 2018). Hexafluoroacetylacetone and acetylacetone exist also as enol tautomers.

According to the reaction of Eq. (3.2), three hfac ligands must react with the oxidized iron atom at the surface. The kinetics of the reaction must be quite complicated. While the detailed reaction path is not known, it is likely that three hfac ligands bond to the Fe atom in a chelate configuration. This means that two oxygen atoms of the same Hhfac bind to the same iron atom in a pincer-like approach. The resulting molecule is large and complex, and it derives its ability to remove the iron atom from the surface from the fact that three Hhfac molecules "wrap" themselves around the iron atom and shield it from the chemical interaction with the surface. This requires complex rearrangements of the molecule. It can be speculated that the surface reaction involving three incoming molecules and a complex rearrangement proceed via a Langmuir–Hinschelwood mechanism involving surface diffusion of Hhfac.

If any atoms are lost in the rearrangement process, the molecule can partially dissociate, and the reaction products can block the surface. In fact, fluorides and carbides were detected on the Fe_2O_3 surface after etching with Hhfac, which is a sign that some of the Hhfac molecules lost fluorine from the CF_3 functional groups (George et al. 1996). Dissociation of reactants or reaction products may require the addition of gases that keep the surface clean or separate cleaning steps. Oxygen plasma has been shown to clean the surface of ZnO after etching with Hacac (Mameli et al. 2018).

Hhfac does not react with metallic iron (George et al. 1996). Equation (3.2) requires iron to be in the fully oxidized 3^+ state. Chelating reactants such as Hhfac and Hacac are not reactive enough to bond to most metal surfaces. Metal bonds are characterized by electrons that move freely between atoms. This gives metals the property to conduct electricity. In the context of etching, however, this is a decremental property because there are no individual bonds for the incoming reactant to attack. Only strongly electronegative elements such as halogens and oxygen have the ability to capture the freely moving electrons from the metal surface and to localize them in covalent bonds.

Once the surface is oxidized, chelating reactants can bond to the surface and complete the etching process as described above. This has two important implications. Firstly, etching with chelating reactants is selective to the oxidation state of metals. It is possible to etch metal oxides selectively to metals. This is very useful for pre-cleaning of metals before deposition of another metal. Secondly, metals can be etched by combining surface oxidation with oxygen or halogens with the removal by chelating reactants. If both reactions have reasonable rates at the same temperatures, an etching process can be designed where the oxidizing gas and the chelating gas are flown simultaneously. If not, both steps can be separated in space or time. If both steps are self-limited, the result is thermal ALE.

Problems

P3.1 Explain why the process window of thermal etching is typically small. Why is this a fortunate situation for our daily life?

P3.2 Assume a given thermal etching process is desorption limited. The highest etching rate on the wafer is 10% higher than the lowest due to gas flow nonuniformities. Use Eq. (2.3) to calculate the temperature delta across the wafer, which is needed to compensate the flow uniformity. Assume that the base temperature is 250 °C and the activation energy 1.2 eV.

P3.3 Why is the repeatability for thermal etching of SiO_2 with HF improved when alcohols are added and the temperature is increased?

P3.4 Why are halogens used in thermal etching?

P3.5 Why do chelating reactants etch metal oxides but not metals?

P3.6 What does ΔG^0 say about the etching rate of a thermal etching process?

References

Dean, J.A. (1999). *Lange's Handbook of Chemistry*, 15e. McGraw-Hill, Inc.

Donnelly, V.M. (2017). Review Article: Reactions of fluorine atoms with silicon, revisited, again. *J. Vac. Sci. Technol., A* 35: 05C202 1–9.

Engstrom, J.R., Nelson, M.M., and Engel, T. (1988). Thermal decomposition of a silicon-fluoride adlayer: evidence for spatially inhomogeneous removal of a single monolayer of the silicon substrate. *Phys. Rev. B* 37: 6563–6566.

George, M.A., Hess, D.W., Beck, S.E. et al. (1996). Reaction of 1,1,1,5,5,5-hexafluoro-2,4-pentadione (H^+fac) with iron and iron oxide thin films. *J. Electrochem. Soc.* 143: 3257–3266.

Helms, C.R. and Deal, B.E. (1992). Mechanisms of the HF/H_2O vapor phase etching of SiO_2. *J. Vac. Sci. Technol., A* 10: 806–811.

Holmes, P.J. and Snell, J.E. (1966). A vapor etching technique for the photolithography of silicon dioxide. *Microelectron. Reliab.* 5: 337–341.

Ibbotson, D.E., Mucha, J.A., Flamm, D.L., and Cook, J.M. (1984). Plasmaless dry etching of silicon with fluorine-containing compounds. *J. Appl. Phys.* 56: 2939–2942.

Mameli, A., Verheijen, M.A., Mackus, A.J.M. et al. (2018). Isotropic atomic layer etching of ZnO using acetylacetone and O_2 plasma. *ACS Appl. Mater. Interfaces* 10: 38588–38595.

Ogawa, H., Arai, T., Yanagisawa, M. et al. (2002). Dry cleaning technology for removal of silicon native oxide employing hot NH_3/NF_3 exposure. *Jpn. J. Appl. Phys.* 41: 5349–5358.

Ruzyllo, J., Torek, K., Daffron, C. et al. (1993). Etching of thermal oxides in low pressure anhydrous HF/CH_3OH gas mixture at elevated temperature. *J. Electrochem. Soc.* 140: L64–L66.

Winters, H.F. and Coburn, J.W. (1979). The etching of silicon with XeF_2 vapor. *Appl. Phys. Lett.* 34: 70–73.

Winters, H.F. and Haarer, D. (1987). Influence of doping on the etching of Si(111). *Phys. Rev. B* 36: 6613–6623.

Winters, H.F., Graves, D.B., Humbird, D., and Tougaard, S. (2007). Penetration of fluorine into the silicon lattice during exposure to F atoms, F_2, and XeF_2: implications for spontaneous etching reactions. *J. Vac. Sci. Technol., A* 25: 96–103.

4

Thermal Isotropic ALE

4.1 Mechanism of Thermal Isotropic ALE

Thermal isotropic atomic layer etching (ALE) uses nonionized gases and thermal energy for etching (Carver et al. 2015). This class of ALE reactions requires spontaneous, sequential, self-limiting thermal reactions to remove with atomic fidelity. Spontaneous means that all reactions are thermodynamically favorable ($\Delta G < 0$). As is the case for all ALE technologies, surface modification and removal are separated into self-limiting steps. These steps can be applied to the wafer in the same reactor chamber in sequence (temporal separation) or in difference chambers (spatial separation). The choice between temporal and spatial implementation of ALE is driven by engineering complexity and the associated cost. In the case of spatial separation, moving the wafers between the modification and removal chambers requires a sophisticated mechanical system to overcome throughput and particle challenges (Roozeboom et al. 2015). The advantage is that both chambers can be optimized for each step.

In the case of temporal separation, process parameters must be changed within seconds or faster. Complete switching between gases is a function of the residence time of the gas. Reactors with small volumes are preferred. Smaller reactors bring the walls closer to the wafer, which increases the sensitivity to wall effects. Finally, ramping wafer temperatures up and down within seconds and faster is an engineering challenge.

The underlying mechanism of thermal isotropic ALE is however independent from the engineering implementation of step separation. A schematic illustration of the process window for thermal isotropic ALE with direct desorption is depicted in Figure 4.1. Similarities and differences compared to the mechanism of thermal etching in Figure 3.1 are obvious. The process window of both etching technologies is determined by the temperature dependence of the adsorption of the reactant gas and the desorption of the reaction products. In contrast to thermal etching, the temperature required to desorb the reaction products does not overlap with the adsorption window. Adsorption and desorption proceed at different temperatures, which requires rapid cycling of the wafer temperature. The reason for the potential existence of such a temperature difference is that the chemical bonds that are formed in the adsorption step are different from those that are broken in the desorption step.

Atomic Layer Processing: Semiconductor Dry Etching Technology, First Edition. Thorsten Lill.

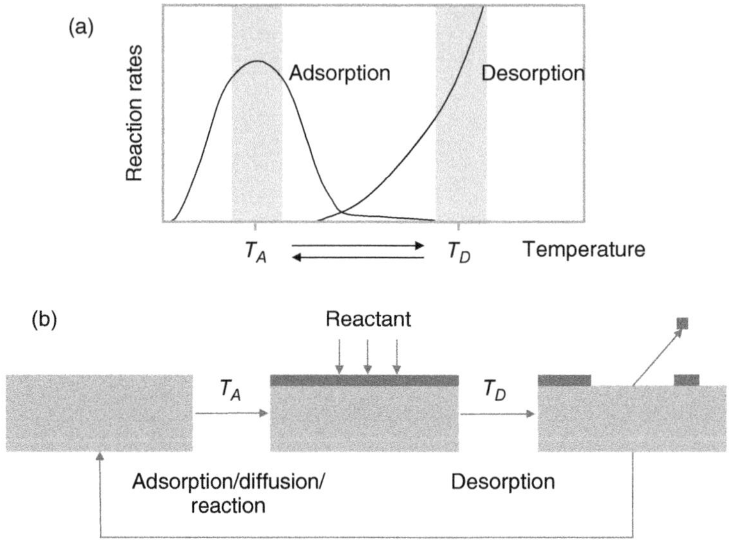

Figure 4.1 Schematic illustration of the temperature process window for thermal isotropic ALE with direct desorption via temperature cycling (a). Processing steps for thermal isotropic ALE (b).

The former are bonds between the incoming molecule and the surface atom while the latter are bonds between the surface atom and the bulk material (see Section 2.3 and Figure 2.5).

Surface modification in thermal isotropic ALE is frequently accomplished by an oxidation reaction, either using oxygen or halogens. When the surface modification step time is increased to reach saturation, adsorbed atoms can diffuse into the bulk material. Diffusion processes slow down as a function of time, but they do not reach complete saturation. In this regard, the Cabrera–Mott (CM) oxidation mechanism, which follows an inverse logarithmic time dependence, provides better saturation or quasi self-limitation than the Deal–Grove (DG) mechanism with its initial linear time dependence followed by a parabolic dependence (see Section 2.7). CM oxidation is observed for metals, while DG oxidation is seen, for instance, in silicon. When quasi self-limited diffusion occurs in the surface modification step, several Angstroms of bulk material can be modified. During the subsequent removal step, this material will be removed via desorption.

Thermal ALE using temperature cycling has been demonstrated using O_2 to oxidize germanium at room temperature followed by desorption with a rapid thermal pulse photon source (Paeng et al. 2019). Shinoda et al. demonstrated thermal isotropic ALE of TiN where surface modification was accomplished using a downstream CHF_3/O_2 plasma at room temperature followed by infrared irradiation for thermal desorption of the modified surface (Shinoda et al. 2019). Miyoshi et al. have applied a similar technique to thermal ALE of Si_3N_4 (Miyoshi et al. 2017).

Temperature cycling of 300 mm wafers, especially if the temperature range is hundreds of degrees, is a complicated engineering challenge. This complication

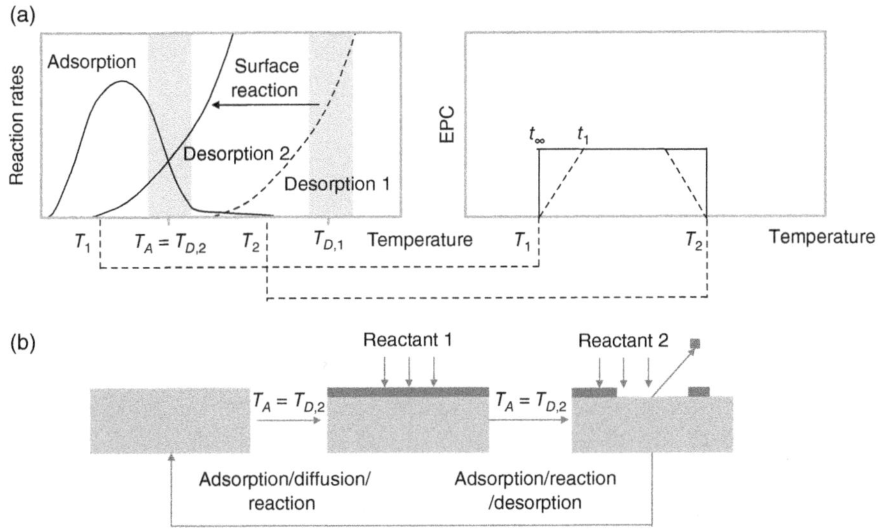

Figure 4.2 Schematic illustration of the temperature process window for thermal isotropic ALE with surface reaction.

can be circumvented by introducing a second reaction, which is designed to reduce the desorption temperature and to create sufficient overlap of the adsorption and desorption temperatures. This embodiment of a removal process is labeled "reaction-assisted desorption" in Table 2.4.

A schematic illustration of the mechanism is presented in Figure 4.2. The graph to the right depicts the so-called ALE window. This window is analogous to the ideal atomic layer deposition (ALD) window (George 2010). The lower limit of the ALE window, T_1, is given by the lowest temperature with appreciable desorption after the second surface reaction. The upper limit, T_2, is given by the disappearance of an appreciable adsorption rate. The amount of material removed in one cycle is called "etch per cycle" (EPC). Theoretically, the EPC jumps from 0 to a finite number at the extreme ends of the window for infinite step times, t_∞. For finite step times, t_1, the boundaries of the process window are sloped.

This representation of the process window for isotropic thermal ALE with reaction-assisted desorption implies that the process is not limited by the surface reaction or dissociation of the reaction products at higher temperatures. In other words, the ALE window for thermal isotropic ALE can also be limited by the temperature dependence of the surface reaction, in which case the upper boundary moves to higher temperatures, or by dissociation reactions, in which case the upper boundary would move to lower temperatures. The energy scan, in the case of thermal isotropic ALE, the measurement of EPC as a function of temperature, is one of three important complementary tests to characterize ALE processes. The two other tests are the synergy test and the measurement of saturation curves (Kanarik et al. 2015).

At least four classes of thermal isotropic ALE with reaction-assisted desorption are known to date: chelation ALE, ligand exchange ALE, conversion ALE, and oxidation/fluorination ALE.

4.1.1 Chelation/Condensation ALE

To illustrate the concept of thermal isotropic ALE with reaction-assisted desorption, let us re-imagine the thermal etching process of Fe_2O_3 with hexafluoroacetylacetone (Hhfac), which was described in Section 3.2, as a cyclic thermal ALE process for metallic iron. The surface of metallic Fe can be oxidized in an oxygen gas or ozone environment. Oxidation is thermodynamically favorable for most metals. The depth of oxidation is quasi limited by the CM oxidation mechanism. The sublimation temperature of Fe_2O_3 is not existent; it decomposes at over 1500 °C. However, a second reaction with Hhfac forms $Fe(hfac)_3$, which desorbs at around 300 °C (George et al. 1996).

Desorption is assisted by a second surface reaction, in this case a chelation reaction. This is why we call this type of ALE thermal isotropic ALE with reaction-assisted desorption. Once all Fe_2O_3 is removed and metallic Fe is exposed, the reaction stops because Hhfac does not etch metallic Fe. Therefore, the removal step is self-limited.

The synergy for any ALE process can be expressed as (Kanarik et al. 2017)

$$\text{ALE synergy \% (S)} = \frac{\text{EPC} - (\alpha + \beta)}{\text{EPC}} \times 100\% \tag{4.1}$$

where the values $\propto$ and β are contributions from the individual modification step A and removal step B. Because Fe_2O_3 cannot desorb, and Hhfac does not react with metallic iron, etching happens only when both steps are applied to the metallic iron surface in sequence. The synergy for this process should therefore be close to 100%.

This class of ALE process has been demonstrated for etching of copper (Mohimi et al. 2018). When the copper surface is oxidized with molecular oxygen at 275 °C, a 3 Å thick Cu_2O layer is formed. The Hhfac gas removes 0.9 Å per cycle. The use of O_3 increases the oxide thickness to 150 Å and EPC to 84 Å. The so-called saturation curves for copper oxidation at 7 mTorr O_2 and 275 °C and for removal of the copper oxide with Hhfac are depicted in Figure 4.3a,b, respectively.

Other metals that have been etched with chelation ALE include cobalt (Zhao et al. 2018; Konh et al. 2019) and iron (Lin et al. 2018). Basher et al. conducted density functional theory (DFT) first principles simulations on metallic nickel and NiO to clarify why enol Hhfac can be adsorbed stably on a metal oxide but not on metal surface. The simulation results show that when Hhfac approaches the NiO surface, its negatively charged oxygen atoms are attracted to the nickel atoms of the NiO surface, where all Ni atoms are positively charged. On a metallic nickel surface, every nickel atom is charge neutral. There are no particular preferential sites to which Hhfac can stably attach. Furthermore, accidental tilting of an Hhfac molecule allows its carbon and fluorine atoms to be bonded with surface Ni atoms, resulting in the dissociation of Hhfac on a metallic surface (Basher et al. 2020). These results explain why Hhfac and Hacac can etch metal oxides but not metals. They also predict the

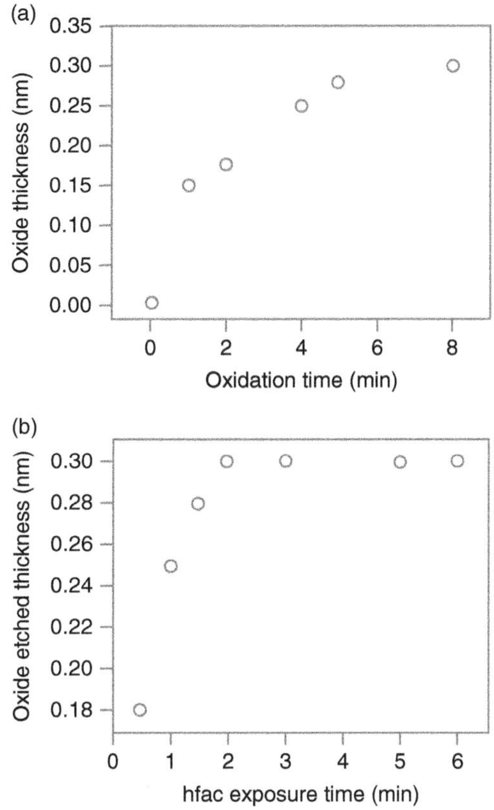

Figure 4.3 Saturation curves for thermal isotropic ALE of Cu via O_2/Hhfac ligand exchange reaction. Source: Mohimi et al. (2018). © 2018, The Electrochemical Society.

effect of precursor dissociation for certain chemistries that could contaminate the surface and cause etch stop or lack of isotropy if not removed.

4.1.2 Ligand Exchange ALE

The archetypical example for ligand exchange ALE is etching of Al_2O_3 with alternating HF (for instance, in the form of HF-pyridine) and $Sn(acac)_2$ gases at around 200 °C (Lee and George 2015). The experimental setup in Lee's work was a hot wall, viscous flow reactor (Elam et al. 2002). The metal oxide surface is converted into metal fluoride by HF in the modification step. This step proceeds via the elementary steps of HF adsorption, reaction with Al_2O_3 to form AlF_3, and desorption of the reaction product H_2O. This surface reaction is exothermic with $\Delta G = -58$ kcal at 200 °C (Lee and George 2015). AlF_3 is formed at the surface as measured by in situ Fourier transform infrared spectroscopy (FTIR) (Lee et al. 2015b). Natarajan and Elliott studied fluorination Al_2O_3 with DFT methods, which predict the existence of chemisorbed HF molecules and fluorine atoms (Natarajan and Elliott 2018).

To achieve fluorination of several monolayers, fluorine must diffuse into the crystal lattice and substitute oxygen. The released oxygen must diffuse to the surface and react with HF to form H_2O and free fluorine radicals, which in turn can diffuse to the AlF_3/Al_2O_3 interface. The resulting depth of fluorination for the reaction of HF

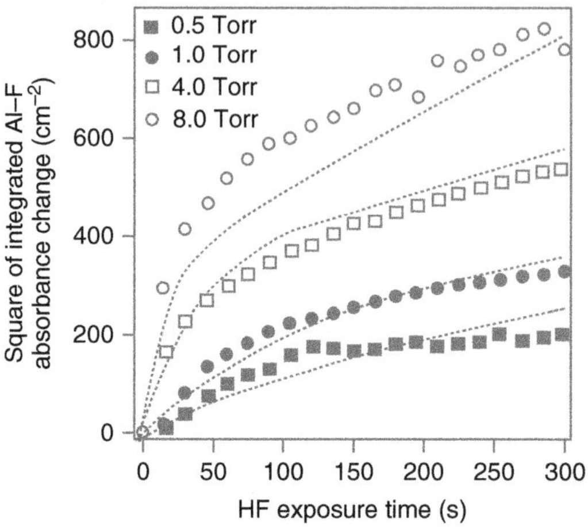

Figure 4.4 Depth of fluorination as a function of HF pressure measured with XPS and fitted with a model based on DG diffusion in the parabolic regime. Source: Based on Cano et al. (2019).

with Al_2O_3 is pressure and time dependent (Cano et al. 2019). Cano et al. developed a model for the depth of fluorination as a function of time and HF pressure. The equation is based on a DG diffusion mechanism in the parabolic regime (see Figure 2.10). The experimental results and best fits are shown in Figure 4.4.

Besides HF, the use of SF_4 as a fluorination reactant has been demonstrated for isotropic thermal ALE of Al_2O_3 and VO_2 via ligand exchange reaction using $Sn(acac)_2$ as the removal reactant (Gertsch et al. 2019). While SF_4 and HF showed similar EPC for Al_2O_3 (0.20 Å for SF_4 and 0.28 Å for HF), SF_4 was more effective for ALE of VO_2 (0.30 Å for SF_4 and 0.11 Å for HF). The increased effectiveness of SF_4 is the result of more favorable thermodynamics compared to HF. F_2 and XeF_2 are predicted to be even stronger fluorination reactants than SF_4 as shown in Figure 4.5. Fluoroform (CHF_3) was found to be not effective as alternative to HF for Al_2O_3 ALE with trimethyl aluminum (TMA) (Rahman et al. 2018).

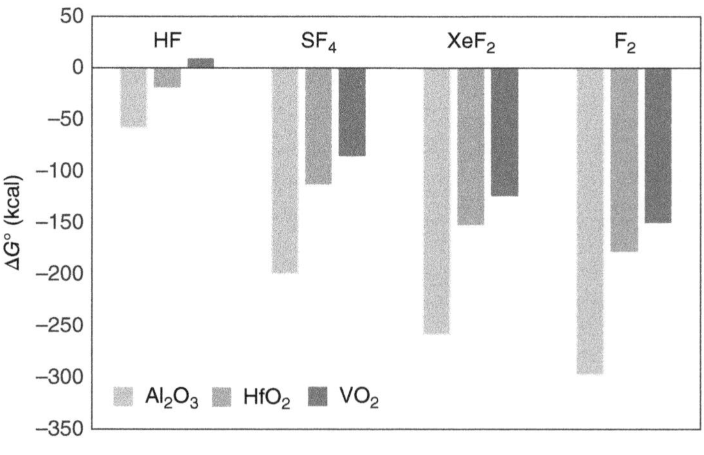

Figure 4.5 Standard Gibbs free energies at 200 °C for fluorination of Al_2O_3, HfO_2, and VO_2. Source: Data from Gertsch et al. (2019).

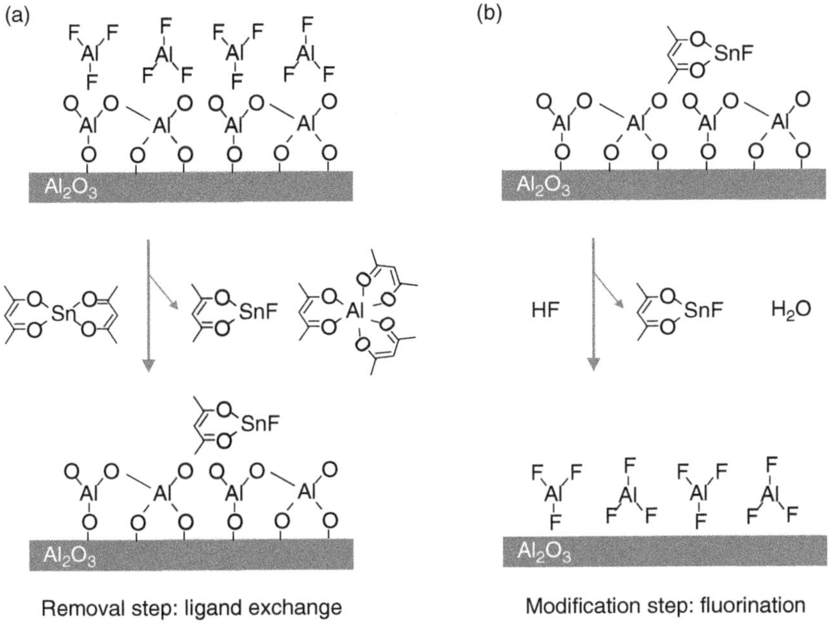

Figure 4.6 Schematic of proposed reaction mechanism for Al_2O_3 ALE showing (a) $Sn(acac)_2$ reaction and (b) HF reaction. Source: Lee and George 2015). © 2015, American Chemical Society.

The removal of aluminum atoms via ligand exchange reaction is illustrated in Figure 4.6. $Sn(acac)_2$ exchanges its acac ligands with the fluorine at the AlF_3 surface. This results in the formation of volatile SnF(acac) and $Al(acac)_3$ reaction products. The term "ligand exchange" comes from this reaction mechanism. Modification and removal steps require three sequential or parallel surface reactions to convert Al_2O_3 first into AlF_3 and then into $Al(acac)_3$. Possibly, diffusion of reactants according to a Langmuir–Hinschelwood mechanism is likely involved in the transport of $Sn(acac)_2$ molecules to the reaction site.

After the AlF_3 layer is etched to saturation via volatile $Al(acac)_3$ and SnF(acac) product formation, SnF(acac) can still be detected on the exposed Al_2O_3 underlayer. Its concentration is lower at higher surface temperatures. This residual SnF(acac) is desorbed subsequently by HF in the next surface modification step (Lee and George 2015; Lee et al. 2015c). Residual SnF(acac) is likely a root cause of poor isotropy performance of this process (Lill et al. 2018).

Mass spectrometry of the reaction products of ligand exchange ALE of Al_2O_3 with HF and TMA ($Al(CH_3)_3$) revealed the existence of dimers containing $AlF(CH_3)_2$ reaction products. They can be found in dimers with itself ($AlF(CH_3)_2 \cdot AlF(CH_3)_2$) or with the precursor molecule ($AlF(CH_3)_2 \cdot Al(CH_3)_3$) (Clancey and George 2017; Clancey et al. 2020).

This finding points to an alternative reaction pathway where the reactant molecule bonds directly with the surface molecule, forms a dimer via a halogen bridge, weakens the bond between the surface molecule and the surface, and desorbs as a dimer.

This would be an Eley–Rideal type of reaction (see Figure 2.6). Ab initio calculations using DFT can illuminate the reaction pathways and the associated energy barriers.

4.1.3 Conversion ALE

Let us explain the mechanism using the archetypical process: etching of SiO_2 via sequential exposure to TMA and HF (DuMont et al. 2017). This process relies on the conversion of SiO_2 into a material that can be etched, for instance, via a ligand exchange reaction. This reaction step gave rise to the term conversion ALE or conversion Etch.

TMA converts SiO_2 into an Al_2O_3/aluminosilicate intermediate via the following reaction:

$$4Al(CH_3)_3 + 3SiO_2 \rightarrow 2Al_2O_3 + 3Si(CH_3)_4 \tag{4.2}$$

In this surface reaction, the electropositive or metallic elements Si and Al are exchanged. Subsequent exposure to HF exchanges the electronegative elements oxygen and fluorine. It converts Al_2O_3 into AlF_3. This reaction is the same as the modification step in ligand exchange ALE of Al_2O_3 with HF/TMA.

After the first two steps, the SiO_2 surface is replaced by AlF_3, which can be etched by ligand exchange with TMA. The ingenuity of the process is that TMA is also the precursor in the first conversion step. Hence, TMA has two functions. Firstly, it is needed to etch AlF_3 via ligand exchange reaction. Secondly, it converts SiO_2 into Al_2O_3 in an exchange of the metallic constituent.

Conversion ALE operates at TMA partial pressures of 1 Torr and higher (see Figure 4.7). The implication is that HF/TMA will etch Al_2O_3 selectively to SiO_2 at pressures below 1 Torr in a ligand exchange reaction. However, selectivity will be lost at pressures of 1 Torr and higher because SiO_2 starts to etch via conversion ALE.

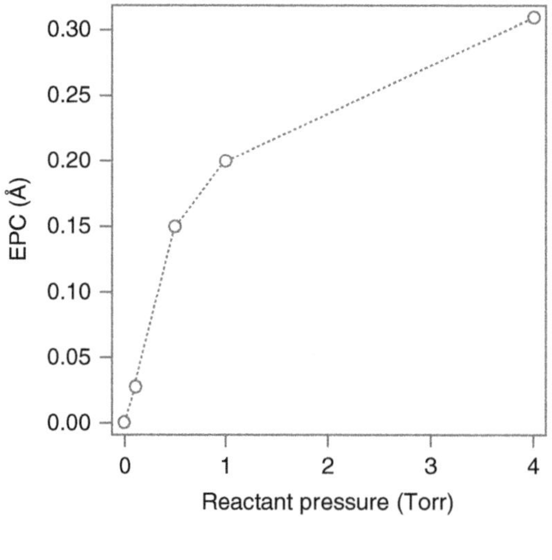

Figure 4.7 EPC as a function of pressure for thermal isotropic ALE of SiO_2 via HF/TMA conversion reaction at 300 °C as measured with X-ray reflectometry. Source: Based on DuMont et al. (2017).

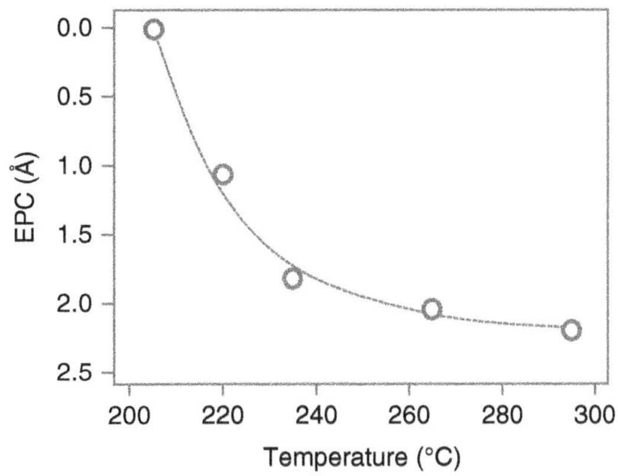

Figure 4.8 Temperature dependence of EPC for ZnO etched with TMA/HF conversion ALE. Source: Zywotko and George 2017). © 2017, American Chemical Society.

Besides SiO_2, conversion ALE has also been used to etch ZnO (Zywotko and George 2017) and WO_3 (Johnson and George 2017). The ZnO etching process uses the same TMA/HF exposure sequence as SiO_2 but at a lower pressure of 60–80 mTorr. First, exposure to TMA converts ZnO into Al_2O_3 according to a reaction similar to Eq. (4.2). The volatile reaction product in this case is $Zn(CH_3)_2$. Al_2O_3 is converted into AlF_3 by HF and subsequently removed by TMA, which also regenerates the Al_2O_3 layer. Quartz crystal microbalance (QCM) measurements showed that the sequential HF and TMA reactions were self-limiting against reactant exposure. EPC ranges from 0.01 Å at 205 °C to 2.19 Å at 295 °C and levels off at higher temperatures (see Figure 4.8).

WO_3 ALE was achieved by an exposure sequence with BCl_3 and HF (Johnson and George 2017) at pressures between 40 and 60 mTorr. BCl_3 converts WO_3 at the surface into B_2O_3 while forming volatile WO_xCl_y products. The conversion depth was shown to be self-limited by means of spectroscopic ellipsometry (SE), most likely due to slowing boron diffusion into the film and tungsten diffusion toward the surface as a function of conversion thickness according to the DG model. Next, HF spontaneously etches the B_2O_3 layer producing volatile BF_3 and H_2O products. This removal step does not involve a ligand exchange. Here, the fluorinated surface has low enough a binding energy to desorb spontaneously.

The conversion of WO_3 to B_2O_3 is required because HF cannot spontaneously etch WO_3 or metallic tungsten. For this reason, the HF step is also self-limited. The synergy is 100% for this reaction. Without HF exposures, BCl_3 alone cannot etch WO_3. Without BCl_3 exposures, HF alone does not etch WO_3. EPC for this process increases from 0.55 Å at 155 °C to 4.19 Å at 207 °C (Johnson and George 2017).

One of the interesting aspects of this process is that material is removed in both steps. Tungsten and some oxygen are removed in the BCl_3 cycle, while the remaining oxygen is removed in the HF step. The thickness loss of 2.99 Å after the BCl_3 exposure leaves a B_2O_3 thickness of 1.30 Å on the surface, which is removed by HF.

This process can also etch metallic tungsten if a third O_2/O_3 plasma step is introduced to form a WO_3 layer. The etch rate for tungsten ALE is 2.5 Å at 207 °C. This process constitutes a plasma-assisted thermal isotropic ALE process. This class of ALE processes will be covered in more detail in Section 4.3.

Table 4.1 Potential conversion ALE reactions based on thermodynamics calculations.

Conversion reaction	ΔG^0 (kcal/mol)
$3ZnO + 2Al(CH_3)_3 \longrightarrow Al_2O_3 + 3Zn(CH_3)_2$	−164.7
$\frac{3}{2}SiO_2 + 2Al(CH_3)_3 \longrightarrow Al_2O_3 + \frac{3}{2}Si(CH_3)_4$	−200.0
$\frac{3}{2}SnO_2 + 2Al(CH_3)_3 \longrightarrow Al_2O_3 + \frac{3}{2}Sn(CH_3)_4$	−229.2
$In_2O_3 + 2Al(CH_3)_3 \longrightarrow Al_2O_3 + 2In(CH_3)_3$	−317.6
$Si_3N_4 + 4Al(CH_3)_3 \longrightarrow 4AlN + 3Si(CH_3)_4$	−362.6

The free energy change values are given for 200 °C.
Source: Zywotko and George (2017). © 2017, American Chemical Society.

Conversion ALE can potentially be applied to a wide range of materials. Table 4.1 lists $\Delta G°$ values for potential conversion ALE reactions (Zywotko and George 2017). Additional reactions have been proposed by Johnson and George (2017).

4.1.4 Oxidation/Fluorination ALE

The etching products in oxidation/fluorination ALE are volatile fluorides. Like thermal etching of silicon and SiO_2 with fluorine (see Chapter 3), oxidation/fluorination ALE leverages fluorine to weaken the bonds between the surface atoms and the bulk.

To design an oxidation/fluorination ALE process, we must find a system that fulfills three requirements: (i) the original surface material does not react with the fluorination reactant. (2) The original surface material can be converted into a material that reacts with the fluorination reactant. (3) The resulting fluoride is volatile.

The archetypical example for ALE oxidation/fluorination ALE is etching of TiN with ozone and HF gases according to the following reaction (Lee and George 2017):

$$TiN + 3O_3 + 4HF \rightarrow TiF_4 + 3O_2 + NO + 2H_2O \tag{4.3}$$

All reaction products are volatile. The EPC is 0.06 Å at 150 °C and 0.20 Å at 250 °C and stays nearly constant for temperatures up to 350 °C. Etching was also accomplished using H_2O_2 and HF with an EPC of 0.15 Å at 250 °C (Lee and George 2017).

The standard free energy for the reaction of TiN with O_3 is thermodynamically favorable with $\Delta G° = -242.1$ kcal/mol at 250 °C. The thermodynamics of the fluorination reaction is less clear. While the standard free energy is favorable at room temperature with $\Delta G° = -6.1$ kcal/mol at 25 °C, the standard Gibbs free energy turns positive at temperatures over 150 °C for entropic reasons (Lee and George 2017). It is possible that ΔG is negative for nonstandard conditions at vacuum pressures where the reaction products are constantly removed from the surface. This example illustrates that thermodynamics calculations are a useful guidance in the design of thermal ALE reactions but may not always be predictive.

Table 4.2 Thermochemistry of a variety of potential oxidation/fluorination ALE processes.

Material	Oxidation	Fluorination
Metal nitride, TiN	$TiN + 3O_3 \rightarrow TiO_2 + NO + 3O_2$	$TiO_2 + SF_4 \rightarrow TiF_4 + SO_2$
	$\Delta G^0 = -241$ kcal/mol	$\Delta G^0 = -62$ kcal/mol
Metal carbide, NbC	$NbC + \frac{7}{2}O_3 \rightarrow \frac{1}{2}Nb_2O_5 + CO + \frac{7}{2}O_2$	$\frac{1}{2}Nb_2O_5 + \frac{5}{4}SF_4 \rightarrow NbF_5 + \frac{5}{4}SO_2$
	$\Delta G0 = -353$ kcal/mol	$\Delta G0 = -74$ kcal/mol
Metal sulfide, WS_2	$WS_2 + 7O_3 \rightarrow WO_3 + 2SO_2 + 7O_2$	$WO_3 + \frac{3}{2}SF_4 \rightarrow WF_6 + \frac{3}{2}SO_2$
	$\Delta G0 = -552$ kcal/mol	$\Delta G0 = -68$ kcal/mol
Metal selenide, $MoSe_2$	$MoSe_2 + 9O_3 \rightarrow MoO_3 + 2SeO_3 + 9O_2$	$MoO_3 + \frac{3}{2}SF_4 \rightarrow MoF_6 + \frac{3}{2}SO_2$
	$\Delta G0 = -519$ kcal/mol	$\Delta G0 = -43$ kcal/mol
Elemental metal, Ta	$Ta + \frac{5}{2}O_3 \rightarrow \frac{1}{2}Ta_2O_5 + \frac{5}{2}O_2$	$\frac{1}{2}Ta_2O_5 + \frac{5}{4}SF_4 \rightarrow TaF_5 + \frac{5}{4}SO_2$
	$\Delta G0 = -323$ kcal/mol	$\Delta G0 = -79$ kcal/mol

The free energy change values are given for 250 °C.
Source: Lee and George (2017). © 2017, American Chemical Society.

Both reaction steps are self-limited. The oxidation step is limited by DG oxidation. Self-limitation of the HF removal step implies that HF etches TiO_2 but not TiN. This can be explained by the nonvolatility of TiF_3 with a boiling point of 1400 °C. In contrast, TiF_4 sublimes at 284 °C for normal conditions (Hall et al. 1958). The reason why this process works is that HF is not strong enough to change the oxidation state from three to four in the reaction with TiN.

This shows that the choice of the fluorination reactant is important for oxidation/fluorination ALE. Some metals and metal compounds will spontaneously etch with fluorination reactants, which are more reactive than HF, for instance, SF_4, F_2, or XeF_2. Etching of other metal nitrides, metal carbides, metal sulfides, metal selenides, and elemental metals should be possible using this oxidation/fluorination ALE (Lee and George 2017). Besides the three requirements mentioned above, the reactivity of the fluorination agent with respect to the original and the oxidized surface are critical. Thermochemistry calculations for other potential oxidation/fluorination ALE reactions are listed in Table 4.2.

Another interesting implementation of oxidation/fluorination ALE is the alternating use of O_2 or O_3 and WF_6 to etch tungsten (Xie et al. 2018). In the first step, tungsten is oxidized. The second step reacts solid WO_3 with gaseous WF_6 to form gaseous WO_2F_2. The process has an EPC of 6.3 Å at a surface temperature of 300 °C. No etching was observed at 275 °C. Thermodynamic modeling reveals that the observed temperature dependence is likely due to the limited volatility of WO_2F_2 (Xie et al. 2018).

Tungsten is unique in that it also forms volatile oxychlorides, which enables thermal ALE of tungsten with alternating O_2 or O_3 and WCl_3 (Xie and Parsons 2020). The EPC is 7.3 and 8.2 Å for temperatures of 205 and 235 °C. This reaction proceeds at lower temperatures than the related O_2/WF_6 ALE, which is the result of the unique thermodynamic properties of tungsten oxyhalides. Other oxidation/chlorination processes may exist.

4.2 Performance Metrics

In this chapter, we will review the performance metrics for thermal isotropic ALE. Like thermal etching, thermal isotropic ALE removes materials in all directions with similar rate and hence metrics that require vertical profiles such as line width roughness (LWR)/line edge roughness (LER) and edge placement error (EPE) do not apply to this class of etching technologies and will not be covered.

4.2.1 Etching Rate (EPC)

ALE is characterized by self-limited modification and removal steps. There is no etching in the modification step. The instantaneous etching rate during the removal step is time dependent – it reaches zero for long enough step times. The total amount of material that is removed after one modification and one removal step is called EPC. EPC is determined by the material, the reactants, and the process conditions. It is largely independent from the engineering implementation. For the same process conditions, EPC should be the same for spatial and temporal step separation.

For industrial applications, the average etching rate is the important metric because it determines the processing cost. The etching rate for an ALE process can be defined as the ratio of the EPC divided by the duration of the process cycle. This definition includes purge steps, which may or may not be needed to avoid parasitic reactions (Zywotko et al. 2018). This means the average etching rate is very much a function of the reactor design, in particular the ability to purge the reactor fast. In the following discussion, we will focus on EPC of different thermal isotropic ALE processes.

Table 4.3 lists EPC for selected ligand exchange reactions that have been reported between 2015 and 2020. This list is by no means complete; new reactions are being reported almost on a weekly basis. A regularly updated compilation is maintained on the internet by members of the Plasma and Materials Processing group at Eindhoven University of Technology (https://www.atomiclimits.com).

Figure 4.9 shows the data from Table 4.3 printed as a function of temperature. Only the data from Prof. George's group at University of Colorado is shown to ensure that the experimental method was consistent. Despite the wide range of values for different materials and removal step precursors, there appears to be a maximum for EPC for temperatures around 250 °C. EPC drops to near zero below 150 °C and above 350 °C. It is possible that this general trend reflects the temperature dependence of fluorination by HF pyridine. The reactions with $Sn(acac)_2$ proceed

Table 4.3 Thermal isotropic etching reactions with ligand exchange reaction.

Data set	Material	Precursor	*T* (°C)	EPC (Å)	References
1	HfO_2	$Sn(acac)_2$	200	0.11	Lee et al. (2015a)
2	HfO_2	$Sn(acac)_2$	150	0.07	
3	AlN	$Sn(acac)_2$	275	0.36	Johnson et al. (2016)
4	HfO_2	$TiCl_4$	300	0.59	Lee and George (2018)
5	Al_2O_3	$Sn(acac)_2$	250	0.61	Lee and George (2015)
6	Al_2O_3	$Sn(acac)_2$	150	0.14	
7	Al_2O_3	$Sn(acac)_2$	200	0.23	Lee et al. (2016a)
8	Al_2O_3	TMA	300	0.45	
9	Al_2O_3	DMAC	250	0.32	
10	ZrO_2	DMAC	250	0.96	
11	HfO_2	DMAC	250	0.77	
12	ZrO_2	$SiCl_4$	350	0.14	
13	ZrO_2	$Sn(acac)_2$	200	0.14	
14	HfO_2	TMA	300	0.10	
15	HfO_2	$SiCl_4$	350	0.05	
16	HfO_2	$Sn(acac)_2$	200	0.06	
17	Al_2O_3	TMA	325	0.75	Lee et al. (2016b)
18	Al_2O_3	TMA	300	1.25	Hennessy et al. (2017)
19	Al_2O_3	DMAC	250	1.40	Fischer et al. (2020a)
20	Al_2O_3	DMAC	250	0.39	Lee and George (2019)
21	HfO_2	DMAC	250	0.98	
22	ZrO_2	DMAC	250	1.33	

at temperatures between 150 and 200 °C, followed by dimethyl aluminum chloride (DMAC) at 250 °C, $TiCl_4$, and TMA at 300 °C, and $SiCl_4$ at 350 °C. DMAC exhibits the highest EPC values with the largest values being 0.96 and 1.33 Å for etching of ZrO_2.

The average EPC per etched material for all tested precursors and temperatures are 0.34 Å for HfO_2, 0.41 Å for Al_2O_3, 0.64 Å for ZrO2, and 0.36 Å for AlN (only one value reported). This is a narrow range given that the etched materials are very different in composition and density. In comparison, reactive ion etching (RIE) etching rates can vary by orders of magnitude.

Let us compare the EPC with the lattice constants of the materials to estimate what fraction of a monolayer is removed per cycle. Crystalline Al_2O_3 has a trigonal structure with lattice constants of 4.8 and 13.0 Å. HfO_2 is monoclinic with lattice constants 5.1, 5.2, and 5.3 Å. ZrO_2 has a cubic structure with a lattice constant of 5.1 Å. This means that less than 10% of one monolayer is removed per cycle in these ALE processes. There are at least two possible explanations why only a small fraction

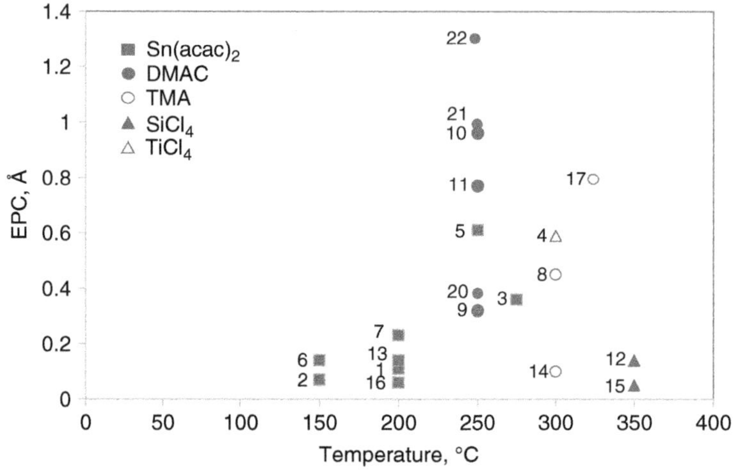

Figure 4.9 EPC as a function of surface temperature for ligand exchange reactions reported by Prof. Steve George's group at University of Colorado. The numbers next to the data points correspond to the data set numbers in Table 4.2.

of one monolayer is etched per cycle: depth of surface fluorination and incomplete removal of the fluorinated metal.

Let us now investigate why the EPC is smaller than one monolayer for these ALE processes. There is experimental evidence that incomplete removal is the main reason for sub-monolayer EPC in Al_2O_3 ALE with HF/Sn(acac)$_2$. FTIR measurements revealed a reduced surface coverage of SnF(acac) at higher temperatures (Lee et al. 2015b). Also, increased mass losses were measured after the complete Sn(acac)$_2$ step at higher temperatures. The combined data suggests that the Al_2O_3 ALE EPC is inversely dependent on the acetylacetonate surface coverage. This indicates that the EPC for Al_2O_3 ALE with HF/Sn(acac)$_2$ is only a fraction of one monolayer because of incomplete AlF_3 removal due to surface reaction site blocking by SnF(acac).

Incomplete fluoride removal due to surface blocking by the precursor or reaction products has important implications for the application of thermal ALE processes to the formation of semiconductor devices. In case of a partial etch where some of the etching material is left in place, the top surface will be fluorinated Al_2O_3. This may impact subsequent deposition steps, and defluorination steps must be developed. Rahman et al. have demonstrated that roughly 90% of the residual fluorine in Al_2O_3 ALE with HF/TMA can be removed with water vapor (Rahman et al. 2018). If the molecule that blocks the etching is a reaction product, this has also important implications for the isotropy of the resulting etch as will be discussed in the section about profile performance.

Surface blocking opens an alternative path to ALE of metal fluorides. Such a process can be designed by combining a ligand exchange or chelation step to remove the top layer until saturation by surface blocking is reached. In a second step, the blocking molecule is removed to "refresh" the surface. The latter can, for instance, be accomplished using plasma (Mameli et al. 2018).

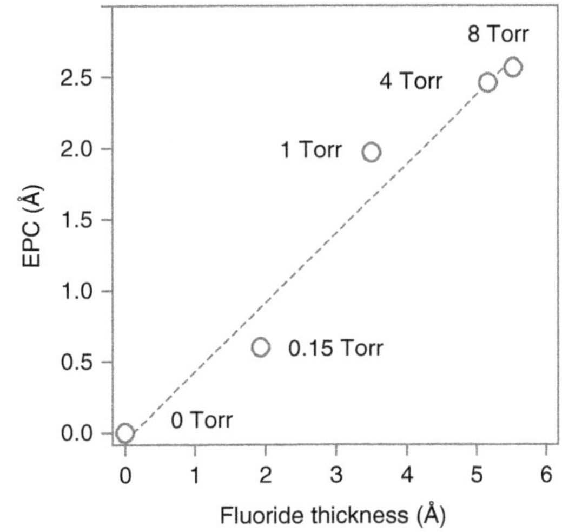

Figure 4.10 Correlation between EPC for thermal ALE of Al_2O_3 with HF/TMA and the depth of fluorination. Source: Cano et al. (2019). © 2019, American Chemical Society.

The EPC for thermal ALE of Al_2O_3 ALE with HF/TMA appears to be limited by the depth of surface fluorination. Figure 4.10 depicts the good correlation between the EPC for HF/TMA ALE and the depth of fluorination as measured with XPS, assuming the composition of the fluorinated layer is Al_2OF_4, which represents a mixture of AlF_3 and various aluminum oxyfluorides (Cano et al. 2019). The experiments were conducted at 300 °C and the change in fluorination depth was achieved by changing the HF pressure at constant exposure time (see Figure 4.4). The EPC is only about 50% of the depth of fluorination because only AlF_3 is removed in the ligand exchange reaction.

EPC is also limited by the depth of fluorination for ALE of Al_2O_3 with HF/DMAC. Fischer et al. demonstrated that dimethylaluminum chloride (DMAC) etches an AlF_3 film spontaneously at substrate temperatures above 180 °C. The thermal etching reaction of AlF_3 with DMAC exhibited no self-limitation and showed a linear dependence on DMAC pressure (Fischer et al. 2020a). However, when Al_2O_3 is fluorinated, DMAC removes the fluorinated layer only partially. Some fluorine remains at the surface. Etching was detected for the HF/DMAC ALE process for temperatures above 170 °C and EPC increases with temperature even though the fluorine concentration after the HF step decreases with temperature. The authors attribute this behavior to an increased reactivity of DMAC with aluminum oxyfluoride. Increases of EPC with temperature have also been reported for HF/DMAC ALE for Al_2O_3, HfO_2, and ZrO_2 (Lee and George 2019).

DFT calculations can shed additional light on the surface modification mechanism of metal oxides with HF (Natarajan and Elliott 2018; Mullins et al. 2019). Natarajan et al. find that HF molecules adsorb on the Al_2O_3 surface by forming hydrogen bonds. They stay intact or dissociate to form Al–F and O–H species. At higher coverages, a mixture of molecularly and dissociatively adsorbed HF molecules in a hydrogen-bonded network is observed. The theoretical maximum of the EPC is calculated to be 0.57 Å (Natarajan and Elliott 2018). This compares to 0.45 and 1.25 Å for HF/TMA at 300 °C (Lee et al. 2016a; Hennessy et al. 2017), and

Table 4.4 Thermal isotropic etching reactions: chelation, conversion, oxidation/fluorination.

ALE process	Material	Reactants	*T*(°C)	EPC (Å)	References
Chelation	Cu	O_2/hfac	275	0.90	Mohimi et al. (2018)
	Co	Cl_2/hfac	185	16.00	Konh et al. (2019)
	Fe	Cl_2/acac	140	5.00	Lin et al. (2018)
Conversion	SiO_2	TMA/HF	300	0.31	DuMont et al. (2017)
	Si	O_2/HF/TMA	290	0.40	Abdulagatov and George (2018)
	Si_3N_4	O_2/HF/TMA O_3/HF/TMA	290	0.25 0.47	Abdulagatov and George (2020)
	ZnO	TMA/HF	295	2.19	Zywotko and George (2017)
	WO_3	BCl_3/HF	207	4.19	Johnson and George (2017)
	W	O_3/BCl_3/HF	207	2.5	Johnson and George (2017)
Oxidation/ fluorination	TiN	O_3/HF	350	0.20	Lee and George (2017)
	TiN	H_2O_2/HF	250	0.15	Lee and George (2017)
	W	O_2/WF_6	300	6.3	Xie et al. (2018)
	W	O_2/WCl_2	235	8.2	Xie and Parsons (2020)

0.32 and 1.30 Å for HF/DMAC C at 250 °C (Lee et al. 2016a; Fischer et al. 2020a). Mullins et al. found a similar adsorption mechanism involving molecular HF and dissociative fluorine adsorption for ZrO_2 and HfO_2 (Mullins et al. 2020).

The variation in reported EPC may have various root causes. Hennessy reported that the EPC for Al_2O_3 ALE with anhydrous HF and TMA was dependent on the wall surface coating (Hennessy et al. 2017). Wall effects were also observed for Al_2O_3 ALE with HF or NF_3 and DMAC (Fischer et al. 2020a). The differences in EPC as reported from different groups could also be related to the crystalline structure of the films. Murdzek and George studied HF/DMAC ALE on HfO_2, ZrO_2, and $HfZrO_4$ films and found that the amorphous films etched faster than the crystalline films for each material (Murdzek and George 2019, 2020). The difference was the most dramatic for the HfO_2 films where the amorphous films etched 8–22 times faster than the crystalline films.

EPC for thermal isotropic ALE via chelation, conversion, and oxidation/fluorination are listed in Table 4.4. The reported EPCs for ALE via chelation are much larger for surface modification with chlorine (Co with Cl_2/Hhfac: 16.00 Å (Konh et al. 2019); Fe with Cl_2/Hacac: 5.00 Å (Lin et al. 2018)) compared to oxygen modification (Cu with O_2/Hhfac: 0.90 Å (Mohimi et al. 2018)). This could be the result of chlorine being more reactive and creating deeper surface modification.

The conversion ALE reactions for ZnO, W, and WO_3 have higher EPC compared to Si, SiO_2, and Si_3N_4. This is because the former are related to fluorination/oxidation ALE and latter to ligand exchange ALE.

Let us now investigate the mechanisms that limit EPC for these ALE processes. For ALE of ZnO with TMA/HF, TMA acts both as the reactant that converts ZnO into Al_2O_3 and as the etchant of AlF_3 that was formed in the reaction of Al_2O_3 with HF. TMA has been found to remove AlF_3 completely and HF fluorinates Al_2O_3 only to less than 1 Å at the given temperature and pressure (Cano et al. 2019). This leads to the conclusion that TMA converts ZnO at least 2 Å deep and that the resulting Al_2O_3 must have a density and/or structure that allows to fluorinate the entire film depth with HF.

ALE of SiO_2 with TMA/HF follows a very similar reaction mechanism, yet the EPC is only 0.31 Å. The removal of AlF_3 by TMA is the same as for ligand exchange ALE of Al_2O_3 with HF/TMA. EPC for the latter at pressures of more than 1 Torr is larger than 2 Å. Therefore, EPC for ALE of SiO_2 with TMA/HF is most likely limited by the SiO_2–Al_2O_3 conversion step.

For ALE of WO_3 with BCl_3/HF, BCl_3 converts the WO_3 surface to a B_2O_3 layer and is believed to form volatile WO_2Cl_2 as a reaction product. HF etches B_2O_3 spontaneously. This mechanism implies that EPC is determined by the conversion of WO_3 into B_2O_3. The process is a diffusion-limited reaction. In conclusion, EPC for the conversion ALE of SiO_2, ZnO, and WO_3 appears to be limited by the depth of the conversion processes.

4.2.2 ERNU (EPC Nonuniformity)

Neutral and ion fluxes must be closely controlled in continuous etching processes to obtain etch rate uniformity across the wafer. Because ALE processes are sequences of self-limiting steps, EPC is flux and dose independent. EPC can vary across the wafer if the ALE steps are not saturated or if parasitic thermal etching exists. The upper panel in Figure 4.11 illustrates a sequence of two modification and two removal steps separated by purge steps. The solid line represents an ideal ALE process and the dashed line a process with continuous thermal etching during the modification step due to incomplete purging. The parasitic etch amount is labeled ΔEPC_1. The parasitic thermal etching process is due to the simultaneous presence of modification and removal reactants. Variation of fluxes across the wafer will introduce nonuniform EPC. This could be caused by nonuniform release of reactants from the reactor wall. Lee and George reported EPC variations depending on where coupons were placed in the reactor for oxidation/fluorination ALE of TiN with O_3/HF oxidation/fluorination (Lee and George 2019). ALE reactors must be designed to eliminate chamber wall effects.

The lower panel shows time dependences of etch or modification depths in one graph. The solid and dashed lines depict ideal and quasi self-saturated steps, respectively. In the case of surface modification, the dashed, quasi saturated curve can, for instance, represent the time dependence of an oxygen or fluorine diffusion process. Time dependences of the oxidation or fluorination depths are shown in Figures 2.10 and 2.11 for DG and CM oxidation, respectively. Even though diffusion depths saturate meaningfully, complete saturation is theoretically never reached.

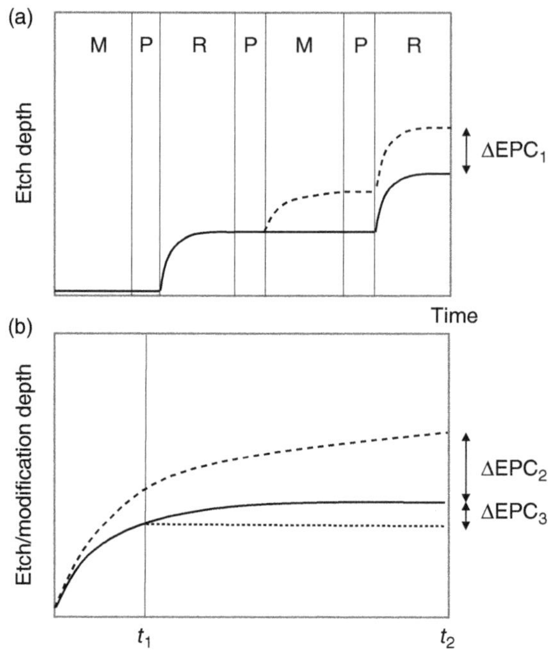

Figure 4.11 Root causes for etch rate nonuniformity for ALE processes.

Diffusion processes are temperature dependent. For a given finite temperature sensitivity of the diffusion process, the wafer pedestal or electrostatic chuck must have a certain temperature uniformity. Deal-Grove (DG) diffusion processes are also pressure dependent. This is typically not an issue in a vacuum chamber where pressures equalize quickly. A nonuniform modification depth leads to a nonuniform removal because the removal step will only remove the modified material. The etch amount per cycle caused by nonideal saturation is labeled ΔEPC_2 in Figure 4.10. It is also possible to imagine an ALE process with ideal saturation of the modification step and quasi self-limited removal step. This is the case when the selectivity is not infinite.

The solid curve in the lower panel of Figure 4.11 shows a perfectly saturated modification or removal step with step times t_1 and t_2 representing incomplete and complete saturation, respectively. Self-limited steps approach saturation asymptotically, which means it can take a long time to reach complete saturation. Similar to ALD, this is not always needed to reach uniformity requirements. The difference in EPC for step times t_1 and t_2 is ΔEPC_3 in Figure 4.11. The nonuniformity across the wafer, which is introduced by reducing the step time, is always smaller than ΔEPC_2 and depends on the flux uniformity. If the flux uniformity is perfect, the process steps must not be saturated.

In contrast to diffusion, adsorption processes are truly self-limited. Once all adsorption sites are occupied, adsorption stops. Still, EPC nonuniformities can arise from adsorption processes. As was discussed in the section 4.1.2, for thermal isotropic ALE of Al_2O_3 with HF/Sn(acac)$_2$, the Sn(acac)$_2$ removal step is limited by adsorption and accumulation of the SnF(acac) remaining on the surface (Lee et al. 2015b). This makes the process temperature sensitive. A uniform temperature across the wafer is needed to achieve EPC uniformity.

In summary, ideal ALE processes are theoretically perfectly uniform across the wafer. ALE processes used in practical applications can deviate from ideal performance due to lack of saturation, lack of synergy, or the existence of secondary reactions. In thermal isotropic ALE processes, lack of saturation can be caused by diffusion processes. For DG diffusion, the depth of modification is a function of temperature and pressure, which have to be controlled across the wafer to achieve EPC uniformity across the wafer. Lack of synergy is caused by lack of selectivity of the removal step or spontaneous etching of the modification step. Here, the choice of the right fluorine gas for modification or removal is very important for ligand exchange and oxidation/fluorination ALE.

Secondary reactions include unintentional mixing of modification and removal of gases, which leads to spontaneous thermal etching. This introduces EPC nonuniformities, which are difficult to control because they depend on temperature and local gas concentrations. Desorption of reactants from reactor walls or incomplete purging are examples for these effects. They must be addressed by the design of the gas injection system and the choice of reactor materials.

Secondary reactions can also be part of the saturation mechanism itself. For instance, blocking of adsorption sites during the removal step can saturate the removal step before all the modified material is removed. Sticking probabilities are a function of temperature and hence wafer temperature across the wafer has to be controlled.

4.2.3 Selectivity

Because all steps in thermal isotropic ALE operate at thermal energies, the amount of "excess energy" is close to zero and selectivities can be infinite. Compared to thermal etching, the selectivity properties are quite similar; however, thermal isotropic ALE offers more flexibility to design selective processes. Selectivity can be obtained when only one of the materials can be modified, for instance, by thermal fluorination. Selectivity is also achieved when only one of the materials permits the removal reaction or only one of the final products desorbs. The choice of the right modification and removal reactants is very important. Temperature can be used as a secondary knob to tune selectivities.

Ligand exchange ALE with HF/DMAC etches Al_2O_3, ZrO_2, and HfO_2 (see Table 4.3). This process however does not etch SiN, SiO_2, or TiN (Lee et al. 2016a). Figure 4.12 shows the etching results for a trench structure with HF/DMAC at 250 °C (Fischer et al. 2019). The selectivity between Al_2O_3 and HfO_2 is 2 : 1 and infinite to SiN, SiO_2, and TiN. The process etches TiO_2 via formation of volatile TiF_4 in the HF step. This example shows that chemistry considerations are critical in the design of processes with several exposed materials.

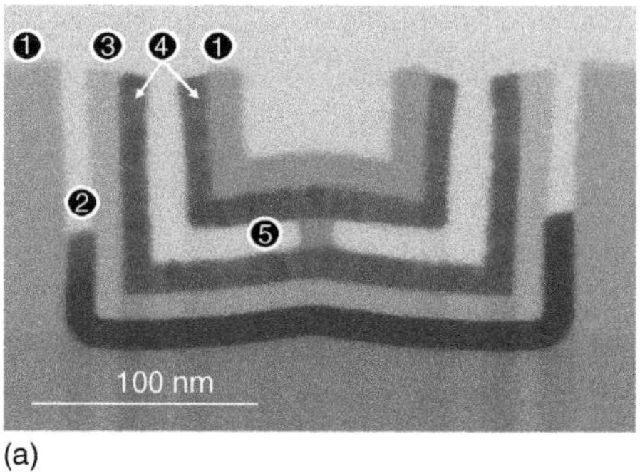

	Material	Etch amount (nm)
1	SiN	Not measurable
2	HfO_2	62.3
3	SiO_2	Not measurable
4	TiN	Not measurable
5	Al_2O_3	111.6

(b)

Figure 4.12 Selective etching of various materials with HF/DMAC thermal isotropic ALE. Source: Based on Fischer et al. (2019).

4.2.4 Profile and ARDE

We will discuss profile performance and aspect ratio dependent etching (ARDE) together because they are closely related for isotropic etches such as thermal isotropic ALE. For a perfectly isotropic etch, the etch amount in normal direction to the surface is the same at any given point on the surface. Figure 4.13 shows schematic illustrations of isotropic profiles for a liner deposited outside a line or pillar, inside a hole or trench, and for a planar film with a trench or hole mask.

Isotropy is defined as the ratio of lateral to vertical etch amount. For the liner structure, isotropy can be measured both on the top and the bottom of the feature: d_1/d_2 (a). ARDE is the difference in the vertical or lateral etch rate between the top and bottom of the feature. It is defined as the difference in the etching amount from the top to bottom divided by the etching amount on the top: d1'/d1 (b). Isotropic etches have the special case of etching of surfaces out of line of sight of the incoming reactants. We call this case shading (c). For complex structures, the term tortuosity is used to quantify this property. Tortuosity is used to describe diffusion and gas flows in porous media.

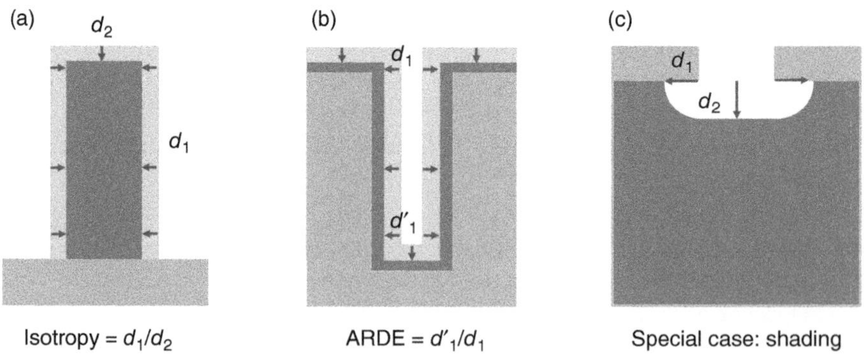

Figure 4.13 Schematic illustrations of perfectly isotropic profiles for a liner deposited outside a line or pillar, inside a hole or trench, and for a planar film with a trench or hole mask.

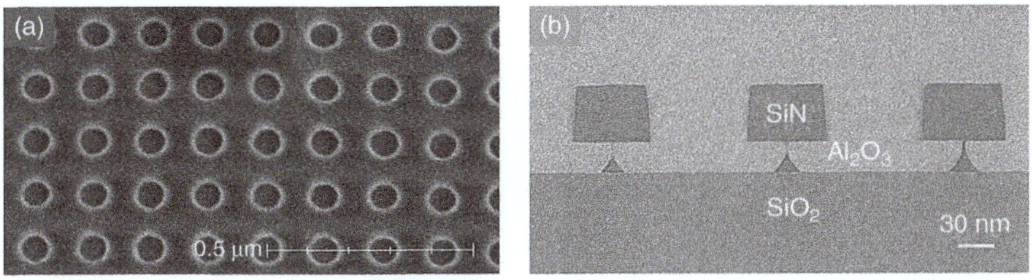

Figure 4.14 Thermal isotropic ALE of Al_2O_3 with HF/DMAC and a Si_3N_4 hole mask: (a) Top-down SEM before mask open etch and resist strip. (b) TEM cross section of the same sample after 1000 ALE cycles. Source: Fischer et al. (2020a).

According to Figure 4.13, using masked planar films to measure isotropy is not entirely correct, because as the feature develops, isotropic etching underneath the mask will be significantly less in the line of sight than the vertical etching component. While these structures are very convenient to measure isotropy, they convolute isotropy and ARDE. They can be used for a rough estimate of isotropy, though.

The fluxes and energy sources in thermal isotropic ALE are largely isotropic if the gas flow is diffusive. This means that the isotropy should be 100%. If an isotropy value of less than 100% is observed, the root cause must therefore be ARDE, i.e. species transport. Because ALE processes use self-limited process steps, theoretically they should not have ARDE. This is not always the case. Next, we will discuss the potential root causes for ARDE in thermal isotropic ALE.

Figure 4.14 shows experimental results for HF/DMAC ALE of Al_2O_3 with a SiN hole mask. The process conditions were 30 mTorr, 250 °C, and 1000 cycles. The isotropy factor for this process is only 44%. ALD learning can be applied to improve isotropy. The analog of "isotropy" in ALD is "conformality." For ALD, higher conformality for high aspect ratio structures is frequently achieved by increasing the doses of the reactants. This allows the precursors to saturate surfaces at the end of high aspect ratio features and surfaces without direct line of sight for the incoming reactant molecules. This approach can also be used for thermal isotropic ALE. The isotropy factor increased to 55% when the pressure was increased from 30 to 140 mTorr. Further process optimization using longer step times increases isotropy to 70% (Fischer et al. 2019).

To achieve saturation at the bottom of high aspect ratio or out-of-line-of-sight features, sufficient precursor molecules must be transported there. The flux of precursors that reaches the bottom of the feature is governed by Knudsen transport and is a function of the transmission coefficient K and the sticking coefficient s (see Eq. (2.17)). The former is a function of the geometry while the latter depends on the surface and precursor molecules. There are fundamental differences between ALD and ALE, which can lead to different conformality or isotropy results even if the same precursors are involved. A direct comparison is possible for a sequence of HF/TMA steps, which etch Al_2O_3 at 300 °C and deposit AlF_3 at 150 °C (DuMont and George 2017). Figure 4.15 illustrates that the transport of reactants into the feature involves the same precursors, TMA or HF. In the case of ALD, the surface is always AlF_3; in

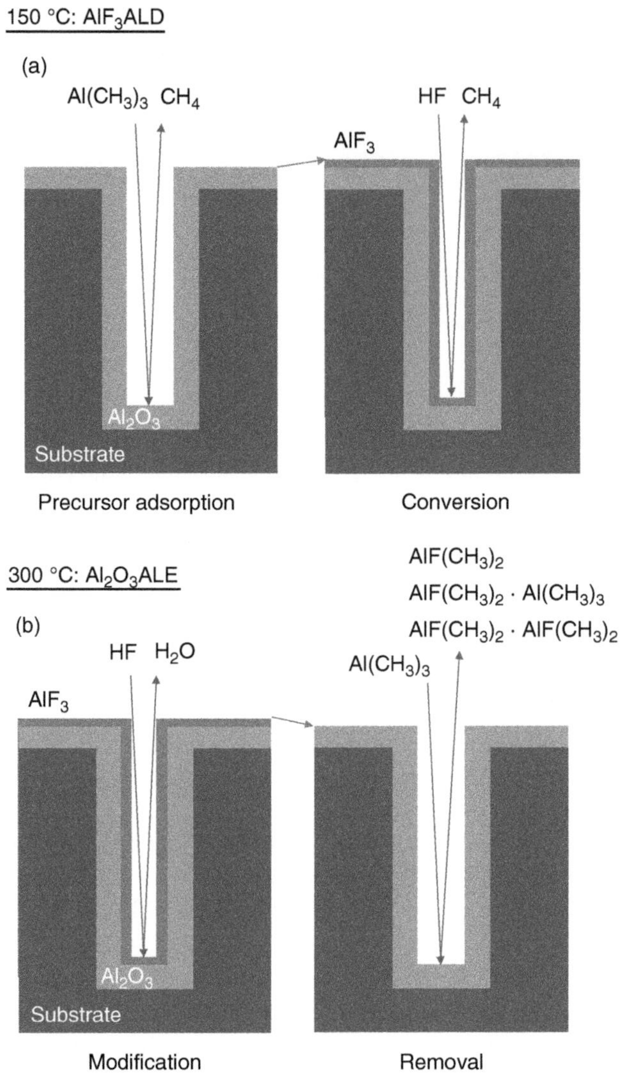

Figure 4.15 Schematic illustration of reactant fluxes in Al_2O_3 ALE and AlF_3 ALD with HF/TMA.

the case of ALE, it changes from Al_2O_3 to AlF_3 and back. This could change sticking coefficients and is the first difference between ALE and ALD.

In the case of ALD, highly volatile CH_4 is transported out of the feature. For ALE, the reaction products are $AlF(CH_3)_2$ and even more complex dimers such as $AlF(CH_3)_2{\cdot}Al(CH_3)_3$ and $AlF(CH_3)_2{\cdot}AlF(CH_3)_2$ (Clancey and George 2017; Clancey et al. 2020). It appears that the reaction products in thermal ALE processes are generally more complex than for ALD, and redeposition and collision-induced dissociation are more likely. This creates another boundary condition for the choice of thermal ALE precursors. In addition to all the other requirements, the

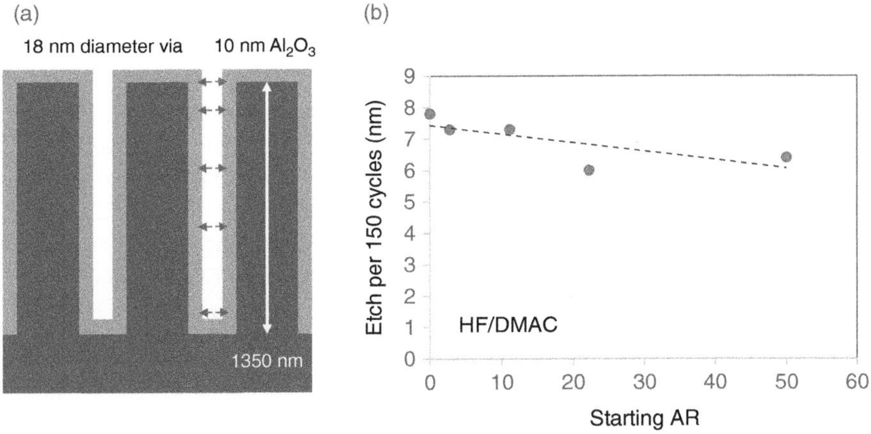

Figure 4.16 ARDE Al_2O_3 ALE with HF/TMA in a high aspect ratio via feature. Source: Lill et al. (2018).

reaction products must be robust enough to survive several surface collisions to be successfully removed from the etching feature.

Fischer et al. proposed an alternative explanation for lack of isotropy in the out of sight feature (Fischer et al. 2020b). The mechanism takes into consideration that HF adsorbs on Al_2O_3 surface in two states, chemisorbed HF and dissociated and chemisorbed F. HF will establish an equilibrium between adsorption and desorption, which will lead to a much lower HF concentration on concealed surfaces because HF must scatter to reach them. If the dissociation of HF is dependent on the surface concentration of adsorbed HF, this would lead to lower concentration of atomic fluorine and less effective fluorination.

Figure 4.16 shows ARDE additional results for Al_2O_3 ALE with HF/DMAC for holes with an initial aspect ratio of 50 (Fischer et al. 2019). The structures were prepared by depositing 10 nm Al_2O_3 ALD into holes with an initial diameter of 38 nm. The final hole diameter was therefore 18 nm. The ARDE was measured to be 18%.

4.2.5 CD Control

Critical dimensions (CDs) in isotropic processes are in most cases the depth of a recess, a remaining thickness of a horizontal layer, or the remaining width of a vertical structure. Isotropic etching does not require a line of sight like ion-assisted etching technologies. Measuring of hidden structures such as an undercut can be complicated if not impossible. Hence many isotropic etches require scanning electron microscopy (SEM) cross sections for verification, which have a limited sampling rate due to its destructive nature.

For thermal etching, the process is controlled by time. A repeatable etching rate requires precise control of the etching and inhibiting species fluxes and of the temperature. This is where thermal ALE has an advantage over thermal etching because if EPC is repeatable, the amount of removed material is a multiple of EPC. The latter is in most cases a function of the depth of modification and is determined by adsorption and diffusion processes as discussed in detail above.

If the CD of an etching process is the remaining thickness or width of a film, it is a function of the amount of material to be removed. For instance, the complete removal of 10 nm with a 1% uniformity across the wafer creates a range of 1 nm across the wafer for the remaining film. For 20 nm removal and 10 nm remaining thickness, the uniformity of the remaining film is 1%. This number would increase to 10% if 100 nm were to be removed. This example shows that recess or etchback applications can be very challenging with respect to uniformity performance. The self-limited nature of ALE is useful to overcome these challenges.

4.2.6 Surface Smoothness

Smoothening of the surface has been reported for thermal isotropic ALE of HfO_2 (Lee et al. 2015a). The initial roughness of about 6 Å was reduced to 3–4 Å after 50 and more ALE cycles. Higher reactivity of convex features at the surface is most likely the underlying mechanism.

4.3 Plasma-Assisted Thermal Isotropic ALE

Plasma can be used to generate radicals that can enhance thermal isotropic ALE processes. This situation is similar to ALD, where plasma-generated ions and radicals are used in the conversion steps. In ALE, radicals can be effectively used in the modification step to eliminate temperature effects of the adsorption steps. For instance, a remote inductively coupled plasma (ICP) plasma was used to generate ozone, which enhanced the oxidation of tungsten in a conversion ALE process (Johnson and George 2017). An ICP plasma in direct contact with the wafer was used to fluorinate Al_2O_3 for subsequent removal with $Sn(acac)_2$ (Fischer et al. 2017). After a few cycles, thin film growth due to Sn accumulation was observed and a H_2 plasma step had to be implemented to remove the tin residues. Apparently, plasma exposure led to the dissociation of residual SnF(acac) on the surface. These residues have been reported for Al_2O_3 ALE with HF/$Sn(acac)_2$ (Lee et al. 2015b). In this case, however, HF was effective in removing SnF(acac) without leaving dissociation products behind.

In AlN ALE with HF/$Sn(acac)_2$, the EPC increased from 0.36 to 1.96 Å when a hydrogen plasma step was inserted after the $Sn(acac)_2$ removal step (Johnson et al. 2016). This increase in EPC was attributed to the removal of Hacac residues. When an argon plasma was used instead of hydrogen, the EPC was 0.66 Å. This leads to the speculation that plasma exposure may create adsorption sites via formation of surface defects.

Another interesting case for the use of plasma is ZnO ALE with Hacac and oxygen plasma (Mameli et al. 2018). In this case, plasma is used to remove Hacac residues. The process relies completely on the self-limiting behavior of the Hacac reaction with ZnO, which is a spontaneous thermal etching process that leaves carbonaceous residues at the surface. When all adsorption sites are blocked by residues, the etching stops and self-limitation is achieved. Oxygen plasma is then used to reset the surface. The uniqueness of this ALE process is that it does not have a modification step, but a plasma reset step.

These examples illustrate that plasma can be a useful tool to remove residues from the surface, which can form as by-products of the removal step. This can be important to implement a wider range of precursors for thermal ALE. Some organic molecules undergo catalytic dissociation on metallic surfaces, which are an important class of materials to be etched (Wang and Opila 2020). This may limit not only the choice of precursors in terms of their stability, but also with respect to the stability of the reaction products, which can be even more complex than the precursor molecules (see Figure 4.14).

Etching selectivity is generally lower when plasma steps are used to enhance thermal isotropic ALE. Remote plasmas greatly reduce or eliminate surface exposure to damaging ions or photons. Still, radicals are by nature very reactive. They can introduce spontaneous etching and eliminate selectivity mechanisms based on selective surface modification.

4.4 Applications Examples

Thermal Isotropic ALE is a new etching technology and applications in the semiconductor industry are still emerging (George and Lee 2016). The characteristic performance attributes such as selectivity, isotropic etching, low ARDE, and their ability to etch metals and metal oxides are very attractive for patterning of advanced semiconductor devices. Many metals and metal oxides, which are challenging to etch with RIE, can be etched isotopically and selectively by thermal isotropic ALE. These materials include Al_2O_3, HfO_2, ZrO_2, cobalt, copper, iron, and others. Coincidentally, these are materials that are also considered for emerging memory devices. While it is difficult to predict all the applications thermal isotropic ALE will be used for, there are two important processing capabilities that can be enabled with this etching technology: area-selective deposition and formation of lateral devices.

4.4.1 Area-Selective Deposition

Patterning by means of etching requires deposition of the material, deposition of a mask film, lithography, and etching. Selective depositing on certain areas on the wafer is one potential approach to reduce cost. Another advantage of area-selective deposition is perfect alignment with the underlying features. One of the techniques to achieve area-selective deposition is the combination of selective ALD and selective ALE.

Selectivity in chemical vapor deposition (CVD) and ALD can be defined as

$$S = \frac{(d_1 - d_2)}{(d_1 + d_2)} \tag{4.4}$$

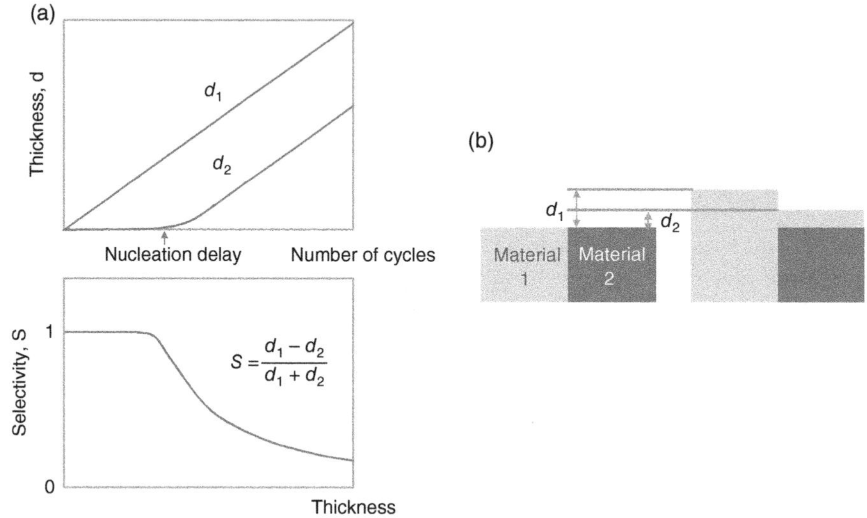

Figure 4.17 Definition of selectivity for deposition processes.

where d_1 and d_2 are the deposited thicknesses of the material on the growth and non-growth areas (Gladfelter 1993; Mackus et al. 2019). Figure 4.17 illustrates that deposition selectivity is the result of a nucleation delay. The growth area in Figure 4.17 is labeled "material 1" and the non-growth area as "material 2." The deposited material is shown as identical to material 1 because this reflects the situation after a few ALD cycles. Non-growth areas will develop nuclei from which the film starts to grow after a certain number of ALD cycles. Once an initial film is formed, the growth rates are the same for the growth and non-growth areas because the surfaces are now identical. Because of this, selectivity in deposition is a function of time. Before nucleation starts in the non-growth area, the selectivity equals 1. After nucleation in the non-growth area, selectivity drops below 1. For an infinite number of ALD cycles, it goes to 0.

ALE selectivity is defined as

$$S = \frac{d_1}{d_2} = \frac{\mathrm{EPC}_1}{\mathrm{EPC}_2} \tag{4.5}$$

where d_1 and d_2 are the depth of removal for material 1 and 2. Typically, the faster etching material is listed in the numerator. Hence, etching selectivities range between 1 and infinite.

Etching selectivity is normally not a function of time because the exposed surfaces do not change over time (Figure 4.18). Exceptions to the rule are etching processes that rely on the buildup of a protective layer on one of the surfaces to achieve selectivity. Owing to nucleation delay of the deposition sub-process, the selectivity in the beginning of the etching process maybe lower (see Figure 6.22).

Selective ALD can be achieved by combining selective deposition and selective etching as shown in Figure 4.19. The ALE process removes grown material in the non-growth area selectively. Thermal isotropic ALE is well suited to be such

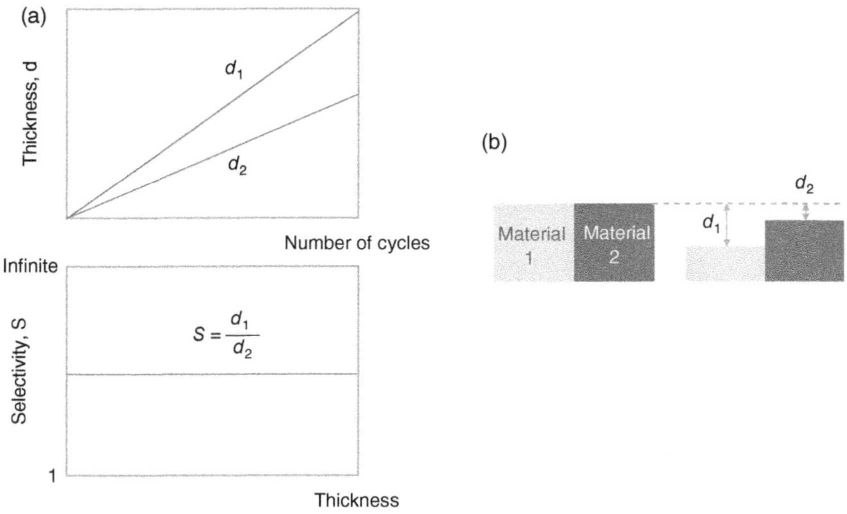

Figure 4.18 Definition of selectivity for etching processes.

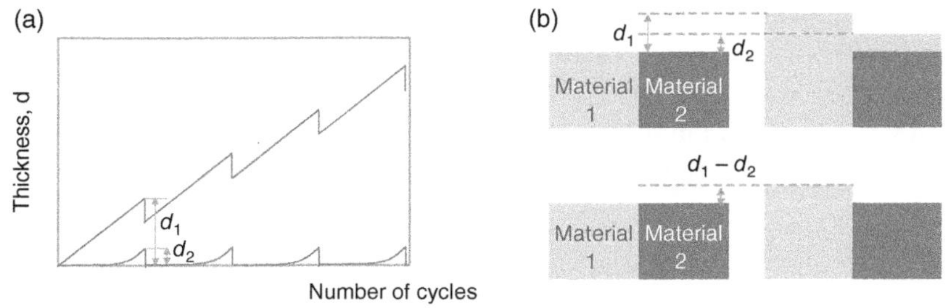

Figure 4.19 Schematic illustration of selective ALD using selective ALE as correction step.

a "repair" step because of its high intrinsic selectivity. Thermal isotropic ALE, however, cannot suppress the so-called "mushroom" growth where material grows on the sidewall of the growing feature. Directional plasma steps reduce the EPC on the top of the material in the growth area, and could be used to remove the sidewall growth with the thermal isotropic ALE step.

4.4.2 Formation of Lateral Devices

Vertical device integration is one of the responses of the semiconductor industry to the performance and cost challenges in traditional scaling. Fin field effect transistor (FinFET) logic devices and 3D NAND flash memory are good examples. The latter represent a very cost-effective way to integrate memory devices. Figure 4.20 shows a high level and generalized integration flow to form device stacks in the lateral direction on the sidewall of a high aspect ratio feature. This approach can in principle also be applied to other emerging memory devices.

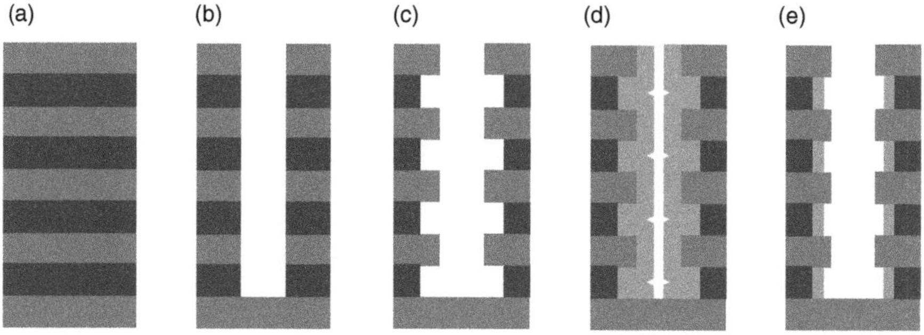

Figure 4.20 Schematic illustration of the formation of horizontal devices using a 3D vertical integration scheme. (a) Stack deposition. (b) High aspect ratio etch. (c) isotropic etch. (d) ALD. (e) ALE.

To form lateral memory devices, a stack of alternating layers is deposited and a high aspect ratio via is etched using RIE and a hardmask. Next, one of the layers is recessed with a selective and isotropic etching technology such as wet etching, thermal etching, thermal isotropic ALE, or radical etching. The recesses layer is conductive to form an electrode. ALD is used to fill the recess conformally. The material can, for instance, be a metal oxide in the case of a resistive random access memory (ReRAM). The deposition thickness must be larger than 50% of the gap height. ALE is then used to recess the layer back to a thickness that is required to form the device.

These ALD and ALE steps can be repeated to form several layers stacked in lateral direction. The device can be connected electrically through the via hole. The number of stacks in the vertical direction can be large. Hundred and more devices can be formed simultaneously in the vertical direction using ALD and ALE in this manner. If thermal isotropic ALE is used, plasma damage is avoided, which is a challenge for forming of sub-10 nm devices with RIE. The critical element of a device formed using ALD and ALE is the interface between the layers. If ligand exchange reaction ALE is used, it is important to remove the remaining fluorine from the surface after etching.

Gate-all-around (GAA) InGaAs fins and nanowires have been fabricated using thermal isotropic ALE (Lu et al. 2018, 2019). The process utilized HF/DMAC at 300 °C, which gave an EPC of 0.24 and 0.62 Å for $In_{0.53}Ga_{0.47}As$ on $In_{0.52}Al_{0.48}As$, respectively. In a first step, fins comprising of $In_{0.53}Ga_{0.47}As$ on $In_{0.52}Al_{0.48}As$ with a width of 28 nm were etched with RIE. Then, thermal isotropic ALE was used to reduce the width of the $In_{0.53}Ga_{0.47}As$ to 4 nm while the $In_{0.52}Al_{0.48}As$ layer underneath was completely removed. Thus, a suspended $In_{0.53}Ga_{0.47}As$ nanowire was etched and subsequently encapsulated with dielectric Al_2O_3. The transmission electron microscopy (TEM) of the final structure is shown in Figure 4.21. The resulting field effect transistor (FET) device was fully functional (Lu et al. 2018, 2019).

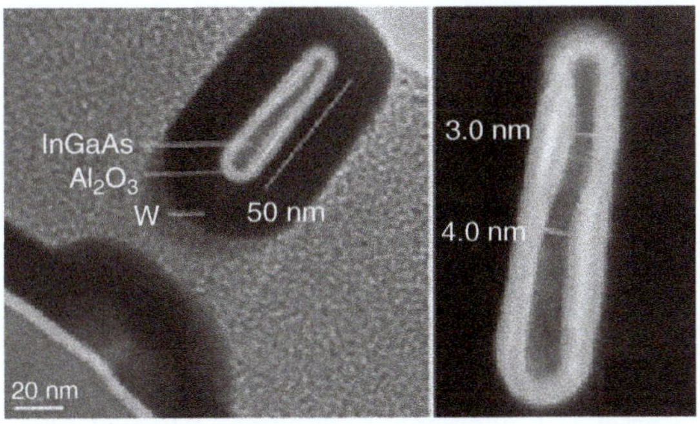

Figure 4.21 TEM cross sections of a 4 nm × 50 nm InGaAs nanowire encapsulated by 3 nm Al_2O_3. Source: Lu et al. (2018).

Problems

P4.1 What is the role of the second surface reaction in thermal isotropic ALE?

P4.2 Name three important tests that are used to identify ALE processes?

P4.3 Explain the role of etching selectivity for the design of thermal isotropic ALE processes with reaction-assisted desorption?

P4.4 Why is the isotropy of Al_2O_3 ALE with HF/Sn(acac)2 less than 100%?

P4.5 How are conversion and ligand exchange ALE related?

P4.6 How does the selectivity to SiO_2 for Al_2O_3 HF/TMA change as a function of pressure?

P4.7 In which step does etching occur on conversion ALE of WO_3 with BCl_3/HF? What are the saturation mechanisms in each step?

P4.8 What are the saturation mechanisms of oxidation/fluorination ALE of TiN with O_3/HF?

P4.9 What are the contributing factors to the throughput of thermal isotropic ALE processes?

P4.10 Explain the relationship of uniformity and throughput for thermal isotropic ALE?

References

Abdulagatov, A.I. and George, S.M. (2018). Thermal atomic layer etching of silicon using O_2, HF, and $Al(CH_3)_3$ as the reactants. *Chem. Mater.* 30: 8465–8475.

Abdulagatov, A.I. and George, S.M. (2020). Thermal atomic layer etching of silicon nitride using an oxidation and "conversion etch" mechanism. *J. Vac. Sci. Technol., A* 38: 022607 1–9.

Basher, A.H., Krstic, M., Takeuchi, T. et al. (2020). Stability of hexafluoroacetylacetone molecules on metallic and oxidized nickel surfaces in atomic-layer-etching processes. *J. Vac. Sci. Technol., A* 38: 022610 1–8.

Cano, A.M., Marquardt, A.E., DuMont, J.W., and George, S.M. (2019). Effect of HF pressure on thermal Al_2O_3 atomic layer etch rates and Al_2O_3 fluorination. *J. Phys. Chem.* 123: 10346–10355.

Carver, C.T., Plombon, J.J., Romero, P.E. et al. (2015). Atomic layer etching: an industry perspective. *ECS Solid State Sci. Technol.* 4: N5005–N5009.

Clancey, J.W. and George, S.M. (2017). In situ mass spectrometer studies of volatile etch products during Al_2O_3 atomic layer etching using HF and trimethylaluminum. *4th International Atomic Layer Etching Workshop, Poster Presentation*, Denver, Colorado, USA (15–17 July 2017).

Clancey, J., Cavanagh, A.S., Smith, J.E.T. et al. (2020). Volatile etch species produced during thermal Al_2O_3 atomic layer etching. *J. Phys. Chem. C* 124: 287–299.

DuMont, J.W. and George, S.M. (2017). Competition between Al_2O_3 atomic layer etching and AlF_3 atomic layer deposition using sequential exposures of trimethylaluminum and hydrogen fluoride. *J. Chem. Phys.* 146: 052819 1–10.

DuMont, J.W., Marquardt, A.E., Cano, A.M., and George, S.M. (2017). Thermal atomic layer etching of SiO_2 by a "conversion-etch" mechanism using sequential reactions of trimethylaluminum and hydrogen fluoride. *ACS Appl. Mater. Interfaces* 9: 10296–10307.

Elam, J.W., Groner, M.D., and George, S.M. (2002). Viscous flow reactor with quartz crystal microbalance for thin film growth by atomic layer deposition. *Rev. Sci. Instrum.* 73: 2981–2987.

Fischer, A., Janek, R., Boniface, J. et al. (2017). Plasma-assisted thermal atomic layer etching of Al_2O_3. *Proceedings of the SPIE 10149*, 101490H 1–5.

Fischer, A., Routzahn, A., and Lill, T. (2019). Characterization of isotropic thermal ALE of oxide films and nanometer-size structures. *66th AVS Symposium*, Columbus, Ohio, October 2019.

Fischer, A., Routzahn, A., Lee, Y. et al. (2020a). Thermal etching of AlF_3 and thermal atomic layer etching of Al_2O_3. *J. Vac. Sci. Technol., A* 38: 022603 1–7.

Fischer, A., Routzahn, A., Wen, S., and Lill, T. (2020b). Causes of anisotropy in thermal atomic layer etching of nano-srtuctures. *J. Vac. Sci. Technol., A* 38: 042601 1–8.

George, S.M. (2010). Atomic layer deposition: an overview. *Chem. Rev.* 110: 111–131.

George, S.M. and Lee, Y. (2016). Prospects for thermal atomic layer etching using sequential, self-limiting fluorination and ligand-exchange reactions. *ACS Nano* 10: 4889–4894.

George, M.A., Hess, D.W., Beck, S.E. et al. (1996). Reaction of 1,1,1,5,5,5-hexafluoro-2,4-pentadione (H^+fac) with iron and iron oxide thin films. *J. Electrochem. Soc.* 143: 3257–3266.

Gertsch, J., Cano, A.M., Bright, V.M., and George, S.M. (2019). SF_4 as the fluorination reactant for Al_2O_3 and VO_2 thermal atomic layer etching. *Chem. Mater.* 31: 3624–3635.

Gladfelter, W.L. (1993). Selective metallization by chemical vapor deposition. *Chem. Mater.* 5: 1372–1388.

Hall, E.H., Blocher, J.M., and Campbell, I.E. (1958). Vapor pressure of titanium tetrafluoride. *J. Electrochem. Soc.* 105: 275–278.

Hennessy, J., Moore, C.S., Balasubramanian, K. et al. (2017). Enhanced atomic layer etching of native aluminum oxide for ultraviolet optical applications. *J. Vac. Sci. Technol., A* 35: 041512 1–9.

Johnson, N.R. and George, S.M. (2017). WO_3 and W thermal atomic layer etching using "conversion-fluorination" and "oxidation-conversion-fluorination" mechanisms. *ACS Appl. Mater. Interfaces* 9: 34435–34447.

Johnson, N.R., Sun, H., Sharma, K., and George, S.M. (2016). Thermal atomic layer etching of crystalline aluminum nitride using sequential, self-limiting hydrogen fluoride and $Sn(acac)_2$ reactions and enhancement by H_2 and Ar plasmas. *J. Vac. Sci. Technol., A* 34: 050603 1–5.

Kanarik, K.J., Lill, T., Hudson, E.A. et al. (2015). Overview of atomic layer etching in the semiconductor industry. *J. Vac. Sci. Technol., A* 33: 020802 1–14.

Kanarik, K.J., Tan, S., Yang, W. et al. (2017). Predicting synergy in atomic layer etching. *J. Vac. Sci. Technol., A* 35: 05C302 1–7.

Konh, M., He, C., Lin, X. et al. (2019). Molecular mechanisms of atomic layer etching of cobalt with sequential exposure to molecular chlorine and diketones. *J. Vac. Sci. Technol., A* 37: 021004 1–8.

Lee, Y. and George, S.M. (2015). Atomic layer etching of Al_2O_3 using sequential, self-limiting thermal reactions with $Sn(acac)_2$ and hydrogen fluoride. *ACS Nano* 9: 2061–2070.

Lee, Y. and George, S.M. (2017). Thermal atomic layer etching of titanium nitride using sequential, self-limiting reactions: oxidation to TiO_2 and fluorination to volatile TiF_4. *Chem. Mater.* 29: 8202–8210.

Lee, Y. and George, S.M. (2018). Thermal atomic layer etching of HfO_2 using HF for fluorination and $TiCl_4$ for ligand-exchange. *J. Vac. Sci. Technol., A* 36: 061504 1–9.

Lee, Y. and George, S.M. (2019). Thermal atomic layer etching of Al_2O_3, HfO_2, and ZrO_2 using sequential hydrogen fluoride and dimethylaluminum chloride exposures. *J. Phys. Chem. C* 123: 18455–18466.

Lee, Y., DuMont, J.W., and George, S.M. (2015a). Atomic layer etching of HfO_2 using sequential, self-limiting thermal reactions with $Sn(acac)_2$ and HF. *ECS J. Solid State Sci. Technol.* 4: N5013–N5022.

Lee, Y., DuMont, J.W., and George, S.M. (2015b). Mechanism of thermal Al_2O_3 atomic layer etching using sequential reactions with $Sn(acac)_2$ and HF. *Chem. Mater.* 27: 3548–3557.

Lee, Y., DuMont, J.W., and George, S.M. (2015c). Atomic layer etching of AlF_3 using sequential, self-limiting thermal reactions with $Sn(acac)_2$ and hydrogen fluoride. *J. Phys. Chem. C* 119: 25385–25393.

Lee, Y., Huffman, C., and George, S.M. (2016a). Selectivity in thermal atomic layer etching using sequential, self-limiting fluorination and ligand exchange reactions. *Chem. Mater.* 28: 7657–7665.

Lee, Y., DuMont, J.W., and George, S.M. (2016b). Trimethylaluminum as the metal precursor for the atomic layer etching of Al_2O_3 using sequential, self-limiting thermal reactions. *Chem. Mater.* 28: 2994–3003.

Lill, T., Kanarik, K.J., Tan, S. et al. (2018). Benefits of atomic layer etching (ALE) for material selectivity. *ALD/ALE conference 2018*, Incheon, Korea.

Lin, X., Chen, M., Janotti, A., and Opila, R. (2018). In situ XPS study on atomic layer etching of Fe thin film using Cl_2 and acetylacetone. *J. Vac. Sci. Technol., A* 36: 051401 1–8.

Lu, W., Lee, Y., Murdzek, J. et al. (2018). First transistor demonstration of thermal atomic layer etching: InGaAs FinFETs with sub-5 nm Fin-width featuring in situ ALE-ALD. *2018 IEEE International Electron Devices Meeting (IEDM)*, pp. 895–898.

Lu, W., Lee, Y., Gertsch, J.C. et al. (2019). In situ thermal atomic layer etching for Sub-5 nm InGaAs multigate MOSFETs. *Nano Lett.* 19: 5159–5166.

Mackus, A.J.M., Merkx, M.J.M., and Kessels, W.M.M. (2019). From the bottom-up: toward area-selective atomic layer deposition with high selectivity. *Chem. Mater.* 31: 2–12.

Mameli, A., Verheijen, M.A., Mackus, A.J.M. et al. (2018). Isotropic atomic layer etching of ZnO using acetylacetone and O_2 plasma. *ACS Appl. Mater. Interfaces* 10: 38588–38595.

Miyoshi, N., Kobayashi, H., Shinoda, K. et al. (2017). Atomic layer etching of silicon nitride using infrared annealing for short desorption time of ammonium fluorosilicate. *Jpn. J. Appl. Phys.* 56: 06HB01 1–7.

Mohimi, E., Chu, X.I., Trinh, B.B. et al. (2018). Thermal atomic layer etching of copper by sequential steps involving oxidation and exposure to hexafluoroacetylacetonate. *ECS J. Solid State Sci. Technol.* 7: P491–P495.

Mullins, R., Natarajan, S.K., Elliott, S.D., and Nolan, M. (2019). Mechanism of the HF pulse in the thermal atomic layer etch of HfO_2 and ZrO_2: a first principles study. ChemRxiv. Preprint. https://doi.org/10.26434/chemrxiv.11310860.v1.

Mullins, R., Natarajan, S.K., Elliott, S.D., and Nolan, M. (2020). Self-limiting temperature window for thermal atomic layer etching of HfO_2 and ZrO_2 based on the atomic scale mechanism. *Chem. Mater. 32: 3414*

Murdzek, J.A. and George, S.M. (2019). Thermal atomic layer etching of amorphous and crystalline hafnium oxide, zirconium oxide, and hafnium zirconium oxide. *2019 International Symposium on VLSI Technology, Systems and Application (VLSI-TSA)*, Hsinchu, Taiwan, pp. 1–2.

Murdzek, J.A. and George, S.M. (2020). Effect of crystallinity on thermal atomic layer etching of hafnium oxide, zirconium oxide, and hafnium zirconium oxide. *J. Vac. Sci. Technol., A* 38: 022608.

Natarajan, S.K. and Elliott, S.D. (2018). Modeling the chemical mechanism of the thermal atomic layer etch of aluminum oxide: a density functional theory study of reactions during HF exposure. *Chem. Mater.* 30: 5912–5922.

Paeng, D., Zhang, H., and Kim, Y.S. (2019). Dynamic temperature control enabled atomic layer etching of titanium nitride. *ALD/ALE 2019*, Bellevue, VA, USA.

Rahman, R., Mattson, E.C., Klesko, J.P. et al. (2018). Thermal atomic layer etching of silica and alumina thin films using trimethylaluminum with hydrogen fluoride or fluoroform. *ACS Appl. Mater. Interfaces* 10: 31784–31794.

Roozeboom, F., van den Bruele, F., Creyghton, Y.Y. et al. (2015). Cyclic etch/passivation-deposition as an all-spatial concept toward high-rate room temperature atomic layer etching. *ECS J. Solid State Technol.* 4: N5076–N5067.

Shinoda, K., Miyoshi, N., Kobayashi, H. et al. (2019). Rapid thermal cyclic atomic-layer etching of titanium nitride in CHF_3/O_2 downstream plasma. *J. Phys. D: Appl. Phys.* 52: 475106 1–9.

Wang, Z. and Opila, R.L. (2020). In operando X-ray photoelectron spectroscopy study of mechanism of atomic layer etching of cobalt. *J. Vac. Sci. Technol., A* 38: 022611.

Xie, W. and Parsons, G.N. (2020). Thermal atomic layer etching of metallic tungsten via oxidation and etch reaction mechanism using O_2 or O_3 for oxidation and WCl_6 as the chlorinating etchant. *J. Vac. Sci. Technol., A* 38: 022605 1–10.

Xie, W., Lemaire, P.C., and Parsons, G.N. (2018). Thermally driven self-limiting atomic layer etching of metallic tungsten using WF_6 and O_2. *ACS Appl. Mater. Interfaces* 10: 9147–9154.

Zhao, J., Konh, M., and Teplyakov, A. (2018). Surface chemistry of thermal dry etching of cobalt thin films using hexafluoroacetylacetone (hfacH). *Appl. Surf. Sci.* 455: 438–445.

Zywotko, D.R. and George, S.M. (2017). Thermal atomic layer etching of ZnO by a "conversion-etch" mechanism using sequential exposure of hydrogen fluoride and trimethylaluminum. *Chem. Mater.* 29: 1183–1191.

Zywotko, D.R., Faguet, J., and George, S.M. (2018). Rapid atomic layer etching of Al_2O_3 using sequential exposures of hydrogen fluoride and trimethylaluminum with no purging. *J. Vac. Sci. Technol., A* 36: 061508 1–11.

5

Radical Etching

5.1 Mechanism of Radical Etching

The mechanism for radical etching can be gleaned from the Lennard-Jones diagram for thermal etching in Figure 2.5. For radical etching, the incoming species are radicals unlike molecules in thermal etching. Hence, the reaction trajectory of the adsorption step is the reverse of the trajectory for dissociative chemisorption. The Lennard-Jones diagram for radical etching is shown in Figure 5.1. If adsorption is thermodynamically favorable, the incoming radicals see a potential energy well and will be attracted to the surface up to a distance with the lowest potential energy. Now, we can compare the energies of the adsorbent to surface bond and the bond between the first and second surface layers. If the latter is lower, etching can occur if the surface temperature is high enough for desorption. The considerations for desorption are the same for thermal and radical etching. The difference between the two is in the adsorption step.

The benefit of radical etching is a larger process window with respect to surface temperature compared to thermal etching. Because radicals are extremely reactive, their activation energy for adsorption is essentially zero. Therefore, radical adsorption is temperature insensitive at least in the range of temperatures considered in the manufacturing of semiconductor devices. This increases the overlap of the adsorption and desorption temperature windows as shown in Figure 5.2.

Radicals also provide higher etching rates because the adsorption step has no activation barrier. Humbird and Graves performed molecular dynamics simulations of reactions between F^*, F_2, Cl^*, and Cl_2 and silicon and found that radicals etch silicon at surface temperatures above 200 K while the molecular species do not etch (Humbird and Graves 2004).

The most common radicals that are used for radical etching are atomic radicals such as F^*, Cl^*, Br^*, O^*, and H^* as well as smaller molecular radicals such as NO^*, OH^*, NH_2^*, CH^*, CH_2^*, CH_3^*, and others. The reason why these radicals contain only a few atoms is the use of high-density plasmas to generate them. High-density plasmas dissociate molecules very efficiently. There are attempts to generate more complex molecules by activation with light, but these are far from implementation.

Atomic Layer Processing: Semiconductor Dry Etching Technology, First Edition. Thorsten Lill.

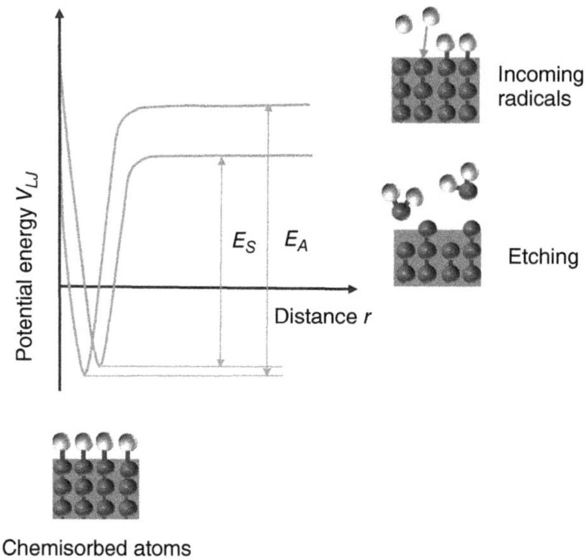

Figure 5.1 Lennard-Jones diagram for radical etching.

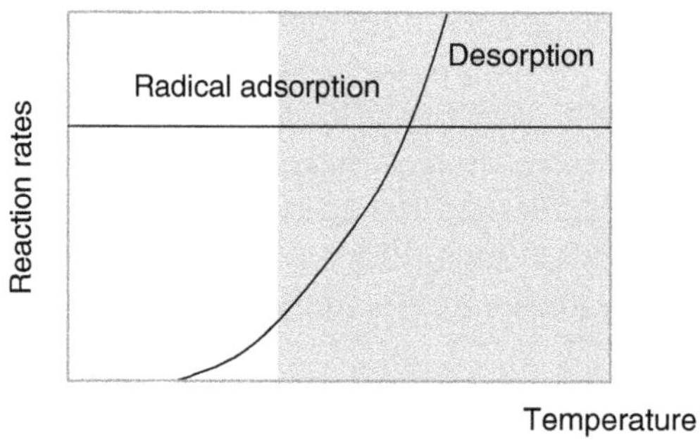

Figure 5.2 Schematic illustration of the temperature process window for radical etching.

5.2 Performance Metrics

5.2.1 Etching Rate and ERNU

Like thermal etching, radical etching also uses only neutral species. In the simplest implementation, only one type of radical is used. This is, for instance, the case for photoresist removal using oxygen radicals generated in a downstream plasma. As for thermal etching, the etching rate is limited by the elementary step with the lowest rate. The elementary process steps of radical etching are the radical transport to the surface, adsorption, surface reaction, and desorption. Because radicals are very reactive, adsorption is unlikely to be the limiting step. If the surface temperature is high enough to facilitate fast surface reaction and desorption, the rate-limiting step will be the rate with which radicals arrive at the surface. Therefore, most radical etching reactors operate at pressures of several 100 mTorr and heat the wafers to elevated temperatures. Under these conditions, the etching rate is driven by the efficiency of the radical source and can be as high as several micrometers per minute.

Etching rate nonuniformity (ERNU) is primarily driven by the flux uniformity to the wafer. The geometry of the reactor and the locations and dimensions of the gas injection points must be optimized to meet ERNU targets. As a rule of thumb, ERNU is more difficult to obtain for radical etching than for atomic layer etching (ALE) with self-limiting steps and ion-flux-limited reactive ion etching (RIE) processes.

Some radical etching processes use more than one kind of neutrals. In some cases, two radicals are used to produce volatile reaction products. This is the case for etching of SiO_2 using a downstream plasma of NH_3 and NF_3 via the formation of volatile ammonium hexafluorosilicate, $(NH_4)_2SiF_6$ (Nishino et al. 1993). This compound sublimates in vacuum at temperatures above 100 °C. We discussed the implementation of this surface chemistry mechanism in the context of thermal etching process and thermal ALE with imperfect self-limitation.

The use of radicals increases the etching rates. Other radical etching processes use etching and passivating radicals and neutrals together to improve the selectivity of the process. In all cases of multiple radical and neutral species, the control of all fluxes to the surface is critical for ERNU. The design of the source and the gas injection system must take this into consideration.

5.2.2 Selectivity

In general, radical etching is used when high selectivity is required; however, its selectivity performance is typically inferior to thermal etching. The fundamental reason is the reactivity of radicals. They carry "excess energy" in the form of chemical energy, which may lead to reactions with more than one material on the wafer surface. Fluorine radicals etch silicon with a high etching rate, but they will also etch Si_3N_4.

The selectivity of radical etching can be boosted using passivating gases. For example, adding oxygen to a N_2/CF_4 or N_2/NF_3 downstream plasma to etch Si_3N_4 can reduce the etching rate of simultaneously exposed silicon and increase the Si_3N_4/Si selectivity. The underlying mechanism is passivation of the silicon surface by means of SiO_2 formation (Kastenmeier et al. 1999).

5.2.3 Profile and ARDE

Radical etching is an isotropic etching technology like thermal etching and thermal isotropic ALE. The root cause for aspect ratio dependent etching (ARDE) in radical etching is the transport of radicals to the etching front. Transport effects can be suppressed by increasing the pressure. This works well for thermal etching with neutral molecules where the adsorption step can be slow enough to become the rate-limiting step at higher pressures. This condition is more difficult or impossible to achieve in radical etching because of the reactivity of radicals. Increasing the pressure will just result in higher etching rates on the top and bottom of the feature without reducing ARDE. Therefore, radical etching has generally a poor ARDE performance.

5.2.4 CD Control

The critical dimension (CD) of radical etching is the remaining film thickness. This can be the thickness of a partially etched layer or the thickness of a stopping layer. In the first case, the remaining film thickness uniformity is driven by ERNU, ARDE, and wafer to wafer repeatability. The remaining thickness of an etch stop layer is determined by the etching selectivity.

5.3 Applications Examples

The most widely used radical etching process is the removal or stripping of photoresist or carbon hardmasks. This task is accomplished using oxygen radicals. A high etching or stripping rate and minimum oxidation of the remaining surface are the critical performance requirements. The latter requirement is most important for resist removal off metal structures where hydrogen radicals are used. The same technology is also deployed for post etch treatments to remove halogen-containing polymers that were generated during the etching. Because exposure of these etching residues to air can lead to condensation and defects, post etch treatment radical etching chambers are frequently integrated with the etching chambers on the same vacuum wafer transfer chamber.

Fluorine radicals are used to etch silicon isotropically and selectively. Advanced logic metal gate structures are formed by etching a sacrificial polysilicon gate. After the formation of the liners and isolation structures, this material must be removed selectively to the gate oxide. In the past, this removal was performed using radical etching to remove the majority of the polysilicon while wet etching removed the rest without damaging the gate oxide. With improvements in radical etching technology damage can be reduced and the entire etch can be performed by radical etching.

Radical etching of silicon nitride is an emerging application for logic and flash memory devices. 3D NAND devices are built inside vertical channels, which are etched into stacks of 100 and more pairs of silicon oxide and silicon nitride. Silicon nitride is removed selectively to silicon oxide and replaced by tungsten to form metal connections between the devices. This removal process is realized today using wet etching. In the future, this application could be performed by radical etching. The chemistry of radical etching of silicon nitride is based on ON^* and F^* radicals (Kastenmeier et al. 1999).

Figure 5.3 illustrates how radical etching can be utilized to form silicon nanowires or nanosheets from a Si/SiGe stack for gate-all-around (GAA) transistors (see Figure 37). Before etching, the sidewall of the line comprised of alternating 10 nm Si and SiGe (30% Ge) was continuous and the goal of the etching process was to recess selectively only SiGe. Selectivity was achieved with a cycling two-step process comprised of a selective silicon surface passivation step (remote He/O_2 plasma) and a silicon germanium etching step (remote $NH_3/NF_3/O_2$ plasma) (Pargon et al. 2019). In this example, the test structures were wider than needed for the final devices to demonstrate the selectivity of the SiGe removal. Even though this process is a cyclic process, it is not ALE because the steps are not self-limited.

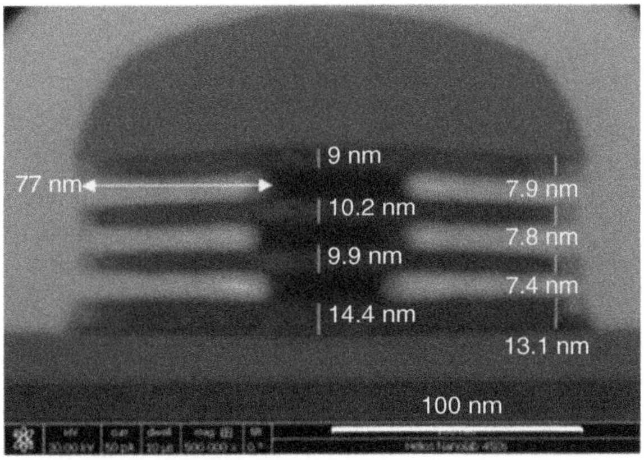

Figure 5.3 Fabrication of stacked silicon nanowires for GAA devices. Source: Pargon et al. (2019).

Problems

P5.1 Why is the process window for radical etching generally larger than for thermal etching?

P5.2 Why is the intrinsic selectivity for radical etching generally lower than for thermal etching?

P5.3 Using Eq. (2.18), explain why ARDE for radical etching is generally worse than for thermal etching?

References

Humbird, D. and Graves, D.B. (2004). Atomistic simulations of spontaneous etching of silicon by fluorine and chlorine. *J. Appl. Phys.* 96: 791–798.

Kastenmeier, B.E.E., Matsuo, P.J., and Oehrlein, G.S. (1999). Highly selective etching of silicon nitride over silicon and silicon dioxide. *J. Vac. Sci. Technol., A* 17: 3179–3184.

Nishino, H., Hayasaka, N., and Okano, H. (1993). Damage-free selective etching of Si native oxides using NH_3/NF_3 and SF_6/H_2O down-flow etching. *J. Appl. Phys.* 74: 1345–1348.

Pargon, E., Petit-Etienne, C., Youssef, L. et al. (2019). New route for selective etching in remote plasma source: application to the fabrication of horizontal stacked Si nanowires for gate all around devices. *J. Vac. Sci. Technol., A* 37: 040601 1–5.

6

Directional ALE

6.1 Mechanism of Directional ALE

The majority of etch processes in the semiconductor industry require directionality. This means that the etching rate must be larger in the vertical than in the horizontal direction. This gave rise to reactive ion etching (RIE) as a replacement for wet etching about 40 years ago. It is also a desired capability for atomic layer etching (ALE) processes.

For ALE to be directional, at least one of the two steps must be directional. Directionality is realized by ion bombardment originating from a plasma in direct contact with the wafer or by energetic beams of ions or neutrals. Electrons and photons could potentially also be used for directional ALE. In this section, we will focus on directional ALE using ions and fast atoms. An overview of electron- and photon-assisted etching and ALE can be found in Chapter 10. Ions and fast atoms can be deployed in the modification step or the removal step or in both steps (see Figure 6.1). The latter case is of less practical interest.

6.1.1 ALE with Directional Modification Step

In the case of a directional surface modification step, a thin modified layer is created by shallow implantation of accelerated ions or atoms (see Section 2.6). This form of ALE is directional because the surface is modified by directional implantation of ions (Chang and Chang 2017; Sang and Chang 2020). The kinetic energy of the ions or atoms must be high enough to penetrate the crystal lattice and to form a modified layer but low enough to prevent too much sputtering to ensure synergy. Depth of modification is self-limited by the projected range of the atoms R_p.

This process is, however, not self-limited with respect to concentration. The concentration at any given depth increases with dose. Eventually, a steady state is reached due to sputtering. At that point, synergy would be compromised. Therefore, dose control by time is needed.

The modified layer is removed by thermal isotropic etching or radical etching, which have been described (see Chapters 2 and 5). The synergy requirement for ALE dictates that the removal steps must have infinite selectivity with respect to

Atomic Layer Processing: Semiconductor Dry Etching Technology, First Edition. Thorsten Lill.

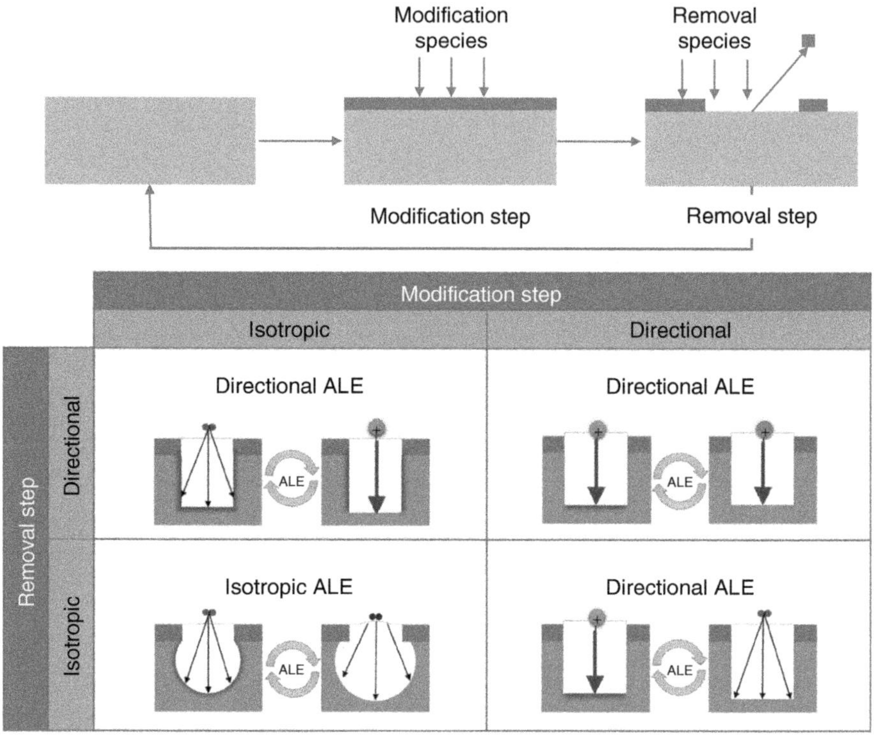

Figure 6.1 Classification of ALE by directionality. Source: Lill et al. (2016).

the unmodified bulk material. In one embodiment of this class of directional ALE, wet processing can be used to remove the modified layer. Engineering challenges of a process time efficient dry/wet integration are, however, a large barrier for practical implementation.

An interesting implementation of this type of ALE is the use of large oxygen gas clusters in an oxidation step (Toyoda and Ogawa 2017). Etching of copper was demonstrated using a 5 keV beam containing 2000–3000 oxygen atoms alternating with a neutral beam of acetic acid, which increased the chamber pressure to 1×10^{-5} Pa. Bulk sputtering was observed for cluster energies above 5 keV or about 2 eV per impinging atom. With exposure to acetic acid, material removal was observed for cluster energies below 5 keV.

The proposed etching mechanism has gas clusters heating the surface locally upon impact. This stimulates the reaction of adsorbed acetic acid with the oxidized Cu surface and leads to the formation of $Cu(CH_3COO)_2$, which is sputtered by the cluster beam. Simultaneously, the surface is freshly oxidized. This mechanism is supported by the observation that removal occurs during the ion cluster bombardment step, not during exposure to acetic acid. Since the ion beam contributes both to the modification and removal, this process actually constitutes ALE with both directional modification and removal steps.

6.1.2 ALE with Directional Removal Step and Modification by Chemisorption and Diffusion

In the case of a directional removal step, the surface is modified by neutrals or radicals. Subsequently, the modified surface layer is removed by low-energy ion bombardment. This approach is the most commonly used implementation of ALE in the semiconductor industry to date. One of the reasons why it finds acceptance in industrial applications is that it can be executed in production-worthy plasma etch chambers with the appropriate hardware to allow fast step to step transitions.

As in thermal isotropic ALE and atomic layer deposition (ALD), the steps in directional ALE can be separated by time or spatially (Roozeboom et al. 2015; Faraz et al. 2015). This classification is not fundamental but a question of engineering implementation. However, it is important that the fluxes of the two steps must be completely separated. Any coexistence will introduce the contribution of parasitic RIE. Step overlap in thermal isotropic ALE introduces parasitic thermal etching. Parasitic RIE or thermal etching diminishes the benefits of ALE because synergy is reduced. In this chapter, we focus on directional ALE with isotropic surface modification via chemisorption, diffusion/conversion and deposition, and removal by ions or fast nonreactive neutrals.

The archetypical system for directional ALE with removal by ion bombardment is ALE of silicon with surface modification by Cl_2 molecules or Cl^* radicals and removal with positively charged Ar ions or fast argon neutrals. In the following, we will denote this process as silicon ALE with Cl_2/Ar^+. This also covers fast argon neutrals because they are accelerated as ions and neutralized before impacting the surface. From an etching mechanism point of view, the charge of the particle is not important because the removal process is by nature an atomic collision process (Wehner 1955). Whether the accelerated atoms carry a charge or not is important only for etching of three-dimensional (3D) features where surface charging can change the trajectory of ions and lead to profile distortions.

Detailed overviews of the history of ALE in general, and silicon ALE with Cl_2/Ar^+ in particular, can be found in review articles by Kanarik (2015) and Oehrlein et al. (2015). The first reports on the use of chlorine in silicon ALE were published by Matsuura et al. (1993) and Athavale and Economu (1996). To explain the mechanism of silicon ALE with Cl_2/Ar^+, we must analyze the modification and removal steps. In the modification step, chlorine is adsorbed at the silicon surface. If chlorination is realized with neutrals, adsorption is a dissociative chemisorption process as shown in Figure 2.5. If chlorination is achieved with radicals, the adsorption is a radical adsorption process as shown in Figure 5.2.

To prevent thermal or spontaneous etching, the surface temperature must be held below the desorption temperature of silicon chloride reaction products. The resulting surface condition is shown in Figure 2.1, which describes the universal surface condition for chemically enhanced etching. The role of chlorine is to form a strong bond to the silicon surface atoms and to weaken the bond between the top silicon layer and the bulk silicon. Under the conditions that E_s is smaller than E_A and E_0, and additional energy can selectively break the bonds between the surface and bulk silicon. This framework for chemically enhanced etching can be extended to surfaces that have been modified by subsurface diffusion. Diffusion can form a thin silicon and chlorine containing mixed layer with weaker bonds than the undisturbed silicon bulk layer.

Surface modification by halogen adsorption and diffusion is common across many etching technologies. If the modification reactant is a molecule and the surface temperature is high enough to stimulate desorption, the result is thermal etching. If the modification reactant is a radical and desorption is possible at the given wafer temperature, the result is radical etching. If the modification reactant is a molecule or a radical but temperature cycling or a second surface reaction is needed, the result is thermal isotropic ALE. In case of surface modification with molecules or radicals and ion bombardment-stimulated desorption, the resulting etching process is directional ALE.

In the case of directional ALE, accelerated ions deliver enough energy to stimulate materials removal. Ion impact increases the surface temperature at the location of the ion impact. As fast ions or atoms impact the surface, a collision cascade is produced. Momentum is transferred from one lattice atom to the next akin to a three-dimensional billiard game. Depending on the relative angle of the colliding atoms, the direction of the collision cascade can change and eventually turn back to the surface where a surface atom can be ejected. The result is physical etching or sputtering (Sigmund 1981).

Let us now apply the collision cascade theory to a chemisorbed surface where E_s is smaller than E_0. According to Eqs. (2.7) and (2.8), the sputtering threshold is lower, and the sputtering yield is higher for the weaker bonded top layer. Figure 6.2 depicts a collision cascade for a silicon surface with chemisorbed chlorine. The sputtering threshold is below 50 eV for this system (Tan et al. 2015) and between 20 and 100 eV for most materials (see Table 2.1). The ratio of ion energy to the energy for breaking a bond is typically around 10 : 1. Thus, 90% of the energy is "excess" removal energy, which is dissipated into the lattice where it can create damage or cause sputtering of the bulk material. Ion-assisted etching is therefore less selective than radical or vapor dry etching. The benefit of using fast ions or atoms for desorption is a greatly enlarged temperature window, especially when radicals are used in the modification step (see Figure 6.3).

The chemically enhanced sputtering mechanism, which is depicted in Figure 6.2, can be realized with neutral and ion beams. One of the seminal experiments in research of plasma etching was conducted by Coburn and Winters using beams.

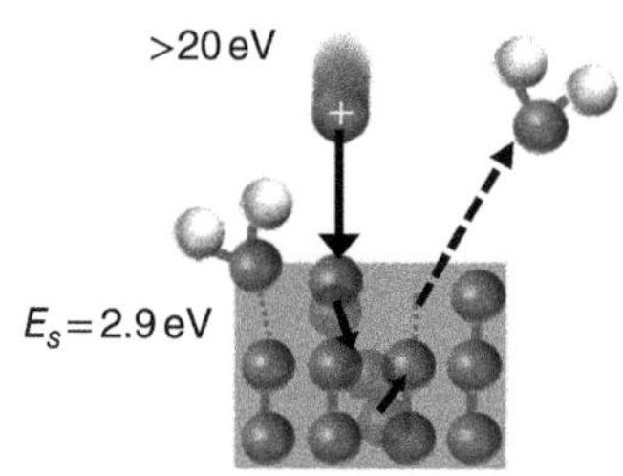

Figure 6.2 Illustration of a collision cascade for an ion impinging a silicon surface with adsorbed chlorine. Over 20 eV of energy is needed to break a bond of 2.9 eV.

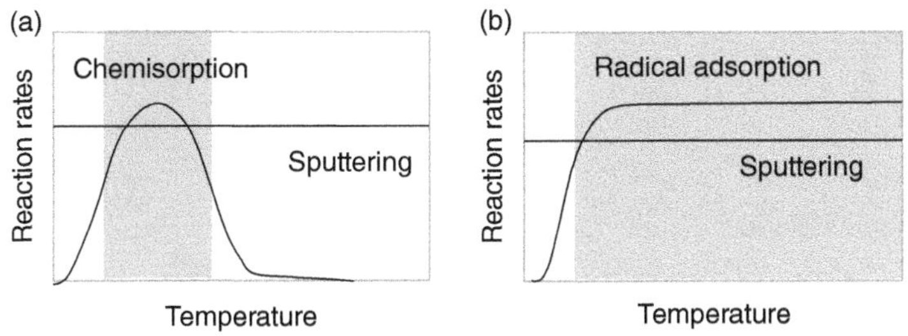

Figure 6.3 Schematic illustration of the process window of directional ALE as a function of surface temperature for etching with (a) neutrals and ions, and for radicals and ions (b).

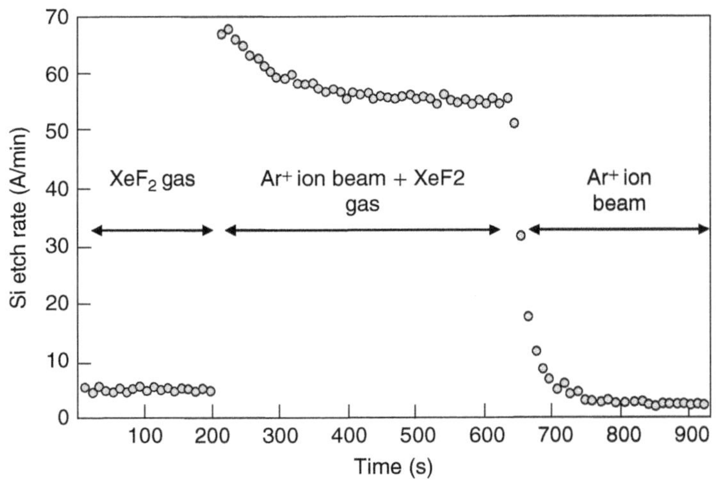

Figure 6.4 Seminal experiment by Coburn and Winters demonstrating neutral – ion synergy. Source: Coburn and Winters (1979). © 1979 AIP Publishing.

They deployed a XeF_2 neutral beam and a 450 eV Ar^+ ion beam to etch silicon (Coburn and Winters 1979). As shown in Figure 6.4, a meaningful etching rate was obtained only when both beams were on simultaneously.

This result illustrates the effect of weakening surface bonds by fluorine adsorption and the resulting enhancement of the sputtering yield. This effect is also called "ion–neutral synergy." Coburn and Winters studied simultaneous neutral and ion beams and discovered RIE. If they would have alternated between these beams and reduced the ion energy below the sputtering threshold of silicon, they would have observed directional ALE. This shows that both directional ALE and RIE leverage ion–neutral synergy.

The removal process in ion-assisted directional ALE can be understood and modeled by applying sputtering theory to directional ALE (Berry et al. 2018). Directional ALE can be achieved when the sputtering thresholds of the modified layer and the bulk material are sufficiently different, and the ion energy has a value between these two thresholds. A functional relationship between the sputtering threshold and the binding energy is given in Eq. (2.8) (Mantenieks 1999). For ion energies far above

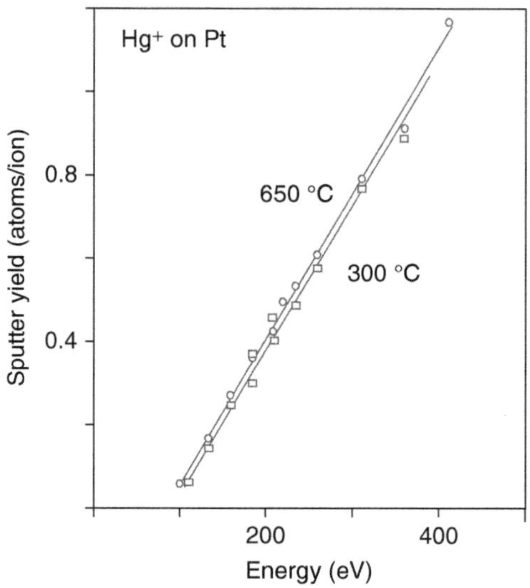

Figure 6.5 Sputtering yield as a function of ion energy for sputtering of platinum with mercury ions. Source: Wehner (1956). © 1956 American Physical Society.

the sputtering threshold, the sputtering yield follows a square root dependence on the ion energy (Sigmund 1981; Steinbrüchel 1989). Chang and Sawin found a square root energy dependence also near threshold sputtering of silicon with chlorine atoms (Chang and Sawin 1997a). However, other dependencies have been proposed for ion energies close to the sputtering threshold (Steinbrüchel 1989; Zalm 1984). Recent molecular dynamics (MD) simulations show a linear energy dependence for very low ion energies (Yan and Zhang 2012).

Figure 6.5 shows as a classic example the sputtering yield of platinum with mercury ions for energies between 100 and 400 eV and for two different temperatures, 300 and 650 °C (Wehner 1956). For this particular system, the sputtering yield is a linear function of ion energy. The graph also shows that the difference between the two target temperatures is small and within the range of possible errors. Therefore, the ideal ALE curve for directional ALE is represented as a function of the ion energy of the removal step.

Let us now combine the concept of sputtering curves of the modified and bulk materials to construct the ideal ALE window for directional ALE. Figure 6.6 depicts on the left side sputtering yield curves for the modified and the bulk layer. For simplicity, we have chosen a linear relationship akin to the yield curves in Figure 6.5. On the right side, the etching per cycle (EPC) for an ALE process with removal by ion or fast atom bombardment is depicted as a function of ion energy. The lowest energy where EPC is observed corresponds to the sputtering threshold of chlorinated silicon. The high-energy branch of the window is given by the sputtering threshold of the bulk silicon.

This curve represents the ideal ALE window for directional ALE. It is the analog to the ideal window for thermal isotropic ALE shown in Figure 4.2, which is a function of temperature. In other words, the x-axis of the ALE window represents the energy provided to initiate removal of the modified layer. In real implementations, the ALE window can be distorted by unintended secondary reactions such as

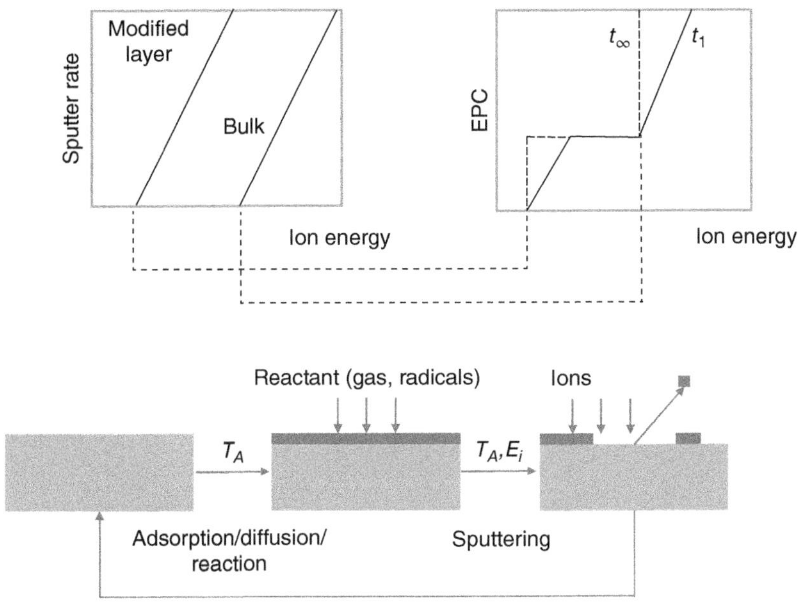

Figure 6.6 Origin of the ideal window of directional ALE as a function of ion energy.

cross-talk between steps (for instance chamber wall effects), spontaneous etching in the surface modification step, or unintended sputtering due to lack of ion energy control.

The link between sputtering yields and ALE window is illustrated in Figure 6.7 where measurements of an ALE curve for silicon ALE with chlorine radical modification and argon plasma removal are shown (Tan et al. 2015). The inserts show results of Monte Carlo simulations of collision cascades for this system (Berry et al. 2018). The plateau of the window starts at about 50 eV and silicon sputtering can be seen from 80 eV. According to the modeling results, for 100 argon ions impacting the surface with 60 eV, only 1 silicon atom is removed from the silicon surface, while 17 silicon atoms are removed from the chlorinated silicon surface. At 200 eV, seven atoms are removed from the silicon surface and 39 from the chlorinated surface. While the removal step is not self-limiting at 200 eV, the removal rate of the modified layer is significantly higher than that of the unmodified layer.

This means that even for energies that are higher than the bulk sputtering threshold, some of the benefits of ALE can be obtained. These benefits are improved uniformity across the wafer, across features with different critical dimension called aspect ratio dependent etching (ARDE), and surface smoothness (Kanarik et al. 2015). Operation under these conditions does provide throughput benefits but the removal step time must be carefully controlled to end when the modified layer is removed (Berry et al. 2018). Throughput and performance must be balanced to achieve the required etch performance. This trade-off is well known in ALD.

The extent and shape of the ALE window shown in Figure 6.8 is determined by the following physical and chemical properties of directional ALE:

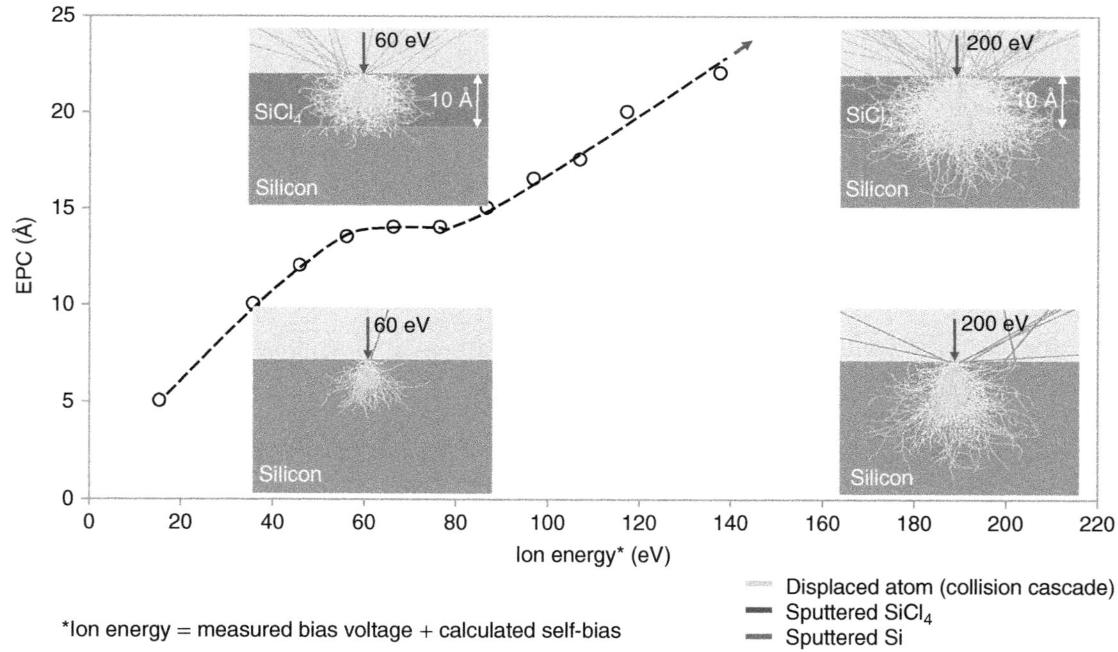

Figure 6.7 Measured ALE window for silicon with chlorine radical modification and argon plasma removal. The inserts show Monte Carlo modeling results for the same system. Source: Tan et al. (2015); Berry et al. (2018).

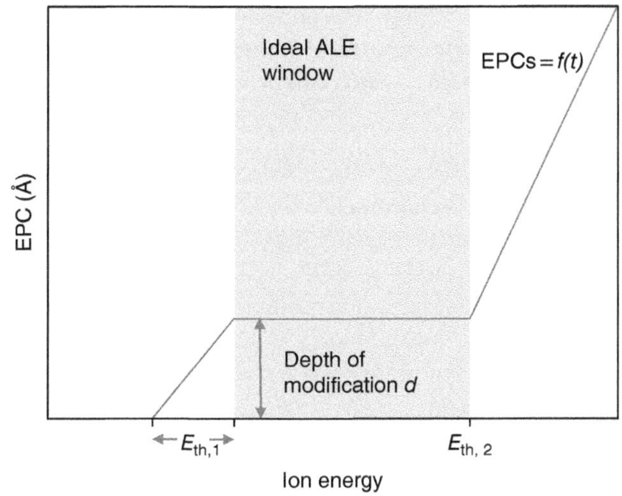

Figure 6.8 Characteristic features of the ideal window for directional ALE.

1) The extent of the ALE window is given by the difference in energy between the sputter thresholds of the modified layer and the bulk material, $E_{th,1}$ and $E_{th,2}$.
2) The upper limit of the ALE window is a physical parameter and is given by the sputter threshold of the bulk material, $E_{th,2}$, the choice of removal ions, and the impact angle.
3) The lower limit of the ALE window is the result of chemical modification. The modification step must be designed to weaken the surface compounds as much as possible without causing spontaneous etching.
4) The slope of the curve for energies higher than the sputter threshold is determined by the duration of the removal step. For infinite step times the slope is vertical.
5) EPC within the ideal ALE window is given by the depth of modification *d*. The depth of modification and the step times determine etch rates for ALE processes.
6) The slope of the curve for ion energies below the ALE window is the result of changing sputtering thresholds as a function of concentration of the modification species. In the case of silicon ALE with Cl_2/Ar^+, it is the concentration of chlorine in silicon. If the surface is single crystal and the modification only adsorption without diffusion, the low-energy slope of the ALE curve will be steeper than for amorphous materials, which are modified by adsorption and diffusion.

In directional ALE, surface modification is accomplished by adsorption and in some cases diffusion and oxidation by oxygen or halogens. The saturation curves can be explained by the same fundamentals as for thermal isotropic ALE, which have been discussed in Chapter 4. In this chapter, we will focus on the saturation curve of removal by sputtering. If the ion energy is above the upper ion energy limit, two sputter events can occur, sputtering of the modified layer and of the bulk material. If sputtering of the bulk material exists, complete saturation of the removal step and therefore 100% synergy cannot be reached.

Let us now explore the functional expression for a non-saturated removal curve. EPC, the amount of removed material, is the integral over time of the instantaneous sputtering rate (SR) during the removal step, which can be expressed as (Gottscho et al. 1992)

$$\mathrm{SR} = \nu\Theta E_i J_i \tag{6.1}$$

Here, ν is the volume removed per unit bombardment energy for a saturated surface ($\mathrm{cm}^3/\mathrm{eV}$), Θ is the surface coverage, E_i is the ion energy (eV), and J_i is the ion flux to the surface ($\mathrm{cm}^{-2}/\mathrm{s}$). In ALE, the ion energy is controlled within a narrow range and ν and E_i can be combined to give the sputtering coefficient k_{sp}, which denotes the volume of material removed per ion impact onto a modified surface site. The sputtering coefficient k_{sp} is equal to the sputtering yield Γ multiplied by the volume per removed atom.

Within the ideal ALE window, only the modified layer is removed. In this case, $k_{\mathrm{sp},1}$ equals the depth of modification d multiplied by the area of material removal A:

$$k_{\mathrm{sp},1} = \nu E_i = dA \tag{6.2}$$

In the case of a surface modified by adsorption, d equals one monolayer. If the surface is modified by adsorption and diffusion, d is the depth of diffusion. The instantaneous sputtering rate of the modified material can be expressed as

$$\mathrm{SR}_1 = \frac{d[N_s]}{dt} = k_{\mathrm{sp},1}[N_0 - N_{\mathrm{sp}}]J_i \tag{6.3}$$

Here, $[N_0]$ is the total number of activated surface sites and N_{sp} is the number of sputtered surface sites. To calculate the total amount of material removed as a function of time, Eq. (6.2) must be integrated:

$$\int_0^{[N_{\mathrm{sp}}]_t} \frac{d[N_{\mathrm{sp}}]}{[N_0] - [N_{\mathrm{sp}}]} = \int_0^t k_{\mathrm{sp},1} J_i dt \tag{6.4}$$

If we normalize $[N_0]$ to 1 and equate $[N_{\mathrm{sp}}]$ to EPC for a give step time of the removal step, the following expression is obtained:

$$[N_{\mathrm{sp}}] = \mathrm{EPC}_1 = 1 - \mathrm{e}^{-k_{\mathrm{sp},1} J_i t} \tag{6.5}$$

Equation (6.5) describes EPC as a function of time for an ideal ALE process without any bulk material sputtering.

Let us now consider an ALE process where the sputter rate of the bulk material is finite. This is the case if the ALE process is operated at ion energies above the upper limit of the ideal ALE window. The instantaneous etching rate, or in this case sputtering rate, is given by the ion flux and the number of sites that have been previously removed by the modified layer times the bulk sputter coefficient $k_{\mathrm{sp},2}$:

$$\mathrm{SR}_2 = k_{\mathrm{sp},2}[N_{\mathrm{sp}}]J_i \tag{6.6}$$

Substituting $[N_{\mathrm{sp}}]$ from Eq. (6.4) and integration over times gives the EPC contribution from bulk sputtering, EPC_2:

$$\mathrm{EPC}_2 = \int_0^t k_{\mathrm{sp},2}(1 - \mathrm{e}^{-k_{\mathrm{sp},1} J_i t}) J_i dt \tag{6.7}$$

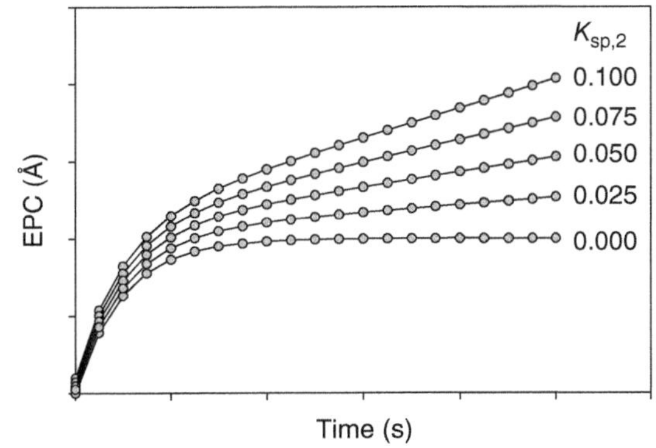

Figure 6.9 Calculated saturation curves of an ion removal step in directional ALE.

Solving the integral results in

$$\mathrm{EPC}_2 = \frac{k_{\mathrm{sp},2}}{k_{\mathrm{sp},1}}(k_{\mathrm{sp},1}J_i t + \mathrm{e}^{-k_{\mathrm{sp},1}J_i t}) \tag{6.8}$$

The total EPC for a nonideal ALE process with significant bulk sputtering equals the sum of EPC_1 and EPC_2:

$$\mathrm{EPC} = 1 - \mathrm{e}^{-k_{\mathrm{sp},1}J_i t} + \frac{k_{\mathrm{sp},2}}{k_{\mathrm{sp},1}}(k_{\mathrm{sp},1}J_i t + \mathrm{e}^{-k_{\mathrm{sp},1}J_i t}) \tag{6.9}$$

Figure 6.9 shows the calculated EPC for ideal and nonideal directional ALE processes using Eq. (6.9) (Lill et al. 2015). Here, $k_{\mathrm{sp},1}$ was normalized to 1 and the $k_{\mathrm{sp},2}$ values are given relative to $k_{\mathrm{sp},1}$. When $k_{\mathrm{sp},2}$ equals zero, this represents the case of ideal ALE without any sputtering of the bulk material, i.e. with complete saturation of the removal process. The values of $k_{\mathrm{sp},2}$ were varied between 0 and 0.1. A value of 0.05 was chosen as the midpoint based on literature data that upon impact on a fluorinated silicon surface, 1 argon ion can remove 25 silicon atoms (Gerlach-Meyer and Coburn 1981). Chang showed that the sputtering yield of silicon does not reach the value of one silicon atom per argon ion below 400 eV (Chang 1997a). Hence values for $k_{\mathrm{sp},2}$ below 0.05 are to be expected for a fluorinated silicon surface. Because the bond energy between silicon and fluorine is stronger than for silicon and chlorine, $k_{\mathrm{sp},2}$ for chlorinated silicon is expected to be larger than 0.05.

For $k_{\mathrm{sp},2} > 0$, EPC does not saturate with time. This case is the equivalent to nonideal saturation shown in Figure 4.11 for thermal isotropic ALE where the contribution of nonideal saturation is denoted as $\Delta\mathrm{EPC}_2$. The root causes for nonideal thermal isotropic ALE include non-saturated diffusion processes in the modification step and thermal etching of unmodified bulk material. The reason for nonideal saturation during the removal step in directional ALE is sputtering of the unmodified bulk material.

Figure 6.9 shows that a meaningful degree of saturation can be achieved if the sputtering yield for the modified material is significantly larger than that of the bulk material. EPC slows down visibly even for $k_{\mathrm{sp},2} = 0.1$, which corresponds to a 10-fold increase of the sputtering yield due to surface modification. For smaller $k_{\mathrm{sp},2}$, the ALE process becomes closer to ideal and uniformity and ARDE performance are improved.

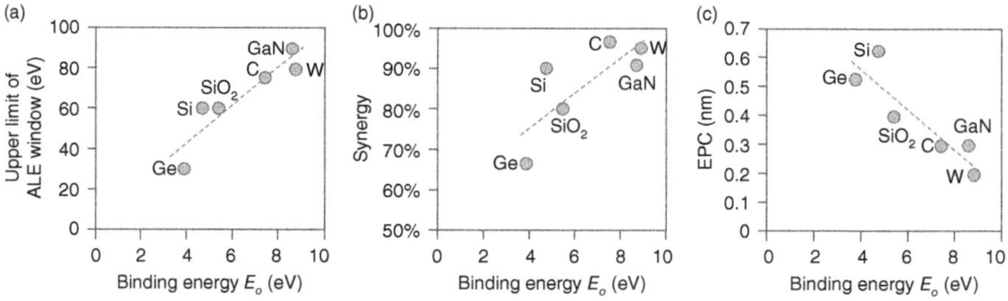

Figure 6.10 Characteristics of directional ALE as a function of bulk binding energy E_O: (a) upper limit of the ALE window, (b) synergy, and (c) EPC. Source: Kanarik et al. (2017).

This improvement in etching performance for smaller $k_{\text{sp},2}$ is the result of increased synergy. The latter is the third critical characteristic of ALE processes besides the existence of an ALE window and saturated or self-limited process steps. The definition of synergy is given in Eq. (4.1). For directional ALE, β represents bulk sputtering for processes with $k_{\text{sp},2} > 0$. Synergy can also be reduced if material is spontaneously removed during the surface modification step, the amount of which is represented in Eq. (4.1) by the term α.

In the case of ALE with directional removal step such as silicon ALE with surface modification by Cl_2 molecules or Cl^* radicals and argon ion removal, α can be the result of thermal or radical etching and can be suppressed by choosing appropriate modification gases and operating at lower temperatures.

Systematic studies of directional ALE of various materials revealed that synergy scales with the binding energy of the bulk material, E_O (Kanarik et al. 2017). This makes sense because E_O determines the sputtering yield Γ and threshold E_{th} for the bulk material (see Eqs. (2.7) and (2.8)). Figure 6.10a depicts experimental results for the upper limit of the ideal ALE window for various directional ALE processes as a function of the binding energy of the bulk material, E_O. Argon ions were used in the removal step in all cases, which allows comparison of the data.

The upper limit of the ideal ALE window is in principle identical to the sputtering threshold of the bulk material and hence the trend should follow the dependency of the sputtering threshold on binding energy given in Eq. (2.7). While the latter predicts a more complicated relationship involving also the atomic mass of the bulk material, the data in Figure 6.10a follows roughly a linear trend for a wide range of materials and atomic masses. The two materials with the largest atomic mass have values for the upper edge of the ALE window below the linear trend line as predicted by Eq. (2.7).

The correlation of the ALE synergy with the binding energy is shown in Figure 6.10b. The ALE synergy is higher for larger E_O, i.e. a higher E_{th} for the bulk material. A higher threshold for bulk sputtering means that it is more difficult to sputter the bulk material. Hence it should be easier to sputter the modified layer selectively. If the modification chemistry can reduce the sputtering threshold to low enough an energy, the ideal ALE window is large and synergy for ion energies within the window is high. Figure 6.10b leads to the somewhat unexpected insight that materials that are harder to etch because of their larger binding energies

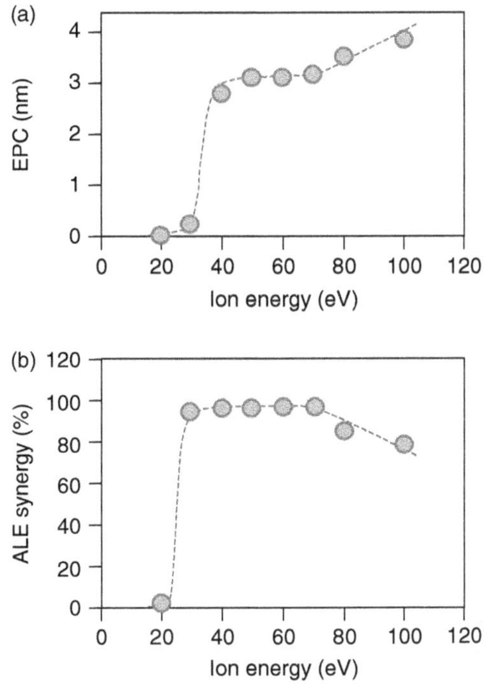

Figure 6.11 Ion energy scan for carbon ALE with O_2/Ar^+ for (a) EPC and (b) synergy. Source: Kanarik et al. (2017).

are better suited for directional ALE with high synergy. Figure 6.10c shows that materials with high synergy exhibit smaller EPC. This implies that the depth of the modified layer depends on E_O as well. Intuitively, this can be understood in terms of the energy needed to break the bonds of the bulk material during modification. Materials with low E_O may also be more prone to thermal or radical etching during modifications, which would increase the value of α.

Figure 6.11 shows EPC and synergy as a function of ion energy for directional ALE of carbon with O_2/Ar^+ (Kanarik et al. 2017). For the range of ion energies below 30 eV, the modified layer is incompletely removed due to insufficient ion energy. Between 30 and 70 eV represents the ideal ALE window and synergy for this system is 100%. Above 70 eV has physical sputtering, which lowers synergy. This analysis shows the importance of staying within the ideal ALE window to obtain the highest synergy. Depending on the system, synergy of 100% may not always be achievable even within the ideal ALE window as discussed above.

It is important to explore synergy for directional ALE at ion energies above the bulk sputtering threshold because of the practical relevance of this process regime. Higher ion energies are beneficial to etch features with higher aspect ratios. Higher ion energies also give higher sputtering rates (see Eq. (2.6)), which reduces the time to saturation and increases etching rates and throughput.

The amount of material removed as a function of time slows down even for high bulk sputtering rates as long as the sputtering rate of the modified material is high enough (see Figure 6.9). This bending of the curves can be explained by the fact that initially the modified layer is sputtered away due to its higher sputtering yield and the fact that the modified layer is on top of the bulk material. Only after the modified layer is removed can sputtering of the bulk occur. This means that at any given point in time during the removal step, a certain amount of modified and bulk material is being removed and an instantaneous synergy can be calculated. By integrating the fluxes of sputtered, modified, and bulk materials, synergies can be calculated as a function of time.

Equation 4.1 can be modified to represented synergy as a function of ALE step time and energy, $S(t, E_i)$ (Berry et al. 2018):

$$S(t, E_i) = \frac{\int_0^t \mathrm{SR}_1(t, E_i)}{d + \int_0^t \mathrm{SR}_2(t, E_i) + \alpha} \times 100\% \tag{6.10}$$

Here, $\mathrm{SR}_1(t, E_i)$ is the time-dependent removal rate of the modified material under ion bombardment normalized to the surface area, $\mathrm{SR}_2(t, E_i)$ the time-dependent removal rate of the bulk material, d the thickness of the modified layer, α the amount of material removed during the chemical modification step, and t the step time. The term $\int_0^t \mathrm{SR}_1(t, E)$ represents the amount of modified material that is removed in one cycle with a step time t and replaces the numerator in Eq. (4.1). For sufficiently large t, $\int_0^t \mathrm{SR}_1(t, E) = d$. The denominator in Eq. (6.10) describes EPC as the sum of the individual components. To achieve 100% synergy, t needs to be large enough that $\int_0^t \mathrm{SR}_1(t, E) = d_m$ while t needs to be small enough that negligible sputtering occurs, i.e. $\int_0^t \mathrm{SR}_2(t, E) \approx 0$. Additionally, α needs to be zero, i.e. no etching occurs during the chemical modification step. If $S(t, E_i) = 100\%$, the ALE process is "ideal."

Using a Monte Carlo sputtering model, the time dependence of modified and bulk material removal can be calculated in a straightforward manner (Berry et al. 2018). Figure 6.12b shows the time dependence of the synergy factor $S(t, E_i)$ for Cl_2/Ar^+ ALE of tantalum for ion energies of 20, 70, and 100 eV. The corresponding experimental ALE curve for this system is shown in Figure 6.12a. The time scale in the calculations is normalized to experimental results.

For an ion energy of 20 eV and all other energies between the lower and upper limits of the ALE window, the time window to achieve ideal ALE is theoretically

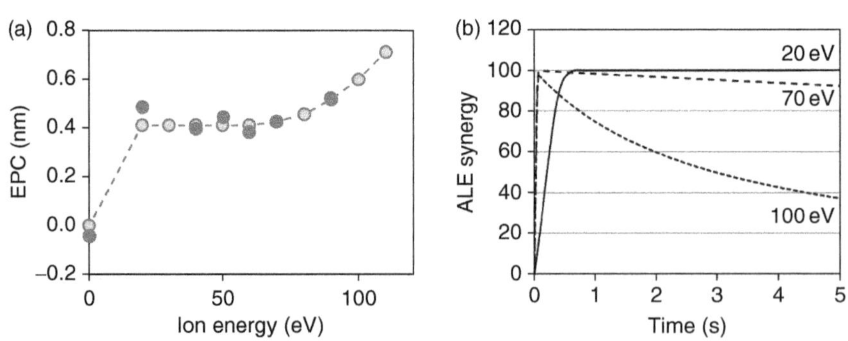

Figure 6.12 Ideal ALE window (a) and time- and energy-dependent synergy (b) for directional ALE of tantalum with Cl_2/Ar^+. Source: Berry et al. (2018).

infinitely long. For an ion energy of 70 eV, which according to Figure 6.12a is just above the upper limit of the ALE window, the synergy reaches 100% after a few milliseconds and is reduced to just above 95% after five seconds. Even for an ion energy of 110 eV, near 100% synergy (98%) can be achieved for very short removal times but the synergy drops below 40% after five seconds sputtering time.

The implication of this analysis is very important for practical applications. Meaningful synergy and the associated benefits for etching rate nonuniformity (ERNU), ARDE, and surface smoothness can be obtained even for ion energies far above the upper ALE energy window. This effect is at the heart of many RIE applications using plasma pulsing and mixed mode pulsing (MMP).

This effect opens the opportunity to reduce the duration of one ALE cycle by leveraging higher sputtering yields and shorter removal step times. Because the optimum cycle time can be very short, the realization of such processes requires new levels of process control. Bias pulsing can increase the duration of the removal step time for the sputtering amount, which has benefits for process control (Tan et al. 2019a).

So far, all examples for directional ALE deployed argon ions or fast argon atoms in the removal step. According to Eqs. (2.7) and (2.8), the sputtering threshold and yield are also a function of the mass ratios of the impinging ion and the surface atoms. This is due to the physics of momentum transfer in Eq. (2.5). It is therefore interesting to explore the effects of using different noble gas ions on the ALE window. Monte Carlo modeling can be used to study the implication of ion mass on directional ALE (Berry et al. 2018). Figure 6.13 depicts calculated ALE curves for two elements with significantly different atomic masses, titanium, and tungsten. To reflect realistic ALE systems, different halogens where modeled for surface modification, chlorine for tungsten and bromine for titanium.

It is important to consider the atomic masses of the ions and the surface atoms to interpret the results in Figure 6.13. They are 48 for titanium, 184 for tungsten, 20 for neon, 40 for argon, and 131 for xenon. In the case of titanium, the impinging ions have similar or higher mass; in the case of tungsten, they have much lower mass than the surface atoms. Figure 6.13 shows that for an ALE process of titanium, the

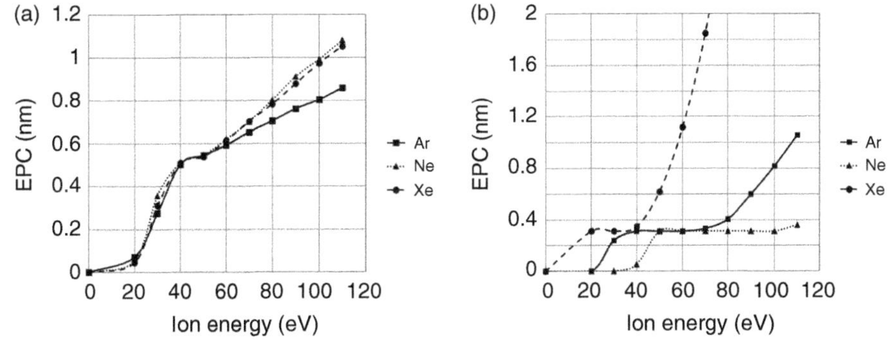

Figure 6.13 Simulated ALE window for Ne, Ar, and Xe ALE of titanium with bromine surface modification (a) and Ne, Ar, Xe ALE of tungsten with chlorine surface modification (b). Source: Berry et al. (2018).

mass of the ions in the removal step does not change the ALE curve significantly. The behavior of tungsten is in stark contrast. Cl_2/Xe^+ ALE of tungsten has a relatively small ALE window of 20 eV starting at 20 eV. Cl_2/Ne^+ ALE of tungsten, however, has a window of 50 eV starting at 50 eV. Based on the results, neon should be a better gas for ALE of tungsten. The ALE window for Cl_2/Ar^+ is between Ne^+ and Xe^+ as is the mass of Ar^+.

Mannequin et al. conducted a study comparing Ar^+ and Kr^+ for ALE of GaN with chlorine activation. They found a much larger and well-defined ALE window for Kr^+ (Mannequin et al. 2020).

6.1.3 ALE with Directional Removal Step and Modification by Reactive Layer Deposition

In directional ALE with modification by a reactive layer, the reactive species are not adsorbed directly but deposited as a constituent of a reactive polymer layer. The archetypical system is SiO_2 ALE with C_4F_8/Ar^+. Other fluorocarbon gases such as CH_3F can also be used to deposit a polymer layer containing fluorine as the reactive element. Argon ions with energies of 10–30 eV sputter the reactive polymer and part of the underlying surface in the removal step. The ion energy is below the sputtering threshold of SiO_2, which is reported to be between 45 eV (Oehrlein et al. 2015) and 65 eV (Kaler et al. 2017). More literature data is listed in Table 2.1. This energy appears sufficient to sputter a mixed layer containing Si, O, C, and F.

Another possible removal mechanism is the formation of volatile SiF_4, CO, CO_2, and COF_2 at the interface between the reactive layer and SiO_2 with subsequent diffusion to the polymer surface and desorption (Huard et al. 2018b). This ALE process was modeled in 2007 and 2009 (Rauf et al. 2007; Agarwal and Kushner 2009) and demonstrated experimentally several years later (Metzler et al. 2014; Hudson et al. 2014). The process was also applied to ALE of silicon (Metzler et al. 2016).

Figure 6.14 depicts the classical experimental results for SiO_2 ALE with C_4F_8/Ar^+ (Metzler et al. 2014). The bottom panel shows the thickness evolution during eight cycles of SiO_2 ALE with C_4F_8/Ar^+ measured with in situ ellipsometry (Metzler et al. 2014). The experiment used a continuous inductively coupled argon plasma at 10 mTorr and the wafer was cooled to 10 °C to accelerate deposition. At the beginning of each cycle, a pulse of C_4F_8 was injected for 1.5 seconds to deposit about 5 Å of a C_xF_y layer. A bias potential of 10 V, which corresponds to an ion energy of about 25–30 eV, was applied to the substrate after the C_4F_8 pulse to sputter the surface. SiO_2 was initially removed fast and the instantaneous removal rate decreased until it eventually stopped after about 10 seconds. At this point, about 2–3 Å of SiO_2 were removed. EPC noticeably increases cycle over cycle in the experiment, which the authors attribute to parasitic contributions from fluorocarbon polymers deposited on the reactor walls.

One of the characteristic features of this type of ALE process is the apparent lack of self-limitation of the modification step because it is a chemical vapor deposition (CVD) process. The thickness of the reactive layer is proportional to the step time. Nevertheless, the combined ALE process is quasi self-limited because excess

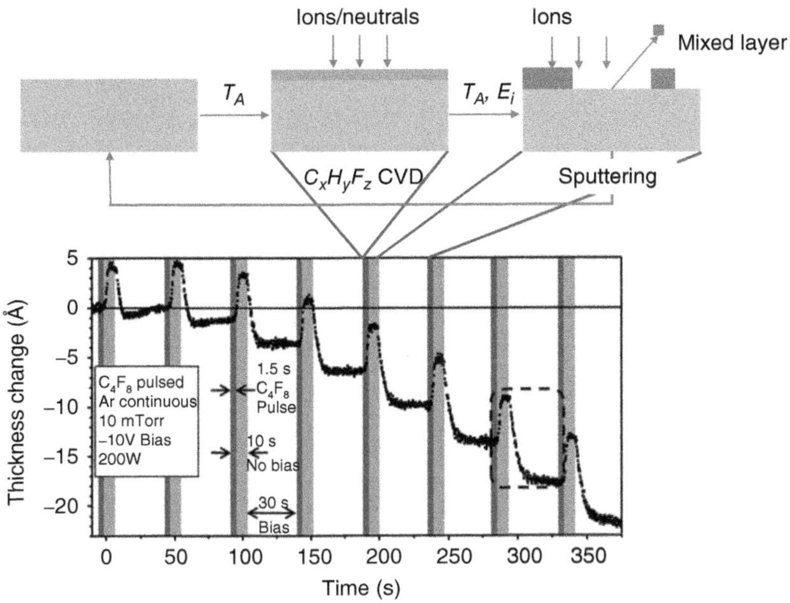

Figure 6.14 Top: Schematic illustration of directional ALE with modification by reactive layer deposition. Bottom: Example of thickness evolution during eight cycles of an SiO_2 ALE process. Source: Metzler et al. (2014). © 2014 AIP Publishing.

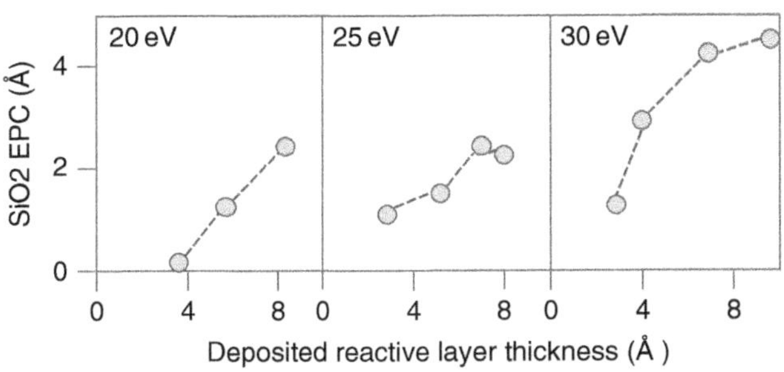

Figure 6.15 EPC for SiO_2 ALE with C_4F_8/Ar^+ for different reactive layer thicknesses and ion energies. Source: Metzler et al. (2014). © 2014 AIP Publishing.

polymer does not contribute to the removal process by sputtering. Figure 6.15 shows that once a critical FC layer thickness on SiO_2 is reached, EPC saturates. If the reactive layer thickness exceeds this critical thickness, which is on the order of the projected range of argon in the reactive layer, any additional reactive polymer will be sputtered by argon ion bombardment with little interaction with the SiO_2 underneath.

The underlying reaction mechanisms have been studied with MD simulations (Hamaguchi et al. 2018a) and voxel-slab models (Kuboi et al. 2019). The calculations show that preferential sputtering plays an important role in this process. Equation (2.9) shows that lighter and weaker bonded species are sputtered preferentially from the surface. In the case of SiO_2 ALE with C_4F_8/Ar^+, preferential sputtering does play an important role as long as only FC polymer is sputtered because the masses of carbon (12 amu) and fluorine (19 amu) are quite similar. If anything, carbon would sputter preferentially, and the layer should become more reactive. Once the SiO_2 surface is reached, however, preferential sputtering of oxygen (16 amu) atoms occurs, which makes the surface more silicon (28 amu) rich. In the presence of an FC layer, this promotes the formation of Si—C bonds. Low-energy Ar^+ irradiation may not be able to remove C atoms completely from the surface. Under such conditions, more carbon atoms may remain on the surface after each ALE cycle and etch stop may eventually occur after several ALE cycles.

MD calculations show that adding small amounts of oxygen and fluorine restores etching by enhancing carbon removal (Hamaguchi et al. 2018a). MD simulations are well suited to study this type of ALE process because it constitutes an extreme case of near threshold sputtering of a mixture of at least four elements. Under these conditions, conventional sputtering theory can still give quantitative guidance but is not accurate enough for qualitative predictions.

Evidence of near or quasi self-limitation during FC film deposition has been found in beam experiments (Kaler et al. 2017). Species from a CF_2-rich plasma were measured to have a 10 times higher sticking coefficient on the pristine SiO_2 surface compared to FC polymer films. A modified Langmuir–Hinshelwood model (see Figure 2.6) was proposed to explain this effect. Once the entire surface is covered with polymer, the polymer deposition rate decreases due to this lower sticking coefficient.

A second characteristic feature of this class of ALE is its selectivity mechanism, which does not rely on intrinsic selectivity but on selective deposition. The definition of intrinsic selectivity was introduced in Section 2.1 as etching selectivity without deposition sub-reactions, relying only on the binding energy of the bulk material. It is the dominant selectivity mechanism for directional ALE with modification by chemisorption (Tan et al. 2015). Per definition, ALE with surface modification by deposition of a reactive layer uses a deposition sub-reaction. This deposition step is used to passivate one surface vs another. Specifically, SiO_2 ALE with C_4F_8/Ar^+ is very selective to Si_3N_4 (Hudson et al. 2014).

Let us now investigate the synergy for this class of ALE processes. Kanarik et al. reported a synergy of 80% for SiO_2 ALE with CHF_3/Ar^+ (Kanarik et al. 2017). The binding energy E_O is about 5 eV, which is the middle of the range for the elements in Figure 6.10. Based on the correlation between synergy and binding energy, the suitability of SiO_2 for high-synergy ALE is average.

Compared to directional ALE with plasma-based halogen chemisorption, deposition-based ALE has the benefit that it is easier to avoid spontaneous etching. This is because polymer deposition suppresses materials removal as soon as the surface is covered. Also, the etching rate for SiO_2 by fluorine radical is low at room temperature. If the ion energy is kept at self-bias levels around 10–20 eV, the polymer deposition step will not etch SiO_2 and the parameter α in Eq. (4.1) will be close to zero.

The contribution of bulk sputtering is in principle zero if the ion energy is below the sputtering threshold of SiO_2. The parameter β will be zero under this condition. However, bulk sputtering can contribute to α indirectly. This mechanism arises from the fact that SiO_2 is a compound material. Oxygen will sputter preferentially when exposed to argon ions because of the lower mass (see Eq. (2.9)).

Todorov and Fossum reported measurable sputtering for energies as low as 40 eV (Todorov and Fossum 1988). Even though the yield for these energies is only 0.004 atoms per impinging ion, this will lead to silicon enrichment of the SiO_2 surface after the FC polymer is cleared. This silicon-rich layer can be etched spontaneously during the modification step in the presence of fluorine radicals. This effect has been predicted by MD simulations (Hamaguchi et al. 2018b). To avoid this mechanism, Ar^+ sputtering should not be extended beyond the time when the mixed layer is cleared.

Synergy is also negatively impacted if the gas exchange is slow or when FC polymer from the reactor walls leads to fluorine species in the Ar plasma during removal. The latter mechanism was evoked to explain the increase in EPC in Figure 6.14 (Metzler et al. 2014). Huard et al. modeled the contribution of residual C_4F_8 from gas lines and wall desorption by adding 100 ppm C_4F_8 to argon in the removal step (Huard et al. 2018b). Synergy was calculated as the ratio between EPC without (ideal EPC) and with C_4F_8 addition (nonideal EPC). This approach agrees with Eq. (4.1) because the term $EPC - (\alpha + \beta)$ in Eq. (4.1) represents ideal EPC.

Figure 6.16 shows that both ideal and nonideal EPC start to saturate for deposition times over one second, which confirms the property of quasi self-limitation discussed above. Synergy starts at zero because in the absence of the deposition step, there is only etching with C_4F_8 contamination in the Ar feed gas. This value is theoretical because without polymer deposition step, there would be no C_4F_8 contamination. When the polymer step time is extended, synergy increases and reaches a maximum of 80%. This is in agreement with the experimental value reported by Kanarik et al. (2017). The reason for the increase in synergy with FC polymer deposition time is the increase in ideal EPC for thicker polymers. Thicker polymers provide a larger source of reactive fluorine. This is an interesting result and can be used to design high-synergy ALE processes. The positive effect of a longer polymer step on synergy is even stronger in practical applications because the time to reach the interface to SiO_2 is longer for a thicker polymer layer. Residual fluorine from reactor wall and gas lines can be flushed out before the interface to SiO_2 is reached.

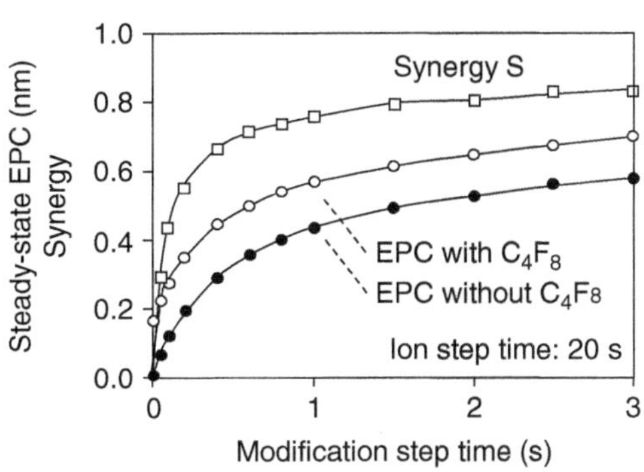

Figure 6.16 EPC with and without C_4F_8 in removal step and resulting ALE synergy as a function of deposition step time for 20 seconds Ar^+ ion removal. Values are shown for ideal and nonideal fluxes. Source: Huard et al. (2018b). © 2018 American Vacuum Society.

Directional ALE of SiO_2 has been demonstrated using C_4F_8 gas without plasma in the modification step (Antoun et al. 2019). The surface had to be cooled down to cryogenic temperatures below −120 °C. The surface modification mechanism is believed to be physisorption of C_4F_8 molecules. The threshold for an observable EPC is quite abrupt. ALE of SiO_2 occurs for a temperature of −120 °C, while no etching was obtained at −110 °C. The EPC is 0.4 nm. The modifications step is self-limited at the given temperature most likely due to the nature of physisorption.

Lin et al. demonstrated selective ALE of HfO_2 over silicon by deposition (Lin et al. 2020). They found that a CH_4/CHF_3 mixture deposits a fluorocarbon film on silicon, whereas it fluorinates the HfO_2 surface with negligible FC deposition. The fluorinated layer was removed by low-energy argon bombardment. They were able to etch through a ~29 Å HfO_2 film with the formation of a passivation layer on Si. The Si thickness loss is less than 13.7 Å before the Si etching stop occurred.

6.2 Performance Metrics

6.2.1 Etching Rate (EPC)

Etching rate for ALE is determined by EPC and the time it takes to complete one cycle. Cycle time can be improved by proper engineering of the reactor. EPC is determined by the mechanism for self-limitation in the modification and/or removal steps. Let us analyze the three directional ALE processes we reviewed above with respect to their fundamental EPC limitations.

In the case of *ALE with directional modification step and isotropic removal*, the depth of modification is determined by the projected range of the implanted ions, which can cover a wide range from several Angstroms to nanometers (see Table 2.2). The depth of modification also depends on the selectivity of the isotropic removal process, which in most cases is a thermal desorption step or chemical reaction with a gas or radicals. If the thermal process requires a precise stoichiometry to enable removal, any areas with lower than critical implant concentration may remain on the wafer and reduce EPC. The distribution of implanted species is roughly a Gaussian distribution with the maximum concentration located at the depth of the projected range R_p. EPC will be equal to the depth of modification where the concentration is higher than the critical value on the falling slope of the distribution. If the concentration on the raising slope is below the critical concentration for isotropic removal, the process may not even start. If very high concentrations are needed, synergy challenges may arise from sputtering effects, which limit the maximum concentration that can be achieved by ion implantation.

In the case of *ALE with directional removal step and modification by chemisorption and diffusion*, EPC is a function of diffusion depth. For pure chemisorption processes without subsequent diffusion, EPC is limited to just one monolayer. Values of 1.36 and 1.57 Å were demonstrated for ALE of (100) Si and (111) Si, respectively (Park et al. 2005a,b; Oh et al. 2007). These values represent precisely one monolayer for these crystallographic orientations of silicon. The surface was modified using neutral chlorine gas without plasma. The silicon surface was held at room temperature in these experiments. Chemisorption can be accelerated using radicals generated by plasma. This approach introduces additional energy in the form of the radicals themselves, residual ions, photons, and electrons, which can stimulate bulk diffusion. Therefore, EPC of silicon ALE with chlorine radicals is over several monolayers. Values between 7 Å (Kanarik et al. 2015) and 14 Å (Tan et al. 2015) have been reported for chlorination in an inductively coupled plasma chamber where the plasma is in contact with the wafer and fast chlorine atoms and ion impinge the surface. This leads to a modified layer, which is several monolayers thick. Computational modeling for germanium ALE with Cl_2/Ar^+ shows that the chlorine penetration depth increases from 0 to 7 Å when the energy per chlorine atom is increased from 0 to 50 eV. The number of chlorine atoms per germanium atom also increases (Zhang et al. 2019). To reduce this effect, remote radical sources can be used.

For *ALE with directional removal step and modification by reactive layer deposition*, EPC is a function of the reactive layer thickness and ion energy as shown in Figure 6.15.

6.2.2 ERNU (EPC Nonuniformity)

As for all ALE processes, uniformity across the wafer is theoretically perfect if the synergy is 100%. Nonideal contributions to EPC can disturb uniformity. In the case of *ALE with directional modification step and isotropic removal*, the modification step is quasi self-limited. The implantation depth is self-limited, but the concentration is not. The isotropic removal step, for instance, thermal desorption, will remove all the material with a certain composition. Any nonuniformity will be revealed by the removal step. Therefore, the modification steps must be very well controlled.

ALE with directional removal step and modification by chemisorption and diffusion can achieve 100% synergy and excellent ERNU. ERNU across the wafer of ±1.5 nm 3σ after etching of 50 nm silicon with a starting uniformity of ±1.4 nm 3σ was reported for silicon ALE with chlorine plasma modification and argon ion removal (Kanarik et al. 2013). This performance can be obtained without many of the uniformity tuning features that have been developed for advanced RIE reactors. If the process has less than perfect synergy, it is important to identify the root cause, which can be material removal in the modification step or in the removal step (components α and β in Eq. (4.1), respectively). The species fluxes in the step that reduce synergy need to be carefully controlled across the wafer similarly to RIE. If the root cause is parasitic RIE due to residual halogens from the chamber wall, purge steps should be extended.

While chemisorption and diffusion rates are a function of temperature, this type of ALE process is normally not so temperature sensitive that a multizone, temperature-controlled electrostatic chuck (ESC) would be needed.

Huard et al. developed a model of an inductively coupled reactor to predict uniformities across the wafer for etching of silicon using comparable RIE and ALE processes (Huard et al. 2018a). The RIE process was continuous with a mixture of 90% Cl_2 and 10% Ar. Directional ALE was modeled by cycling between exposure to the fluxes produced by the chlorine plasma without a bias to passivate the surface

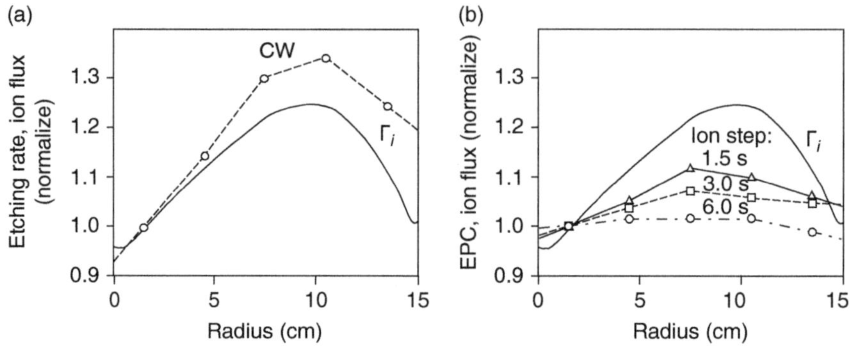

Figure 6.17 Etch rate and ion flux to the wafer as a function of radius for RIE and ALE. Source: Modified from Huard et al. (2018a).

with chlorine radicals and to the argon plasma with a radio frequency (RF) bias providing moderately energetic ion bombardment to etch the passivated surface. The argon removal step contained 10 ppm of chlorine to account for imperfect chlorine removal from the reactor in the purge step. The results are shown in Figure 6.17. The ERNU for continuous RIE is slightly worse than the ion flux uniformity across the wafer. The ERNU for directional ALE is better for all conditions and improves as the argon ion removal step is extended to reach saturation. The residual nonuniformity can be attributed to the addition of small amounts to the argon, which causes RIE type of etching.

ALE with directional removal step and modification by reactive layer deposition is quasi self-limited and EPC depends on the thickness of the reactive layer as shown in Figure 6.15. Therefore, uniformity of the polymer deposition step must be carefully controlled to obtain good uniformity across the wafer. If fluorine is released from the chamber walls or gas lines during the sputtering step, it will introduce parasitic RIE and nonuniformities. Purge steps and longer polymer deposition times can be used to increase synergy and to improve the intrinsic uniformity.

6.2.3 Selectivity

Directional ALE has fundamentally lower intrinsic selectivity than thermal isotropic ALE because the ions involved carry excess kinetic energies of tens of electron volts (see Figure 6.2). Selectivity mechanisms are different for the three classes of directional ALE discussed above. For *ALE with directional modification step and isotropic removal*, selectivity can in principle be obtained but the underlayer that is not supposed to be etched will be damaged by implantation. The depth of implantation depends on the material properties but implantation into this stopping layer will occur. This damage depends on the chosen projected range of the implant species, which in turn determines the profile and EPC. In some cases, damage to the stopping layer is not important or a sacrificial layer can be implemented.

Selectivity of *ALE with directional removal step and modification by chemisorption and diffusion* can be explained using the framework of the ideal ALE window

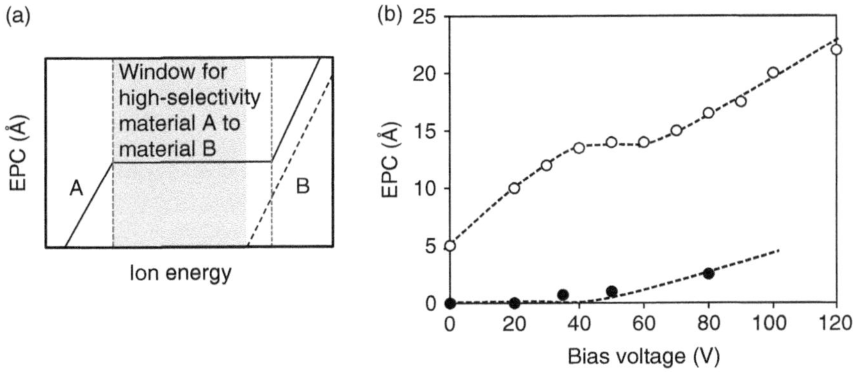

Figure 6.18 (a) Schematic of EPC for material A (e.g. silicon) and material B (e.g. silicon oxide) as a function of ion energy. Hypothetically, infinite etch selectivity can be reached in the energy range that etches material A and not material B. (b) Measured EPC for silicon (open circles) and silicon oxide (solid circles) as a function of RF bias. Source: Tan et al. (2015).

discussed in Section 6.1.2. In Figure 6.18a, material A exhibits an ALE window while material B does not. In the case of material B, the bonding energy of the adsorbed layer is significantly lower than for the bulk material. In this case, the adsorbed species would be removed as atomic species (EPC equals zero) and the removal of the bulk material realized only if the energy reaches the energy needed to sputter the bulk material. If this sputter threshold energy is higher than at least part of the energy range for ideal ALE of material A, high selectivity can be obtained. Figure 6.18b shows the actual, measured EPC for single crystal silicon and thermal silicon oxide as a function of RF bias voltage (Tan et al. 2015). Silicon has an ALE process window between 40 and 60 V of RF bias. This translates to 60–80 eV average ion energy. The EPC for thermal silicon oxide is essentially zero for RF bias voltages of up to 50 V or ion energies of 70 eV. Based on these results, silicon ALE can be achieved with high selectivity to oxide for ion energies up to 60 eV.

The selectivity benefit of ALE compared to continuous processing is illustrated in Figure 6.19 where silicon to SiO_2 selectivity results are compared for chlorine plasma/Ar^+ ALE and continuous RIE with a plasma of a chlorine and argon mixture (Tan et al. 2015). For ALE, no etching was measured for silicon oxide for RF bias voltages up to 40 V or about 60 eV ion energy. This region is labeled "infinite selectivity." In contrast, the selectivity decreases immediately for continuous processing under similar conditions for bias voltages above 0 V.

Infinite selectivity as measured by thin film metrology does not necessarily mean that the stopping layer is not damaged by ion bombardment especially if the stopping layer is only a few nanometers thin. To investigate the effects at a truly atomic level, transmission electron microscopy (TEM) is a very useful tool. Figure 6.20 compares TEM micrographs for a polysilicon/gate oxide structure before and after ALE with 30 nm overetch. In this experiment, a xenon plasma without bias power was used in the removal step to minimize surface damage. No gate oxide loss is observed, and only minimal if any damage of the bulk silicon underneath the gate oxide is visible.

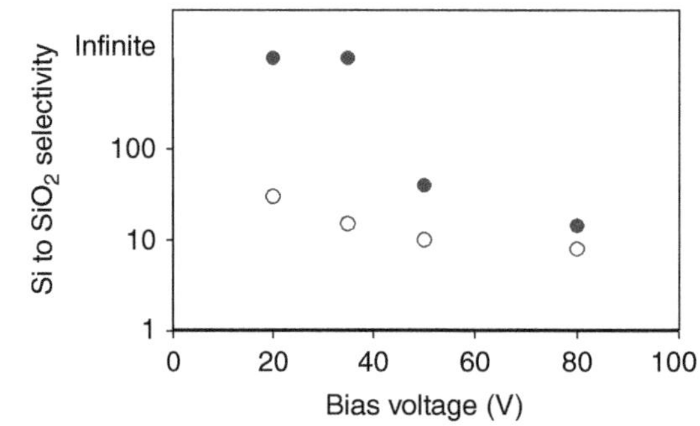

Figure 6.19 Selectivity of single crystal silicon to thermal silicon oxide as a function of RF bias voltage. ALE results (full circles) and continuous process results (open circles) are compared. Source: Tan et al. (2015).

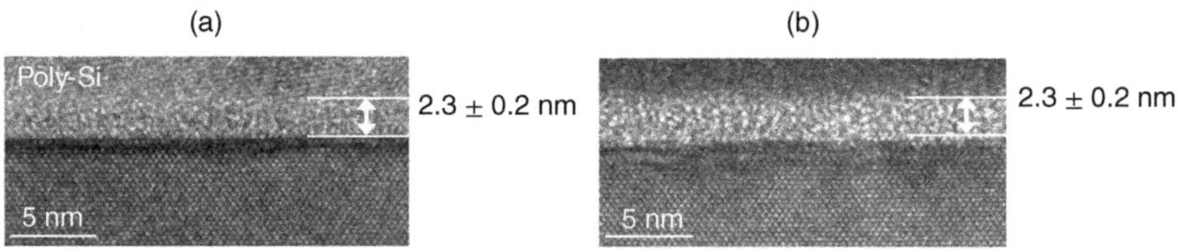

Figure 6.20 TEM micrographs of gate oxide before (a) and after (b) ALE etch with xenon plasma desorption and 0 V bias voltage. The overetch amount was 30 nm. Source: Tan et al. (2015).

The fact that the darker regions extend to a greater silicon depth than in the pre-etch TEM indicates that there is potentially some lattice damage (Tan et al. 2015).

For an equivalent time of 70 nm polysilicon, some gate oxide loss was observed. The gate oxide integrity becomes compromised due to secondary processes such as preferential sputtering of oxygen and ALE removal of silicon atoms. Conceivably, the ALE process removes some of the silicon-rich surface layer that forms during argon ion bombardment and eventually the gate oxide layer will be thinned to a critical thickness.

Other explanations are also possible. Petit-Etienne et al. reported that for chlorine plasma with 100 eV ion energy, chlorine atoms penetrate through 2.5 nm thick SiO_2 and accumulate as $SiCl_x$ at the interface causing silicon recess and accelerated gate oxide breakthrough (Petit-Etienne et al. 2011). It is possible that a chlorine penetration/accumulation mechanism plays a role during the final stages when the gate oxide is sufficiently thin.

The silicon damage for ALE compares very favorably with published results for processes in continuous mode. Figure 6.21 shows a comparison of Cl_2/Xe ALE (Lill et al. 2018) and conventional HBr/O_2 RIE polysilicon gate overetches (Vitale and Smith 2003). In the case of ALE, the gate oxide thickness was about 2 nm and in the case of RIE, it was 4 nm. The ALE process shows no gate oxide or substrate loss. The RIE experiments exhibit classic silicon recess behavior. The RIE process relies on oxygen implantation to extend the thickness of the gate oxide to cope with 100 eV ion energy and more. This oxide is removed during wet cleans, which causes the recess, which has negative impact on device performance because the overlap between the transistor channel region with source and the drain is compromised.

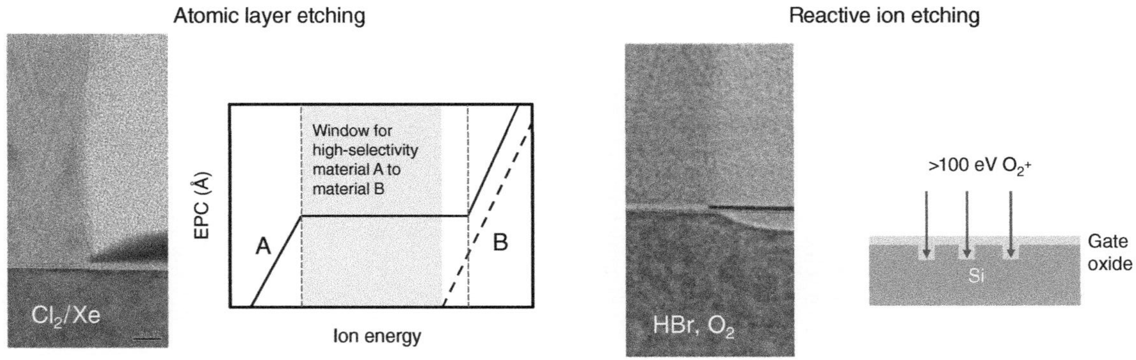

Figure 6.21 Comparison of gate oxide loss for Cl_2/Xe ALE and conventional HBr/O_2 RIE polysilicon gate overetches. Source: Vitale and Smith (2003); Lill et al. (2018).

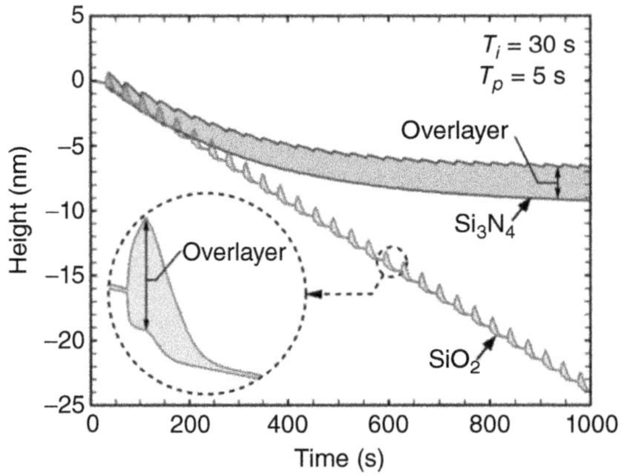

Figure 6.22 Calculated thickness evolution for ALE of SiO_2 and Si_3N_4. Source: Huard et al. (2018b). © 2018 American Vacuum Society.

ALE with directional removal step and modification by reactive layer deposition is characterized by a deposition-based selectivity mechanism. Intrinsic selectivity only matters for the short period of time until a protective polymer layer is formed on the stopping layer. Deposition-based etching selectivity in SiO_2 ALE with C_4F_8/Ar^+, for instance, relies on the buildup of polymer to levels that terminate etching of Si_3N_4 and silicon, a sequence that requires several pulses to occur as shown in Figure 6.22 (Huard et al. 2018b). Significant etching of Si_3N_4 stops after approximately 15 cycles when the polymer layer has reached a thickness of 2 nm. The argon ions had a modeled distribution ranging from 25 to 45 eV, which is below the sputtering threshold of SiO_2 (see Table 2.1).

The reason why there is a polymer buildup on Si_3N_4 but not on SiO_2 is the presence of oxygen in the latter. This creates a pathway for the removal of carbon via the formation of CO or CO_2,which does not exist in Si_3N_4. As Hamaguchi et al. have shown, SiC can be formed in the absence of oxygen (Hamaguchi et al. 2018a). The same arguments can be applied to the case of SiO_2 ALE with C_4F_8/Ar^+ selectively to silicon.

Figure 6.22 shows that by the time Si_3N_4 is fully passivated, 8 nm of Si_3N_4 is lost. This can be too much for advanced etching applications. One solution to increase selectivity is to use the lowest possible ion energies with a narrow energy distribution when the Si_3N_4 is not yet passivated. RF waveform shaping can be used to achieve this goal (Wang and Wendt 2000). This technique will be discussed in more detail in Section 9.4. Another approach is to leverage selective deposition of carbon-rich precursors onto Si_3N_4 to enhance the formation of a passivation layer (Gasvoda et al. 2019). If the initial loss is eliminated, this class of ALE can theoretically achieve infinite selectivity, which is of great interest for practical applications.

6.2.4 Profile and ARDE

ALE with directional modification step and isotropic removal can, in principle, create near vertical profiles by implanting modification species at the bottom of the feature and not into the sidewall. The deeper the projected range of the implanted ions, the more the effect of straggling will lead to modification also in the lateral direction, which would compromise the generation of a perfectly vertical profile (see Table 2.1). Scattering of ions from the mask can also cause an isotropic etching component. Vertical profiles have been demonstrated for Si_3N_4 spacer etching by He ion bombardment and isotropic removal of the modified layer with diluted HF acid (Posseme et al. 2014). Near vertical profiles were reported for etching of cobalt with oxygen ion implantation at 200 eV followed by removal of the cobalt oxide with formic acid vapor (Chen et al. 2017b). The mask material in this experiment was TiN.

ALE with directional removal step and modification by chemisorption and diffusion is used in the semiconductor industry, for instance, as a highly selective overetch process for FinFET gates. The polysilicon gate profile shown in Figure 6.21 illustrates the ability of this type of ALE to generate vertical profiles. The reason for this etching behavior is that the flux toward the bottom of the feature is much higher than that toward the wall and the ions impacting the wall are predominantly scattered (see Figure 2.16). To achieve perfect profiles without ARDE, secondary effects must be considered. When Cl_2/Ar^+ ALE is applied to a silicon layer with a SiO_2 hardmask of narrow lines and spaces, re-entrant or bowed profiles are observed even when the ion energy is within the ALE window (Ranjan et al. 2016).

Figure 6.23 illustrates that sputtering of the sidewall by non-vertical ions is amplified by lowering of the sputtering threshold for oblique ion incidence. The modeling results in Figure 6.23 show that the more oblique incidence angles shift the ALE window to smaller energies and reduce its width. This negative effect can be mitigated by making sure that the hardmask has a perfectly vertical profile since this reduces the number of ions scattered form the hardmask.

Another approach is to operate the argon sputtering step at pressures of a few millitorrs or lower. This gives narrower angular distributions of the incoming ions. A third approach is to increase the ion energy to several 100 eV which also decreases the width of the ion angular distribution. This will of course lead to bulk sputtering and reduce the synergy of the process. However, if the time of the sputtering step is very short, the moment of maximum synergy can be achieved as shown in Figure 6.12.

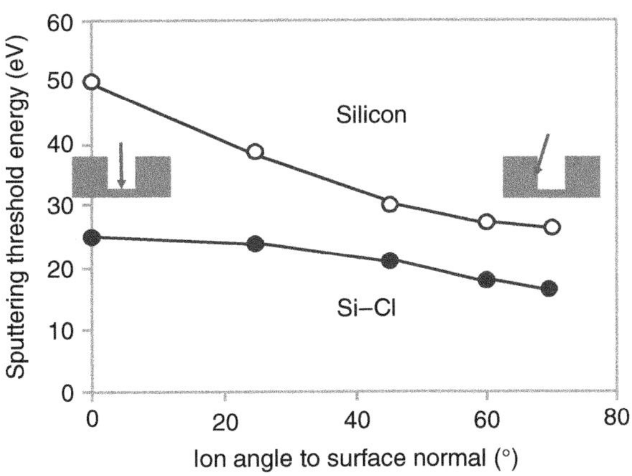

Figure 6.23 Calculated ALE window as a function of incident ion angle for Cl_2/Ar ALE of silicon. Source: Berry et al. (2018).

The step time for this approach will typically have to be well below one second. To achieve such short sputter step times and for better process control, plasma pulsing (for instance bias pulsing) with duty cycles of 10% and lower can be used. Silicon gate profiles without bowing and vertical ruthenium metallization lines have been demonstrated with this so-called high-energy directional ALE approach (Tan et al. 2019a).

Another complication for this class of directional ALE arises from the absence of "volatility" of the reaction products. Etching reaction products are called "volatile" when they desorb from the surface with thermal energies as in thermal or radical etching. Etching products generated by thermal desorption have low probability to adsorb back onto the surface. This makes them well suited for removal from the bottom of high aspect ratio structures.

In directional ALE, this etching mechanism is absent by design. Thermal etching during the modification step must be eliminated to achieve 100% synergy with its benefits for uniformity and ARDE. The reaction products are sputtered from the surface at the endpoint of a collision cascade where they overcome the desorption barrier thanks to their kinetic energy. This means the molecule or fragment that leaves the surface can undergo the reverse process and redeposit onto a surface location in the line of sight. Redeposition becomes more probable the higher the aspect ratio of the etched feature.

The effects of this sidewall redeposition are shown in Figure 6.24, which depicts 2D Monte Carlo sputter model simulations of silicon fin structures etched with 200 cycles of a Cl_2/Ar ALE. The sticking coefficient of the sputtered $SiCl_x$ species was varied between 0.0 and 0.6 to illustrate the effect of redeposition on the profile and ARDE. As the sticking coefficient is increased in the model, the profile becomes less bowed and eventually tapered. This is the result of sidewall protection by the etching products. At the same time, ARDE emerges. These modeling results show that the nonvolatility of the etching products can help create vertical profiles but can also give rise to ARDE despite perfect synergy. The sticking coefficient and ARDE can be reduced by increasing the wafer temperature. Bowing can be improved by using higher ion energies and low ion fluxes as discussed above.

An optimized ALE process is shown in Figure 6.25. for germanium ALE with Cl_2/Ar^+. The aspect ratio of the small trench is about 4 and hence the requirement is not as demanding. Also, the synergy of this process is only 66% due to isotropic etching. The sticking coefficient of the reaction germanium chloride products is low and hence vertical profiles without any ARDE are possible (Kanarik et al. 2017).

ALE with directional removal step and modification by reactive layer deposition has the important advantage of built-in sidewall passivation by the reactive polymer. As we discussed in the section about the underlying mechanism, bulk material is not removed in the sputtering step until the remaining polymer layer is thinner than the projected range of the ions. On the sidewall with oblique ion incidence, the ion flux is much lower and if the ion angular distribution is narrow, most impinging ions are scattered. This allows creating a process regime where the sidewall is always protected by the polymer. However, if polymer modification step is CVD it is also

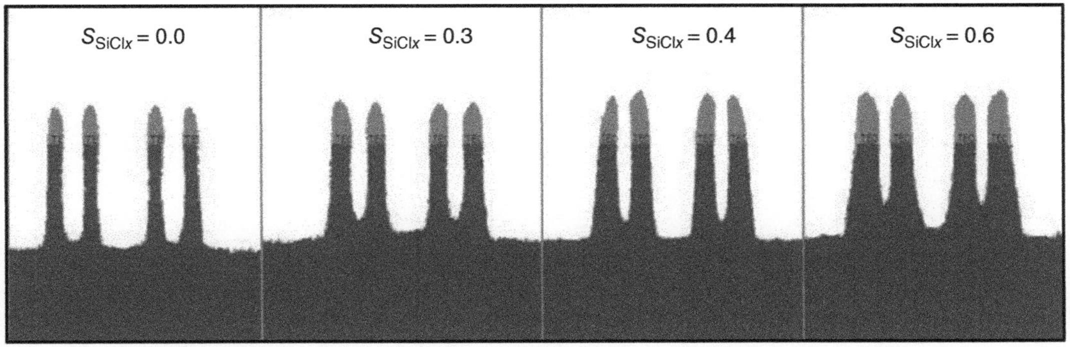

Figure 6.24 Calculated 2D simulation of 200 cycles of a Cl_2/Ar ALE of silicon and varying sticking coefficient of the sputtered species. Source: Berry et al. (2018).

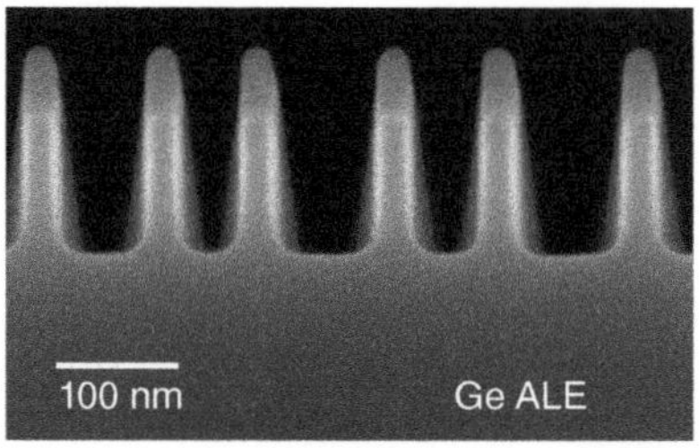

Figure 6.25 Germanium ALE on patterned test wafer, showing flat etch front and aspect ratio independence of the etch depth for 110 nm etch depth in 55 and 77 nm trench widths. Source: Kanarik et al. (2017).

not conformal, which means small features can close on the top and the reactive layer thickness on the bottom of smaller features can be thinner compared to larger features. This can be prevented by inserting a polymer removal step. This type of ALE is used to etch logic contact structures with aspect ratios of 10 and higher.

6.2.5 Surface Smoothness and LWR/LER

Smoothening of the etched surface has been observed for several directional ALE processes with ion bombardment in the removal step (Kim et al. 2013; Kanarik et al. 2018). Figure 6.26 shows experimental results for ruthenium and silicon. Surface smoothening is a desirable property because rough surfaces cause variability of device dimensions and electrical properties of the device, which can hurt performance. Kim et al. demonstrated a reduction of contact sheet resistance when ALE was applied to remove a damaged layer from the bottom of a contact hole etched with RIE (Kim et al. 2013).

The smoothening properties of directional ALE can be utilized to improve the local CD uniformity. In critical pattern-transfer applications, smoothness of less than 1 nm is required. The effect of ALE on line width roughness (LWR) was demonstrated, for instance, during etching of ruthenium lines with SiO_2 hardmask.

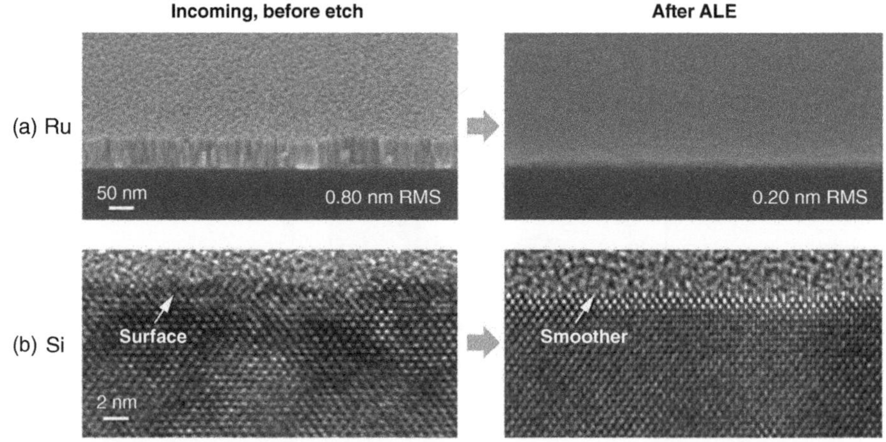

Figure 6.26 Surface before and after directional ALE in (a) tilt scanning electron microscopy (SEM) with Ru ALE, 100 cycles and (b) side-view HR-TEM with Si ALE, 50 cycles. Source: Kanarik et al. (2018).

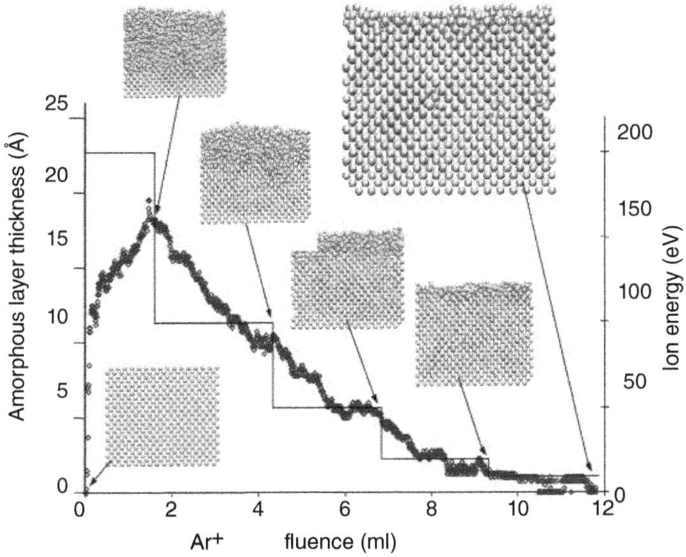

Figure 6.27 Modeled amorphous layer thickness vs argon ion fluence with changing ion energy. Source: Graves and Humbird 2002). © 2002 Elsevier.

Directional ALE maintained a line edge roughness (LER) of 2.8 nm while an RIE process increased LER to over 4 nm (Tan et al. 2019a).

One of the surface smoothening mechanisms for directional ALE is that the surface atom mobility is enhanced during the sputtering removal step. The smoothening effect of argon ion bombardment has been predicted in MD simulations (Graves and Humbird 2002; Humbird and Graves 2002). Graves and Humbird demonstrated that bombardment of silicon surfaces with argon ions induces silicon mixing and creates an amorphous layer in the top part of the simulated slab (Figure 6.27). The thickness of this layer depends on the ion energy, ranging from 1 to 2 Å at 20 eV to 18 Å at 200 eV. The simulation starts with the highest energy and the energy was lowered while the surface was continuously bombarded. The results demonstrate that it is possible to reverse the amorphization damage by gradually decreasing the ion energy. The recrystallized regions are, however, not completely free of defects. It appears that ion bombardment with sufficiently low energies provides just enough kinetic energy to the surface to stimulate the diffusion of atoms to regroup in an energetically favorable crystalline state.

Because surface diffusion is the dominant mechanism behind the smoothening effect of argon ion bombardment, pure metals are especially suited for smoothening with directional ALE. This is of great practical interest because many metals tend to deposit with rough surfaces. Surface smoothening by directional ALE has also been recorded for nonmetals such as carbon and binary compounds such as GaN (Kanarik et al. 2018). Surfaces as smooth as the freshly grown samples were obtained for GaN and AlGaN (Aroulanda et al. 2019). A 65% improvement of the surface roughness of tungsten after 60 directional ALE cycles has been observed experimentally and modeled with a Monte Carlo sputtering model (Gottscho 2018).

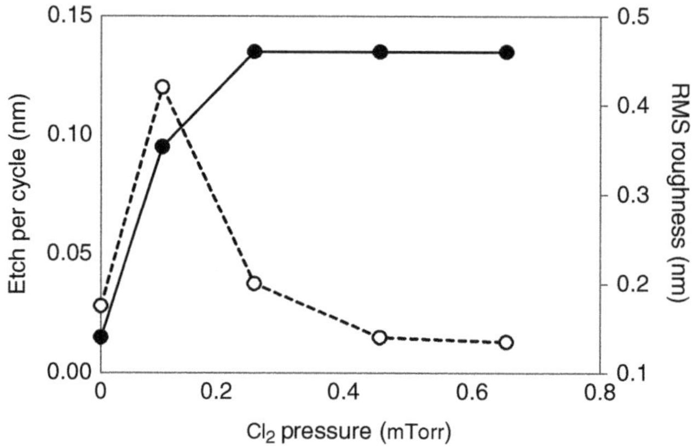

Figure 6.28 EPC and surface roughness as a function of chlorine pressure for directional ALE of (100) oriented silicon with neutral chlorine modification and neutral argon beam removal. The fast argon atom energy was 50 eV and the sputter step time 780 seconds. The chlorine exposure time was 20 seconds. Source: Park et al. (2005c).

It has been postulated that the surface smoothening effect correlates with the synergy of the ALE system and hence with the bulk binding energy, E_O, of the material (Kanarik et al. 2018). Other mechanisms that may contribute to the smoothening effect are a higher reactivity of convex surfaces in the modification step, higher sputtering yields at oblique surfaces, and sputtering redeposition from surface peaks to valleys. Computational studies of SiO_2 ALE with C_4F_8/Ar^+ predict that the SiO_2 surface roughness will increase if the ion energy is above 100 eV due to bulk sputtering (Agarwal and Kushner 2009).

To obtain smooth surfaces with directional ALE, it is important that the process steps are self-limited. If the modification step is incomplete, only modified areas at the surface are removed by argon ion bombardment. While ion-stimulated surface diffusion may smoothen out this effect, the surface may roughen over time. This effect has been reported by Park et al. Their results are shown in Figure 6.28 (Park et al. 2005c).

In contrast to directional ALE, RIE often creates rough surfaces (Nakazaki et al. 2014). This can be attributed to the formation of a several nanometer thick disordered and mixed surface, which is called a selvage layer. The main reason for the formation of this selvage layer in RIE is the effect of reactive ions (Feil et al. 1993). Feil et al. conducted MD simulations of silicon sputtering with 200 eV argon ions in a chlorine environment. The surface was saturated with chlorine after each argon ion impact to simulate RIE with silicon with a chlorine-rich argon/chlorine plasma. This scenario is very different from the situation of directional ALE of silicon with Ar^+/Cl_2, where the chlorine on the surface is removed in the argon sputtering removal step. The simulations produce a rough surface with tower-like structures of silicon atoms with all silicon bonds passivated with chlorine atoms. Closer observation of the collision cascades revealed that the formation of the surface roughness is caused by recoil atoms moving toward the surface. In the absence of chlorine,

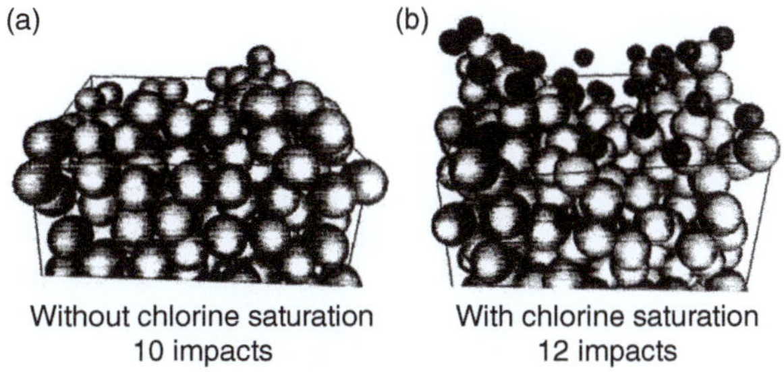

Figure 6.29 MD simulation results for 200 eV argon ion bombardment of a (111) oriented silicon surface without (a) and with (b) chlorine saturation after each argon ion impact. Source: Feil et al. (1993). © 1993 American Vacuum Society.

the atoms diffuse across the surface resulting in smoothening. With chlorine, strong Si—Cl bonds are formed, which suppresses surface diffusion and "freezes" convex features in place. Figure 6.29 shows the MD modeling results comparing the surface after argon ion sputtering with and without the presence of chlorine (Feil et al. 1993).

Based on these modeling results, smoothening in directional ALE requires a completely saturated sputtering removal step. In the absence of this condition, chlorine or other halogens will remain at the surface suppressing surface diffusion and causing surface roughness.

6.3 Applications Examples

6.3.1 ALE with Directional Modification Step

This type of directional ALE has been applied to Si_3N_4 gate spacer etching. Vertical profiles have been demonstrated for Si_3N_4 spacer etching by hydrogen or ion bombardment and isotropic removal of the modified layer with diluted HF acid (Posseme et al. 2014). Posseme et al. demonstrated less than 6 Å recess of the SiGe channel layer. Micrographs of the resulting spacer structures are shown in Figure 6.30.

The diluted HF acid used in this work etched pristine Si_3N_4 with 1 nm/min. This means that the removal step was not self-limited. Removal with HF also lacks selectivity to SiO_2. Finally, the implementation of a dry or vacuum process for surface modification and a wet etching technique for removal makes the implementation of several cycles impractical. To address these challenges removal by a capacitively coupled fluorine plasma with SF_6 or NF_3 feed gas instead of diluted HF acid has been proposed and demonstrated (Sherpa and Ranjan 2017).

Nakane et al. demonstrated ALE of Si_3N_4 via a capacitively coupled plasma (CCP) hydrogen plasma followed by exposure to fluorine radicals generated by a downstream plasma (Nakane et al. 2019). After the hydrogen plasma exposure, Si—H bonds were present near the surface, while N—H bonds were mainly located deeper into the film. The etching thickness can be controlled by adjusting the Si–H rich layer thickness, for example, by changing the bias power to control the hydrogen ion energy (Nakane et al. 2019).

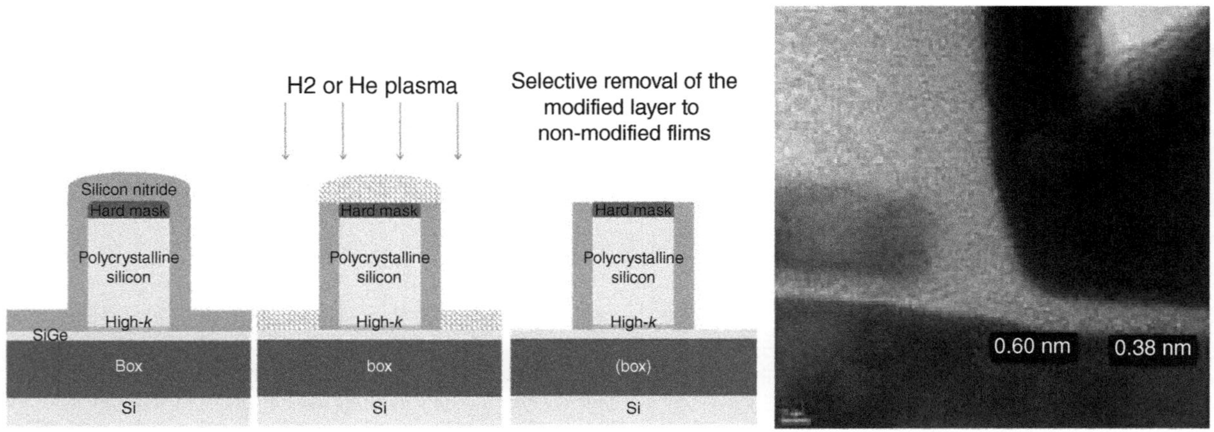

Figure 6.30 (a) Schematic of nitride spacer etching directional ALE with H2 or He implantation and removal with diluted HF acid. (b) TEM pictures after silicon nitride etching using H_2 plasma and 60 seconds 1% HF dip. Source: Posseme et al. (2014). © 2014 AIP Publishing.

ALE with directional modification step is a promising candidate for etching of materials that are difficult to etch with RIE such as copper, cobalt, nickel, iron, palladium, platinum, and other transition metals (Sang and Chang 2020). These materials are used or are candidates for the formation of advanced IC interconnects and emerging memories including magnetic random access memory (MRAM) devices. Some of these metals are difficult to oxidize, either with oxygen or with halogens. Oxidation is the prerequisite for removal by organic reactants. Using shallow ion implantation is an effective way for oxidation (Chen et al. 2017a). Tapered profiles were reported for etching of cobalt with oxygen ion implantation at 200 eV followed by removal of the cobalt oxide with formic acid vapor (Chen et al. 2017b). The mask material in this experiment was TiN.

6.3.2 ALE with Directional Removal Step and Modification by Chemisorption and Diffusion

This class of ALE is finding its way into IC manufacturing for etching of logic silicon FinFET gates. This application requires low damage and high selectivity to stop on a thin stopping layer. At the same time, long overetching is needed to clear the corners at the base of the fins. High selectivity and low ARDE are the performance advantages of directional ALE of silicon with Cl_2/Ar^+, which come to bear in FinFET gate etching.

Huard et al. modeled a Cl_2/Ar^+ ALE FinFET gate etch process (Huard et al. 2017). The geometry used for this theoretical study consists of a periodic array of vertical crystalline silicon fins, each with a width of 10 nm and height of 42 nm. The pitch, which is the sum of the line width and space between the lines, is 42 nm. The fins are covered with a 1 nm SiO_2 etch stop layer on the sides and 10 nm on the top to prevent damage to the fin. The process was modeled as an RIE process until the top of the fins was exposed. The rest of the process was modeled either as ALE or as RIE for comparison as shown in Figure 6.31. The ALE process is able to clear the corners in a three-dimensional FinFET transistor. There is little or no silicon left on the side of the fins whereas silicon did persist on the sidewalls in the RIE case. This is the result of the flat etch front for ALE and reduced ARDE. As expected, the process time for ALE is longer compared to RIE.

To increase EPC and reduce process time, Tan et al. proposed a high-ion-energy directional ALE process (Tan et al. 2019a). It is a promising process for patterning of silicon and metal structures with aspect ratios of 1 to at least 5. We introduced the mechanism for this type of ALE in Section 6.1.2.

Directional ALE can be applied to carbon materials such as amorphous carbon hardmasks, polymers, and resists. The process used surface modification with O_2 gas and removal with argon ions (Vogli et al. 2013; Kanarik et al. 2017).

Other potential applications for this class of directional ALE are etching of non-volatile materials, for instance, MRAM structures (Tan et al. 2019b), and the removal of damaged layers after contact hole etching into SiO_2 stopping on silicon (Kim et al. 2013). Removal of HfO_2 gate oxide selective to a thin SiO_2 underlayer with ALE comprised of BCl_3 gas/surface modification, and fast argon atom removal has been

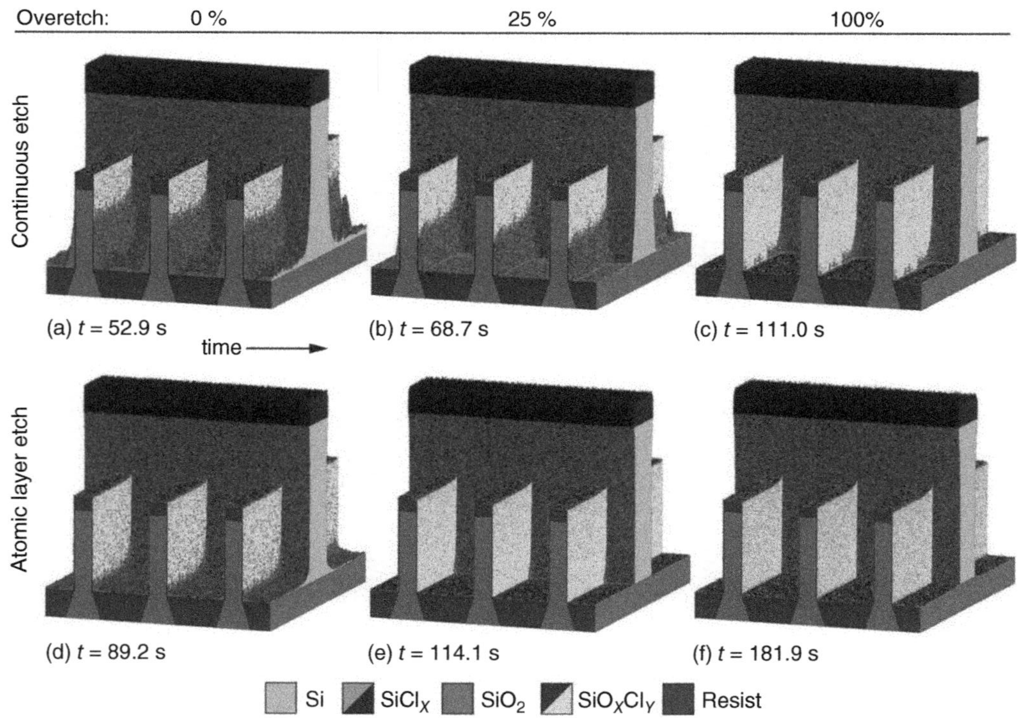

Figure 6.31 Profiles resulting from etching the gate structure with a (a–c) continuous etching process, and (d–f) the optimized ALE process. Frames are taken at equal overetch (as a percentage of the time required to expose the bottom SiO_2), not at equal etch times. The etch times listed for the ALE process (d–f) are active (plasma on) times and ignore any purge or dwell times necessary for a functional ALE process. Source: Huard et al. (2017). © 2017 American Vacuum Society.

demonstrated (Park et al. 2009). When metal oxide semiconductor field effect transistor (MOSFET) devices were fabricated with an ALE process, a 70% increase in drain current and a lower leakage current were observed compared to devices etched with RIE. This improved performance is attributed to reduced structural and electrical damage. This ALE application works only for planar devices where the HfO_2 gate oxide is parallel to the wafer surface. For FinFET devices with vertical sidewalls, thermal isotropic ALE must be used (see Chapter 4). Surface smoothening applications outside the IC industry have also been proposed (Kanarik et al. 2019).

6.3.3 ALE with Directional Removal Step and Modification by Reactive Layer Deposition

This embodiment of directional ALE was the first to be used in volume IC manufacturing. The introductory application is the formation of high aspect ratio, high selectivity logic contact holes for the 10 nm and below logic device nodes. These

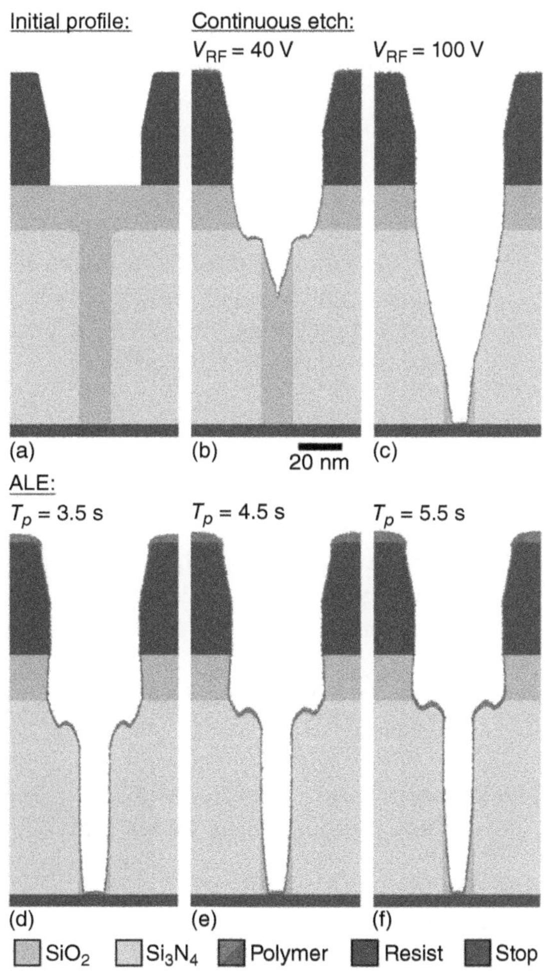

Figure 6.32 Self-aligned contact etch profiles for continuous etching and ALE. (a) Initial profile. Continuous etch using the FC gas mixture with 1200 W inductively coupled plasma (ICP) power for (b) a bias voltage $V_B = 40$ V and (c) $V_B = 100$ V. ALE profiles after 165 cycles using ion bombardment time of 20 seconds and modification step times of (d) 3.5, (e) 4.5, and (f) 5.5 seconds. Source: Huard et al. (2018b). © 2018 American Vacuum Society.

contacts must land on the fin structures between gates lines (see Figure 6.31). Because the space is too narrow for the contacts to land between the gates, the so-called self-aligned contacts are implemented. In this technology, the contact hole is printed large enough to overlap partially two neighboring gates and the space between them. As the etch process reaches the top of the gate lines, it must stop with high selectivity in them while continuing into the intra-gate space.

This is illustrated in Figure 6.32, which compares modeling results for ALE and RIE (Huard et al. 2018b). The gates are represented as Si_3N_4 blocks while in practice the gate is made of metal layers and encapsulated by Si_3N_4 liners. Highly selective SiO_2 ALE with C_4F_8/Ar^+ is used to stop in the Si_3N_4 liners. The mechanism for this ALE process was discussed in Section 6.1.3.

The modeling results show the superior selectivity of ALE compared to RIE with the same plasma chemistry. This is attributed to the lower ion energies. In addition, ALE enables more control over the polymerization of this process compared to RIE. For similar amounts of polymer deposition, ALE will continue etching while RIE may encounter etch stop.

This type of ALE has also been proposed for selective etch back of cobalt. Here, BCl_3 is used as the polymerizing gas (Yang et al. 2018). Another potential application is patterning of SiON transfer layers selectively to extreme ultraviolet (EUV) resists with fluorocarbon polymer deposition and argon ion removal (Metz et al. 2017).

Problems

P6.1 Use the information in Table 2.1 to discuss why it impossible to realize a directional ALE process for silicon by forming a silicon oxide surface layer and sputtering with noble gas ions. How is this different for the same approach when applied to carbon?

P6.2 Apply the ideal ALE window concept to directional ALE with directional modification by ion implantation and removal with neutrals and radicals. Consider both steps in the design of the window.

P6.3 Explain the relationship between thermal isotropic ALE with ligand exchange and directional ALE with surface modification by adsorption and diffusion?

P6.4 What is the fundamental reason for the plateau in the ideal ALE window for directional ALE?

P6.5 Use Eq. (2.8) to explain why there is synergy for directional ALE for energies above the sputtering threshold of the bulk material. Discuss the dependence of synergy on step time and ion energy under these conditions?

P6.6 What is the mechanism that enables directional ALE to improve surface smoothness in many cases?

References

Agarwal, A. and Kushner, M.J. (2009). Plasma atomic layer etching using conventional plasma equipment. *J. Vac. Sci. Technol., A* 27: 37–50.

Antoun, G., Lefaucheux, P., Tillocher, T. et al. (2019). Cryo atomic layer etching of SiO_2 by C_4F_8 physisorption followed by Ar plasma. *Appl. Phys. Lett.* 115: 153109 1–4.

Aroulanda, S., Patard, O., Altuntas, P. et al. (2019). Cl_2/Ar based atomic layer etching of AlGaN layers. *J. Vac. Sci. Technol., A* 37: 041001 1–4.

Athavale, S.D. and Economu, D.J. (1996). Realization of atomic layer etching of silicon. *J. Vac. Sci. Technol., B* 14: 3702–3705.

Berry, I.L., Kanarik, K.J., Lill, T. et al. (2018). Applying sputtering theory to directional atomic layer etching. *J. Vac. Sci. Technol., A* 36: 01B105 1–7.

Chang, J. and Chang, J.P. (2017). Achieving atomistic control in materials processing by plasma-surface interactions. *J. Phys. D* 50: 253001.

Chang, J.P. and Sawin, H.H. (1997a). Kinetic study of low energy ion-enhanced polysilicon etching using Cl, Cl_2, and Cl^+ beam scattering. *J. Vac. Sci. Technol., A* 15: 610–615.

Chen, J.K.C., Altieri, N.D., Kim, T. et al. (2017a). Directional etch of magnetic and noble metals. I. Role of surface oxidation states. *J. Vac. Sci. Technol., A* 35: 05C304 1–6.

Chen, J.K.C., Altieri, N.D., Kim, T. et al. (2017b). Directional etch of magnetic and noble metals. II. Organic chemical vapor etch. *J. Vac. Sci. Technol., A* 35: 05C305 1–8.

Coburn, J.W. and Winters, H.F. (1979). Ion- and electron-assisted gas-surface chemistry – an important effect in plasma etching. *J. Appl. Phys.* 50: 3189–3196.

Faraz, T., Roozeboom, F., Knoops, H.C.M., and Kessels, W.M.M. (2015). Atomic layer etching: what can we learn from atomic layer deposition? *ECS J. Solid State Sci. Technol.* 4: N5023–N5032.

Feil, H., Dieleman, J., and Garrison, B.J. (1993). Chemical sputtering of silicon related to roughness formation of a Cl-passivated Si surface. *J. Appl. Phys.* 74: 1303–1309.

Gasvoda, R., Verstappen, Y.G.P., Wang, S. et al. (2019). Surface prefunctionalization of SiO_2 to modify the etch per cycle during plasma-assisted atomic layer etching. *J. Vac. Sci. Technol., A* 37: 051003 1–10.

Gerlach-Meyer, J.W. and Coburn, E.K. (1981). Ion-enhanced gas-surface chemistry: the influence of the mass of the incident ion. *Surf. Sci.* 103: 177–188.

Gottscho, R.A. (2018). Rethinking the art of etch. *Keynote Presentation at the 2018 Dry Processing Symposium*, Nagoya, Japan.

Gottscho, R.A., Jurgensen, C.W., and Vitkavage, D.J. (1992). Microscopic uniformity in plasma etching. *J. Vac. Sci. Technol., B* 10: 2133–2147.

Graves, D.B. and Humbird, D. (2002). Surface chemistry associated with plasma etching processes. *Appl. Surf. Sci.* 192: 72–87.

Hamaguchi, S., Okada, Y., Isobe, M. et al. (2018a). Surface reaction analysis by molecular dynamics (MD) simulation for SiO_2 atomic layer etching (ALE). *AVS Symposium 2018*, Long Beach, CA, USA.

Hamaguchi, S., Okada, Y., Isobe, M. et al. (2018b). Atomistic simulations of plasma-surface interaction for ALD and ALE processes. *APS Annual Gaseous Electronics Meeting Abstracts.*

Huard, C.M., Zhang, Y., Sriraman, S. et al. (2017). Atomic layer etching of 3D structures in silicon: self-limiting and nonideal reactions. *J. Vac. Sci. Technol., A* 35: 031306 1–15.

Huard, C.M., Lanham, S.J., and Kushner, M.J. (2018a). Consequences of atomic layer etching on wafer scale uniformity in inductively coupled plasmas. *J. Phys. D: Appl. Phys.* 51: 155201 1–8.

Huard, C.M., Sriraman, S., Paterson, A., and Kushner, M.J. (2018b). Transient behavior in quasi-atomic layer etching of silicon dioxide and silicon nitride in fluorocarbon plasmas. *J. Vac. Sci. Technol., A* 36: 06B101 1–25.

Hudson, E., Srivastava, A., Bhowmick, R. et al. (2014). Highly selective atomic layer etching of silicon dioxide using fluorocarbons. *AVS Symposium*, paper PS2+TF-ThM2, November 2014.

Humbird, D. and Graves, D.B. (2002). Ion-induced damage and annealing of silicon. Molecular dynamics simulations. *Pure Appl. Chem.* 74: 419–422.

Kaler, S.S., Lou, Q., Donnelly, V.M., and Economou, D.J. (2017). Atomic layer etching of silicon dioxide using alternating C_4F_8 and energetic Ar^+ plasma beams. *J. Phys. D: Appl. Phys.* 50: 234001 1–11.

Kanarik, K.J., Tan, S., Holland, J. et al. (2013). Moving atomic layer etch from lab to fab. *Solid State Technol.* 56: 14–17.

Kanarik, K.J., Lill, T., Hudson, E.A. et al. (2015). Overview of atomic layer etching in the semiconductor industry. *J. Vac. Sci. Technol., A* 33: 020802 1–14.

Kanarik, K.J., Tan, S., Yang, W. et al. (2017). Predicting synergy in atomic layer etching. *J. Vac. Sci. Technol., A* 35: 05C302 1–7.

Kanarik, K.J., Tan, S., and Gottscho, R.A. (2018). Atomic layer etching: rethinking the art of etching. *J. Phys. Chem. Lett.* 9: 4814–4821.

Kanarik, K.J., Tan, S., Lill, T. et al. (2019). ALE smoothness: in and outside semiconductor industry. US Patents 9,984,858 and 1,030,4659.

Kim, J.K., Lee, S.I., Kim, C.K. et al. (2013). Atomic layer etching removal of damaged layers in a contact hole for low sheet resistance. *J. Vac. Sci. Technol., A* 31: 061302 1–7.

Kuboi, N., Tatsumi, T., Komachi, J., and Yamakawa, S. (2019). Insights into different etching properties of continuous wave and atomic layer etching processes for SiO_2 and Si_3N_4 films using voxel-slab model. *J. Vac. Sci. Technol., A* 37: 051004 1–13.

Lill, T., Kanarik, K.J., Tan, S. et al. (2015). Divide et impera: towards new frontiers with atomic layer etching. *ECS Trans.* 69 (7): 259–268.

Lill, T., Kanarik, K.J., Tan, S. et al. (2016). Directional atomic layer etching. In: *Encyclopedia of Plasma Technology*, 1e (ed. L. Sohet). CRC Press. 1–10.

Lill, T., Kanarik, K.J., Tan, S. et al. (2018). Benefits of atomic layer etching (ALE) for material selectivity. *ALD/ALE conference 2018*, Incheon, Korea.

Lin, K.Y., Li, C., Engelmann, S. et al. (2020). Selective atomic layer etching of HfO_2 over silicon by precursor and substrate-dependent selective deposition. *J. Vac. Sci. Technol., A* 38: 032601 1–18.

Mannequin, C., Vallée, C., Akimoto, K. et al. (2020). Comparative study of two atomic layer etching processes for GaN. *J. Vac. Sci. Technol., A* 38: 032602 1.

Mantenieks, M.A. (1999). Sputtering Threshold Energies of Heavy Ions. *NASA report NASA/TM-1999-209273.*

Matsuura, T., Murota, J., Sawada, Y., and Ohmi, T. (1993). Self-limited layer-by-layer etching of Si by alternated chlorine adsorption and Ar^+ ion irradiation. *Appl. Phys. Lett.* 63: 2803–2805.

Metz, A.W., Cottle, H., Honda, M. et al. (2017). Overcoming etch challenges related to EUV based patterning. *SPIE Proceedings Vol. 10149, Advanced Etch Technology for Nanopatterning VI*, 1014906.

Metzler, D., Bruce, R.L., Engelmann, S. et al. (2014). Fluorocarbon assisted atomic layer etching of SiO_2 using cyclic Ar/C_4F_8 plasma. *J. Vac. Sci. Technol., A* 32: 020603 1–4.

Metzler, D., Li, C., Engelmann, S. et al. (2016). Fluorocarbon assisted atomic layer etching of SiO_2 and Si using cyclic Ar/C_4F_8 and Ar/CHF_3 plasma. *J. Vac. Sci. Technol., A* 34: 01B101 1–10.

Nakane, K., Vervuurt, R.H.J., Tsutsumi, T. et al. (2019). In situ monitoring of surface reactions during atomic layer etching of silicon nitride using hydrogen plasma and fluorine radicals. *ACS Appl. Mater. Interfaces* 11: 37263–37269.

Nakazaki, N., Tsuda, H., Takao, Y. et al. (2014). Two modes of surface roughening during plasma etching of silicon: role of ionized etch products. *J. Appl. Phys.* 116: 223302 1–20.

Oehrlein, G.S., Metzler, D., and Li, C. (2015). Atomic layer etching at the tipping point: an overview. *ESC J. Solid State Sci. Technol.* 4: N5041–N5053.

Oh, C.K., Park, S.D., Lee, H.C. et al. (2007). Surface analysis of atomic-layer-etched silicon by chlorine. *Electrochem. Solid-State Lett.* 10: H94–H97.

Park, S.D., Lee, D.H., and Yeom, G.Y. (2005a). Atomic layer etching of Si(100) and Si(111) using Cl_2 and Ar neutral beam. *Electrochem. Solid-State Lett.* 8: C106–C109.

Park, S.D., Min, K.S., Yoon, B.Y. et al. (2005b). Precise depth control of silicn etching using chlorine atomic layer etching. *Jpn. J. Appl. Phys.* 44: 389–393.

Park, S.D., Oh, C.K., Lee, D.H., and Yeom, G.Y. (2005c). Surface roughness variation during Si atomic layer etching by chlorine adsorption followed by an Ar neutral beam irradiation. *Electrochem. Solid-State Lett.* 8: C177–C179.

Park, J.B., Lim, W.S., Park, B.J. et al. (2009). Atomic layer etching of ultra-thin HfO_2 film for gate oxide in MOSFET devices. *J. Phys. D: Appl. Phys.* 42: 055202 1–4.

Petit-Etienne, C., Darnon, M., Vallier, L. et al. (2011). Etching mechanisms of thin SiO_2 exposed to Cl_2 plasma. *J. Vac. Sci. Technol., B* 29: 051202 1–8.

Posseme, N., Pollet, O., and Barnola, S. (2014). Alternative process for thin layer etching: application to nitride spacer etching stopping on silicon germanium. *Appl. Phys. Lett.* 105: 051605 1–4.

Ranjan, A., Wang, M., Sherpa, S.D. et al. (2016). Implementation of atomic layer etching of silicon: scaling parameters, feasibility, and profile control. *J. Vac. Sci. Technol., A* 34: 031304 1–13.

Rauf, S., Sparks, T., Ventzek, P.L.G. et al. (2007). A molecular dynamics investigation of fluorocarbon based layer-by-layer etching of silicon and SiO_2. *J. Appl. Phys.* 101: 033308 1–9.

Roozeboom, F., van den Bruele, F., Creyghton, Y.Y. et al. (2015). Cyclic etch/passivation-deposition as an all-spatial concept toward high-rate room temperature atomic layer etching. *ECS J. Solid State Technol.* 4: N5076–N5067.

Sang, X. and Chang, J.P. (2020). Physical and chemical effects in directional atomic layer etching. *J. Phys. D: Appl. Phys. 183001 1–12*

Sherpa, S.D. and Ranjan, A. (2017). Quasi-atomic layer etching of silicon nitride. *J. Vac. Sci. Technol., A* 35: 01A102 1–6.

Sigmund, P. (1981). *Sputtering by Ion Bombardment: Theoretical Concepts*, Chapter 2 in: Topics in Applied Physics, vol. 47 (ed. R. Behrisch), 9. Berlin: Springer-Verlag.

Steinbrüchel, C. (1989). Universal energy dependence of physical and ion-enhanced chemical etch yields at low ion energy. *Appl. Phys. Lett.* 55: 1960–1962.

Tan, S., Yang, W., Kanarik, K.J. et al. (2015). Highly selective directional atomic layer etching of silicon. *ECS J. Solid State Technol.* 4: N5010–N5012.

Tan, S., Yang, W.' Kanarik, K.J. et al. (2019a). Atomic layer etching – advancing its application with a new regime. *ALD/ALE Conference 2019*, Bellevue, WA, USA.

Tan, S., Kim, T., Yang, W. et al. (2019b). Dry plasma etch method to pattern MRAM stack. US Patents 9,806,252 and 1,0374,144.

Todorov, S.S. and Fossum, E.R. (1988). Sputtering of silicon dioxide near threshold. *Appl. Phys. Lett.* 52: 365–367.

Toyoda, N. and Ogawa, A. (2017). Atomic layer etching of Cu film using gas cluster ion beam. *J. Phys. D: Appl. Phys.* 50: 184003 1–5.

Vitale, S.A. and Smith, B.A. (2003). Reduction of silicon recess caused by plasma oxidation during high-density plasma polysilicon gate etching. *J. Vac. Sci. Technol., B* 21: 2205–2211.

Vogli, E., Metzler, D., and Oehrlein, G.S. (2013). Feasibility of atomic layer etching of polymer material based on sequential O_2 exposure and Ar low-pressure plasma-etching. *Appl. Phys. Lett.* 102: 253105 1–4.

Wang, S.B. and Wendt, A.E. (2000). Control of ion energy distribution at substrates during plasma processing. *J. Appl. Phys.* 88: 643–646.

Wehner, G.K. (1955). Sputtering of metal single crystals by ion bombardment. *J. Appl. Phys.* 26: 1056–1057.

Wehner, G.K. (1956). Controlled sputtering of metals by low-energy Hg ions. *Phys. Rev.* 102: 690–704.

Yan, C. and Zhang, Q.Y. (2012). Study on low-energy sputtering near the threshold energy by molecular dynamics simulations. *AIP Adv.* 2: 032107 1–15.

Yang, Y., Zhou, B., Shen, M. et al. (2018). Cobalt etch back. US Patent 9,8708,99.

Zalm, P.C. (1984). Some useful yield estimates for ion beam sputtering and ion plating at low bombarding energies. *J. Vac. Sci. Technol., B* 2: 151–152.

Zhang, S., Huang, Y., Tetiker, G. et al. (2019). Computational modelling of atomic layer etching of chlorinated germanium surfaces by argon. *Phys. Chem. Chem. Phys.* 21: 5898–5902.

7

Reactive Ion Etching

7.1 Reactive Ion Etching Mechanisms

So far, we have reviewed etching technologies where only one kind of species is involved in the etching (thermal etching and radical etching) or where different species are separated by space or time (thermal isotropic and directional atomic layer etching (ALE)). In reactive ion etching (RIE), many species such as neutrals, radicals, ions, electrons, and photons impact the surface simultaneously and continuously. The resulting etching process is orders of magnitude more complex and cannot be easily described. This is why RIE has also been called a "black box" (Winters et al. 1977; Gottscho et al. 1999).

In this chapter, we will apply the insights from analyzing simpler etching technologies to RIE. In addition, we will introduce mechanisms that are unique to RIE. These specific mechanisms stem from the fact that in RIE, a chemically reactive plasma is in direct contact with the etching surface. These RIE-specific mechanisms are etching rate dependence on simultaneous fluxes, chemical sputtering, the formation of a mixed surface layer, and redeposition of etching products.

7.1.1 Simultaneous Species Fluxes

One of the main features of RIE is the synergy of ion and neutral fluxes, which gives rise to greatly enhanced etching rates. This effect was demonstrated by Coburn and Winters by measuring silicon etching with an argon ion beam, with a XeF_2 neutral beam and a combination of the two beams (Coburn and Winters 1979a) (see Figure 6.4). Ion-neural synergy was also demonstrated in beam experiments with molecular chlorine and argon ion beams (Coburn 1994b).

Gottscho et al. developed a model that accounts for ion-neutral synergy based on the assumption that the yield per ion is proportional to ion energy times the surface coverage of the chemically assisting neutral species. In this case, the etching rate is given by Gottscho et al. (1992)

$$\mathrm{ER} = \nu \Theta E_i J_i \tag{7.1}$$

This is the same equation as Eq. (6.1), which we used in the discussion of the removal step in directional ALE. Instead of the sputtering rate, Eq. (7.1) describes the etching

Atomic Layer Processing: Semiconductor Dry Etching Technology, First Edition. Thorsten Lill.

rate. As in Eq. (6.1), v is the volume removed per unit bombardment energy for a saturated surface (cm^3/eV), Θ is the surface coverage, E_i is the ion energy (eV), and J_i is the ion flux to the surface (cm^2/s). In this model, etching requires the surface to be covered by reactive species. The reactive sticking probability of neutrals is proportional to the number of open sites on the surface. The neutral etching rate can be expressed as (Gottscho et al. 1992)

$$\mathrm{ER} = v_n s(1 - \Theta)J_n \tag{7.2}$$

Here, v_n is the volume removed per reacting neutral (cm^3), s is the sticking probability of the neutral species (molecules or radicals) on the bare surface, and J_n is the neutral flux to the surface. Equating Eqs. (7.1) and (7.2), the following expression is obtained for surface coverage as a function of the ion energy flux to neutral flux ratio (Gottscho et al. 1992):

$$\Theta = \frac{1}{1 + vE_iJ_i/(v_n sJ_n)} \tag{7.3}$$

Substituting Eq. (7.3) into Eq. (7.1), the etching rate can be written as a function of the ion and neutral flux (Gottscho et al. 1992):

$$\mathrm{ER} = \frac{vE_iJ_i}{1 + vE_iJ_i/(v_n sJ_n)} \tag{7.4}$$

Equation (7.4) is the mathematical expression of the Coburn–Winters synergy experiment, which laid the foundation of RIE (Coburn and Winters 1979a). If the neutral flux is close to zero, the etch rate vanishes. When the ion energy flux is negligible, the etch rate goes to zero. The total etch rate with both neutrals and ions is greater than the etching rates with either species alone. This model considers only purely synergistic etching. Equation (7.4) describes the synergistic effect for RIE with simultaneous fluxes and can be viewed as analogous to Eq. (6.9), which describes synergy in the sputter removal step in directional ALE (see Section 6.1.2).

The meaning of some of the values in Eq. (7.4) can be interpreted using the simple etching framework we presented in Section 2.1. In this framework, chemically active species are delivered to the surface through the gas phase and adsorb on the surface. The role of the adsorbents is to weaken the bonds of the surface atoms to the bulk material as shown in Figure 2.1. We call this process surface modification in ALE but it also exists as a simultaneous process in RIE. The depth of this modification d and the modified surface binding energy E_s determine v, the volume removed per unit bombardment energy in Eq. (7.3). The parameter v is related to the sputtering yield of the modified layer assuming that bulk sputtering can be neglected. The parameter v_n is the volume removed per reacting neutral and describes the reach of the adsorbed species in terms of how many bulk atoms are impacted by bond weakening. This parameter depends on the adsorption mechanism such as the number of bonds it forms and the strength of these bonds. It is important to point out that v_n is not related to the neutral etching yield of a thermal reaction.

Thermal etching, physical sputtering, or radical etching are not considered in Eq. (7.3). These etching pathways are, however, present in many implementations of RIE. For instance, high aspect ratio etching processes operate with ion energies of several thousand electron volts, which will clearly lead to bulk sputtering. These limitations notwithstanding, Eq. (7.4) provides very important insights into the effects of etching with simultaneous, synergistic etching species fluxes. Etching species fluxes must be carefully balanced to maintain repeatable etching rates. This balance ought to be maintained at the bottom of each feature, across the wafer, and wafer to wafer. The sensitivity of the etching rate to the ion or neutral fluxes is not linear. It is smaller when both fluxes are similar or one of the fluxes becomes very small. We will discuss the impact on RIE performance metrics in Section 5.2.

Joubert et al. expanded this model to RIE of silicon oxide, which relies on the deposition of reactive fluorocarbon polymers for etching (Joubert et al. 1994). When these polymers become too thick to be penetrated by ions for a given ion energy, etching can shut down. This effect can be observed as "etch stop" in high aspect ratio dielectric etching. The inclusion of passivating species makes a given RIE process even more reliant on careful flux balancing. The analog of this type of RIE process among ALE processes is directional ALE with surface modification by reactive layer deposition. There, the thickness of the reactive polymer must also be carefully controlled (see Section 6.1.3).

Cyclic RIE has been introduced into device manufacturing over the last two decades to overcome the trade-offs that arise from simultaneous fluxes and the need for flux balancing. Process windows can be increased meaningfully by introducing rapidly alternating process steps, which have different properties. A special case of this idea is directional ALE with surface modification and removal steps. In directional ALE, these steps are self-limited. Without the property of self-limitation, the process is cyclic or pulsed RIE.

The boundaries between cyclic RIE and directional ALE are fluent. As discussed in Chapter 6, not all directional ALE processes are perfectly self-limited or have 100% synergy and can be considered cyclic RIE. However, even in the absence of perfect self-limitation, ALE benefits such as improved etching rate non-uniformity (ERNU) across the wafer, reduced aspect ratio dependent etching (ARDE), and surface smoothness can be obtained. This is one of the reasons why cyclic RIE is increasingly used in manufacturing of semiconductor devices. Especially, the reduction of ARDE is a very important attribute of some cyclic RIE processes.

Cyclic RIE techniques can be classified by the method of step modulation into plasma pulsing and mixed mode pulsing (MMP). In plasma pulsing, the parameter that is periodically changed is the power delivered to the plasma. Gas flow and pressure remain unchanged step to step. The benefit of this method is that the cycle times can be very short, down to milliseconds and below. The switching from one step to another is almost instantaneous, which is beneficial for tool throughput and cost. Chemical variability between the steps, however, is limited because the feed gas mixture remains the same. This constraint is absent in MMP where gases and pressures are changed from step to step. The drawback of MMP is that depending on the reactor volume and gas pressure, it can take several hundred milliseconds or even seconds to exchange the gases and to switch between steps. MMP can be combined with plasma pulsing to leverage the effects of both techniques.

In *mixed mode pulsing*, the individual steps are designed to achieve different objectives. There are few limits to the creativity of the process engineers. Frequently, a step that deposits a protective layer on the sidewall or on top of the mask is paired with a reactive etching step. This passivation can be a fluorocarbon polymer, which is deposited using in situ CVD with $C_xF_yH_z$ feed gases. High aspect ratio deep silicon etching for MEMS applications is a widely used embodiment of this type of cycling RIE (Wu et al. 2010). Directional ALE of SiO_2 with C_4F_8/Ar^+ was implemented in an RIE reactor using MMP (Hudson et al. 2014). In other use cases, the passivation can also be in situ ALD or a simple oxidation step. *Plasma pulsing* found widespread use in the semiconductor industry in the middle of the 2000s when the aspect ratios of devices such as isolation trenches and contact became so severe that the ARDE benefits of plasma pulsing started to become clear.

Let us now discuss how plasma pulsing can reduce ARDE and how the underlying mechanism is related to directional ALE. In this chapter, we will focus on the chemical effects of plasma pulsing on the wafer surface. A more in-depth discussion of the physics of plasma pulsing can be found in Section 9.5.

Advanced RIE tools have at least two radio frequency (RF) power delivery systems, one of which is used to drive plasma density and hence ion flux density. This process parameter is frequently called source power. The other RF power delivery system is used to accelerate ions and the process parameter is called bias power. When the source power is on, ions and radicals are generated with a density that increases with the amount of delivered power. However, in the absence of applied bias power, ions in the plasma are only accelerated by a voltage called self-bias (see Section 9.1). This self-bias is typically not high enough to create chemical or chemically enhanced sputtering with a meaningful rate. Such a plasma is quite useful to modify the surface by depositing reactive fluorocarbon layers or to adsorb halogens at the etching surface. Therefore, plasmas generated by source power only are used in the modification step of directional ALE as discussed in Section 6.1.2.

When bias power is added, positive ions will be extracted from the plasma and accelerated to the surface with an energy that increases with higher bias power. These accelerated ions impact the surface and cause chemical or chemically assisted sputtering. A third possibility to operate a system with source and bias power is to run bias power only. In this case, a plasma can either not be operated in a stable mode or, if the frequency of the RF signal is high enough, a low-density plasma with comparably high ion energies is created.

The plasma mode where the source power is constant and the bias power is turned on and off is called bias pulsing. In the first state without bias power, the surface is modified and in the second state, the modified layer is sputtered. This is very similar to directional ALE. The difference is that in directional ALE, the ions are nonreactive noble gas ions with energies below the sputtering threshold of the bulk material. In bias pulsing, the ions can have energies of several hundred or even thousand electron volts, the ions are reactive and nonreactive (if noble gases are added to the feed gas), and the surface modification continues while the surface is sputtered.

Therefore, state 2 in a bias pulsing process is not self-limited. But because the surface is modified in the first state, the instantaneous removal rate in the second state

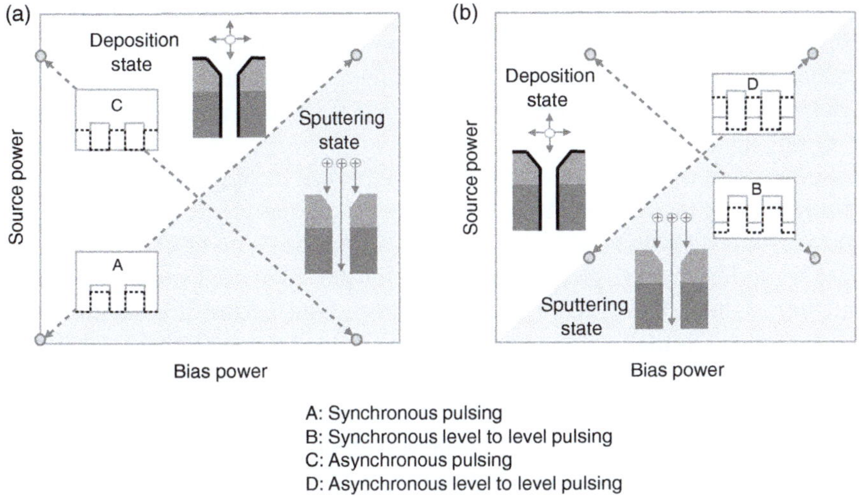

Figure 7.1 Schematic illustration of synchronous and asynchronous plasma pulsing techniques with one (a) and two power levels (b).

levels off as the modified layer is sputtered away. This slowdown, even though it does not go to zero, gives the ARDE benefit similar to nonideal ALE. It is of course also possible to pulse the source and keep the bias constant. This would introduce large variations in ion flux and modulate the ion energy between two levels. This approach is not widely used.

It is also possible to pulse source and bias power simultaneously. Some of the options are shown in Figure 7.1. If source and bias power are pulsed with the same phase, the plasma is turned on and off and the main effect is a slower net etching rate. This so-called synchronous pulsing has not found many practical uses because it does not generate additional process conditions. There is a secondary effect that negative ions can be generated immediately after the power is turned off and these negative ions can be extracted and be used to neutralize positive charges on the wafer. This would help maintain narrow ion angular distributions inside high aspect ratio features. We will discuss plasma pulsing and negative ions in Section 9.5.

The case where source and bias power are pulsed out of phase is called asynchronous pulsing. This technique is similar to bias pulsing except the source power is off during the bias power-on phase. This means that even higher ion energies can be achieved.

Advanced pulsing techniques mix and match different pulsing states to achieve the desired process results. This so-called multistate pulsing is used in high aspect ratio etching to control how deep into the feature proactive polymers are being deposited. The mechanism of how this works is based on ions with different starting ion energies to deliver passivating and reactive species to different locations inside a high aspect ratio structure.

Equation (7.4), which describes the synergy effect in RIE, says that ions and neutrals are required to facilitate etching. These neutrals are molecules or radicals with thermal energies, i.e. species that have not been accelerated in the plasma. They reach the bottom of the feature by means of Knudsen transport as described in Eq. (2.18). The flux of neutrals attenuates much faster than for ions because they have a broader angular distribution than ions. The flux of these neutrals to the bottom of features with aspect ratios above 10 : 1 is negligible in high aspect ratio dielectric etching (Huang et al. 2019).

Therefore, neutralized ions are frequently used to deliver species to the bottom of features with very high aspect ratios. Fortunately, these species are generated in situ by glancing angle surface scattering of ions. We discussed the mechanisms of ion surface scattering and charge exchange in Section 2.8.2. For each surface scattering event, the ion energy is reduced. This means that neutralized, fast atoms with energy depending on the initial ion energy and the number of sidewall surface collisions are generated. If the kinetic energy of the fast atoms is low enough, they will stick to the surface. In that manner, reactive (for instance, fluorine-rich fluorocarbon-rich polymers) and passivating species (for instance, carbon-rich fluorocarbon polymers) can be distributed at various depths of the feature and on the bottom.

This approach can be implemented by using multistate pulsing where each state creates molecular ions with different starting energies and somewhat modulated composition. It is important to point out that the starting ion energies are not discrete but have a distribution, which we will discuss in Section 9.2. This does not change the basic idea of using multistate pulsing in high aspect ratio etching.

7.1.2 Chemical Sputtering

As described in chapter 7.1.1, the main mechanism of RIE is *chemically assisted sputtering*, where both ions and halogen atoms/molecules etch the surface synergistically (Coburn and Winters 1979a). However, when a reactive plasma is in direct contact with the etching surface, chemically reactive ions also impact the surface. This is different from directional ALE where noble gases are used in the removal step. This is a prerequisite to achieve self-limitation, which is the hallmark of ALE.

The most common sources of chemically reactive ions are halogen-, oxygen-, and hydrogen-containing gases, which are dissociated and ionized in the plasma. Ionic fluorocarbon species such as CF_2^+ are very important for etching of SiO_2. When reactive ions hit the surface, physical energy and chemical reactivity are provided simultaneously and the result is an enhancement of the sputtering rate compared to pure physical sputtering by means of noble gas ions. This effect is utilized in reactive ion beam etching (RIBE), which we will cover in Chapter 8. It is also one of the etching mechanisms in RIE, and depending on the process conditions, it can be more or less important. Modeling of high aspect ratio etching of SiO_2 has shown that chemically enhanced sputtering plays a very important role beyond aspect ratios of 10 : 1 (Huang et al. 2019).

Tachi et al. measured chemical sputtering yields of silicon under bombardment by mass-selected reactive F^+, Cl^+, and Br^+ ions as well as nonreactive Ne^+, Ar^+, and Kr^+ ions (Tachi and Okudaira 1986). Since ions with nearly equal mass were compared, there should no difference in the sputtering yield because the sputtering yield is dependent on the mass and energy of the impinging ion (see

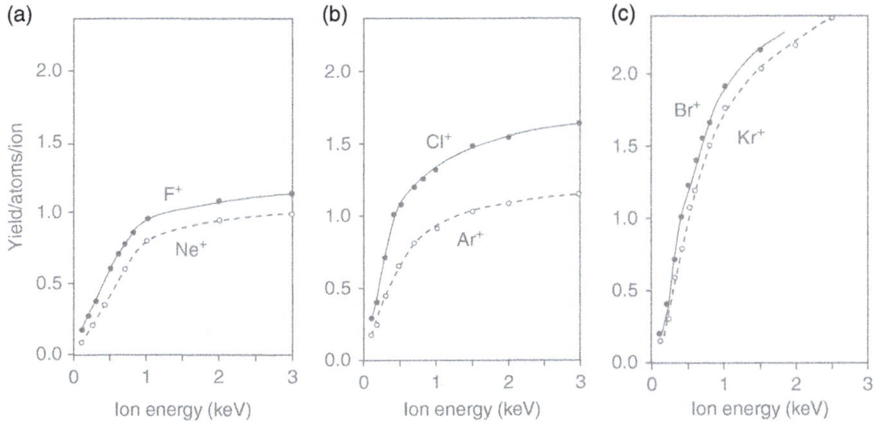

Figure 7.2 Total sputtering yield of silicon by reactive and nonreactive ions with similar mass. Source: Tachi and Okudaira (1986). © 1986 AIP Publishing.

Eq. (2.8)). However, the experimental results in Figure 7.2 show that in comparison with their nonreactive neighbors in the periodic table, the sputtering yields for the reactive ions were higher. The chemistry effect is larger for chlorine than fluorine and nearly absent for bromine. The effect is also more pronounced for lower ion energies. These results can be explained by implantation of halogen ions and the formation of strong bonds with silicon. This will lower the binding energy E_O of the modified layer and lead to higher sputtering yields (see Eq. (2.8)).

This effect has been called chemical sputtering (Tachi and Okudaira 1986). The term "*chemical sputtering*" has also been applied to the enhancement of the sputtering yield by the presence of halogens and other reactive species due to adsorption (Coburn 1994a). In the light of recent developments in directional ALE, we will use the term "*chemical sputtering*" to describe strictly the effect of reactive ions. This effect is present in RIE and RIBE but not in directional ALE. Here, noble gas ions are used in the removal step. This mechanism can be best described as physical sputtering of a chemically weakened surface or *chemically assisted sputtering*. Of course, this mechanism is also present in RIE and is in many cases the main etching mechanism because of the typically large fluxes of neutrals and radicals. Interestingly, chemical sputtering of metals with oxygen can also reduce the sputtering yield. This is an effect utilized in RIBE to increase selectivities.

Chemical reactivity of impinging ions also changes the angular dependence of the sputtering rate. The typical sputtering rate maximum for silicon sputtering with argon ions is absent for chlorine ion sputtering of silicon. The yield is flat from normal incidence to about 40° and then drops gradually to zero for glancing incidence (Chang and Sawin 1997; Chang et al. 1997; Mayer et al. 1981; Guo and Sawin 2009). It is likely that reactive ions are implanted deeper under normal incidence, which weakens the material to a greater depth and enhances the

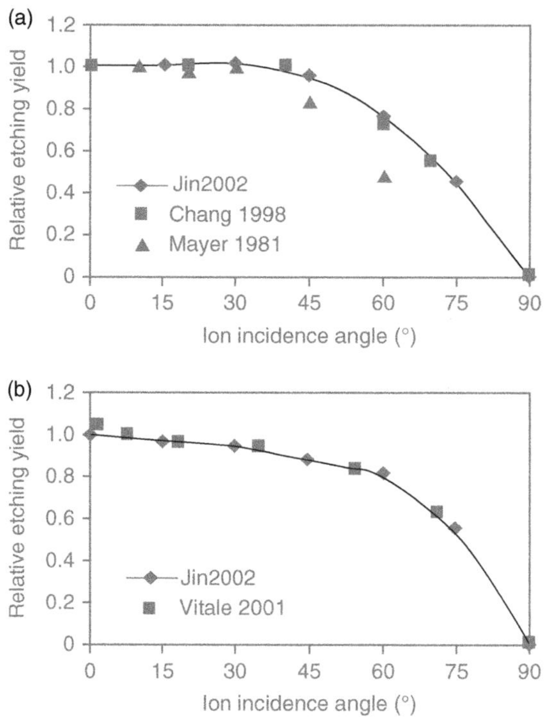

Figure 7.3 Etching yield of polysilicon as a function of ion incident angle. Ion energy = 150 eV, and saturated surface coverage. (a) In Cl_2 plasma and (b) in HBr plasma. Source: Jin et al. (2002). © 2002 AIP Publishing.

sputtering rate, which results in a flattened angular dependence for chemical sputtering (Mayer et al. 1981). The surface may also get roughened by reactive ion bombardment, which could be an additional effect causing a flatter ion angular dependence.

Figure 7.3 shows etching yields of polysilicon as a function of ion incident angle for chlorine and hydrogen bromide plasmas at an ion energy of 150 eV (Jin et al. 2002). The results match well with values reported by Mayer et al. (1981), Chang et al. (1998), and Vitale et al. (2001).

Incident-angle-dependent etching and sputtering yields for SiO_2 are shown in Figure 7.4 (Mayer et al. 1981). A maximum of the sputtering yield around an incidence angle of 60° can be seen for nonreactive argon ions at an energy of 600 eV. The behavior for reactive CF_3^+ ions is quite different. The yield peaks at normal incidence for ion energies of 200 and 400 eV. The maximum shifts to around 30° when the ion energy increases to 600 and 1000 eV. Cho et al. reported deviations of actual RIE rates from fluorocarbon beam studies, which they attribute to the formation of a passivation layer (Cho et al. 2000).

These experimental results illustrate that the incidence-angle-dependent etching yield varies strongly as a function of etching materials, the nature of the ions and the ion energy, and the formation of passivation layers. In addition, the yield will change as a function of surface roughness, which changes the apparent incidence angle depending on the impact location. This behavior must be considered when modeling profile evolution during RIE.

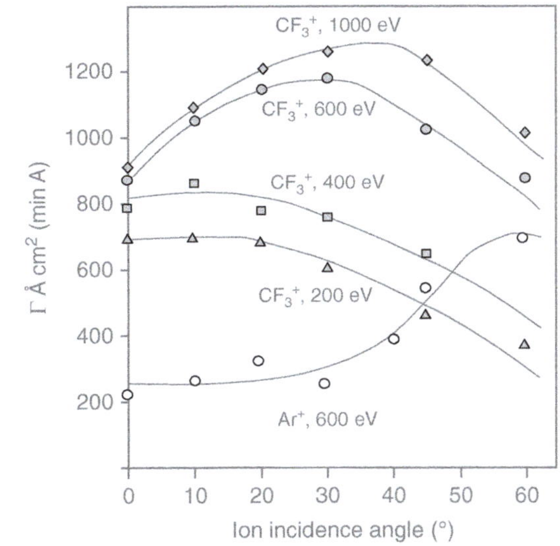

Figure 7.4 Etching and sputtering yields of SiO_2 as a function of angle of incidence and energy in CF_3^+ and Ar^+ beams. Source: Mayer et al. (1981). © 1981 AIP Publishing.

7.1.3 Mixed Layer Formation

In RIE, intermixing of implanted atoms into the bulk leads to the formation of a layer that is also referred to as a "selvage layer." This layer contains reactive species such as halogens. Graves and Humbird conducted molecular dynamics (MD) simulations for the interaction of argon ion as well as fluorine molecules and ions with silicon (Graves and Humbird 2002). They found that the mixed layer is formed mainly due to direct implantation of fluorine ions into the amorphous silicon surface. However, fluorine adsorbed on the silicon surface does not mix into the layer under argon ion impact. Closer examination shows that fluorine ion impacts increase the silicon surface area by creating crevices and cracks, and that the fluorine remains mainly on the surface of this layer (Graves and Humbird 2002). Feil et al. came to the same conclusion for etching of silicon with chlorine (Feil et al. 1993).

If these findings can be generalized, it means that the formation of a mixed layer in RIE is caused by reactive ions. These reactive ions are not present in directional ALE, which uses noble gas ions such as Ar^+ in the removal step to achieve self-limitation. This has important consequences for surface smoothness and selectivity as will be discussed in section 7.2. Another reason why a mixed layer is present in RIE but not in directional ALE is that in the latter, the reactive species are removed completely in each removal step. If there was any intermixing after the modification step, that layer is removed in the following step.

The thickness of the mixed layer has been simulated with MD for the case of silicon etching in a chlorine plasma (Gou et al. 2010). The calculated thickness of the mixed layer is shown in Figure 7.5 as a function of ion energy for a chlorine exposure of 45 monolayers, which was enough to reach steady state. The results are in good agreement with experimental data (Layadi et al. 1997). The thickness is calculated from the full width at half maximum and is found to be 10 and 22 Å deep when the ion energy increases from 25 to 150 eV. The slope of the curve indicates that there is

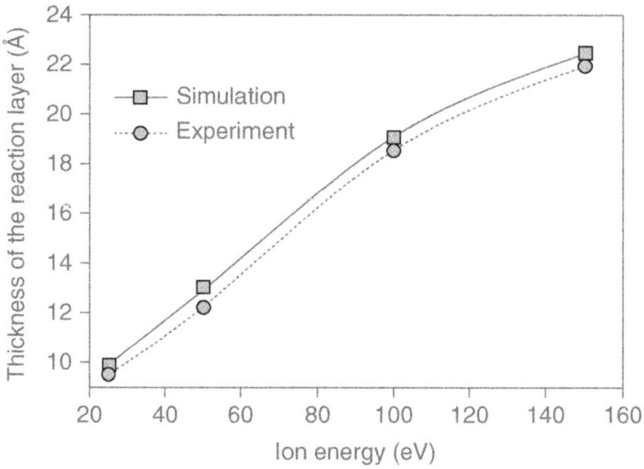

Figure 7.5 Calculated thicknesses of the reaction layer as a function of incident energy, in comparison with experimental data (Layadi et al. (1997). Source: Gou et al. (2010). © 2010 AIP Publishing.

saturation at higher energies because of simultaneous etching. In general, the mixed layers in RIE are a few nanometers deep, which was good enough when structures to be etched had dimensions in the order of several tens of nanometers. Advanced devices, however, have critical dimensions that are below 10 nm. This is one of the reasons why ALE is making inroads into the etching of semiconductor devices.

The composition of mixed layers is not uniform across depth. Depending on chlorine concentration, different $SiCl_x$ species are formed. For an ion energy of 50 eV, Gou et al. find a stoichiometry distribution where $SiCl:SiCl_2:SiCl_3$ is equal to 1.0 : 0.12 : 0.005 (Gou et al. 2010). No $SiCl_4$ was found in the simulations. These species are nonvolatile, and will not desorb thermally at the temperatures used in RIE. Hence ion bombardment is needed to etch. This is exactly the mechanism described in Eq. (7.4). The composition of chlorinated species that are ejected by ion bombardment of the mixed layer reflects the stoichiometry of the mixed layer as shown in Figure 7.6.

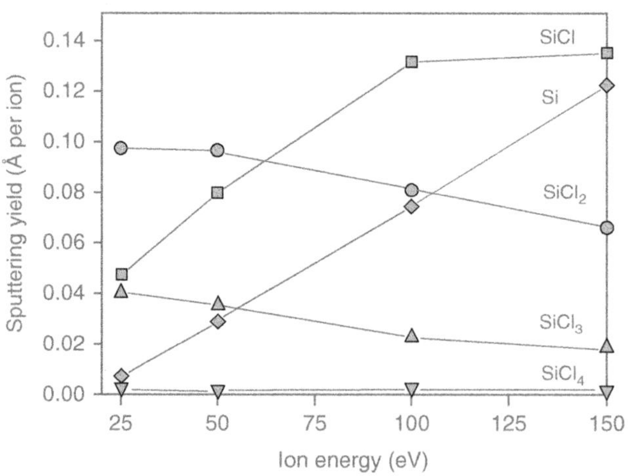

Figure 7.6 Calculated product yields of $SiCl_x$ ($x = 0, 1, 2, 3, 4$) species as a function of incident energy. Source: Gou et al. (2010). © 2010 AIP Publishing.

The fact that all etching products are "nonvolatile" as in "are desorbed by sputtering" has important consequences for etching profiles. If these species are ejected in the direction of a feature sidewall, they will redeposit there. This leads to sidewall passivation and can be used to obtain vertical profiles. As we will show below, direct redeposition is not the only source of passivating species. The latter can also be formed in the plasma when etching species are dissociated and ionized in the plasma.

Similar mixing layers are formed when etching silicon with fluorine- or bromine-containing species. Silicon fluoride is very volatile and thermal etching is possible. Therefore, fluorocarbon passivation gases are used to passivate sidewalls when etching silicon with fluorine-rich gases such as SF_6 or NF_3.

For RIE of SiO_2 and Si_3N_4, the mixed layer is comprised of a fluorocarbon layer, the surface of the silicon compound, and a zone of intermixing. This is very similar to the case of directional ALE with surface modification via deposition of a reactive layer. The thickness of this layer has to be precisely controlled as a function of ion energy to avoid etch stop (Joubert et al. 1994). The ion energy has to be high enough for the ions to penetrate the selvedge layer and to activate the formation of volatile etching products at the interface.

7.1.4 Role of Etching Products

In RIE, volatile and sputtered etching products re-enter the plasma generation region where they can be ionized and dissociated. This leads to the formation of species that contain material from the wafer and that participate in the etching process, either as fluxes directly to the wafer or indirectly via deposition to the reactor walls. This effect is absent in thermal and radical etching and ALE with removal steps that do not involve plasma. In directional ALE with removal by sputtering, the effect is present but to a much lesser degree because the flux of products from the wafer diminishes as the removal reaches self-limitation, which leads on average to lower product fluxes from the wafer.

Redeposition of reaction products onto the etching surface influences RIE etching results greatly. Some etching processes such as silicon gate and isolation trench etch processes rely on redeposition of silicon-containing species for sidewall passivation to achieve vertical etching profiles. Nonuniform redeposition can lead to etching rate and critical dimension (CD) nonuniformities across the wafer. Redeposition onto chamber walls leads to accumulation of reaction products, which requires periodic plasma cleans. Because etching products are in the neutral state when they leave the wafer, neutral transport determines if and where ionization and dissociation by electron impact as well as redeposition will occur (Kiehlbauch and Graves 2003).

The amount of etching species in the plasma that contain atoms from the surface in the plasma is surprising. Quantitative mass spectroscopy of silicon etching plasmas comprised of HBr, Cl_2, and O_2 revealed that under certain experimental conditions, up to 50% of the ion flux was $SiCl_xBr_y^+$ (Cunge et al. 2002). Of course, the concentration is reduced if a larger area on the wafer is covered by a mask. But even in this case, etching products from the mask, assuming the selectivity is not infinite, will participate in the etching process, which complicates matters even more.

The MD simulation results in Figure 7.6 show that as the ion energy is increased, etching products are less saturated with halogen, which makes the products less volatile and increases the sticking probability (Gou et al. 2010). This is a general trend. The participation of reaction products in the process is very important and frequently overlooked in the discussion of RIE mechanisms.

7.2 Performance Metrics

In this section, we discuss etching performance in general terms and link them to mechanisms RIE has in common with simpler etching technologies as well as to RIE-specific mechanisms. There are of course a wide range of different RIE processes, and they have quite specific performance characteristics. A Fin Field Effect Transistor (FinFET) gate etching process, for instance, is very different than a high aspect ratio dielectric etching process. We will discuss these specific cases in Sections 7.3.2 and 7.3.3.

7.2.1 Etching Rate

The etching rate in RIE is the sum of the physical sputtering rate, thermal (spontaneous) etching rate, and synergistic ion-neutral etching rate (Gerlach-Meyer and Coburn 1981; Chang et al. 1997; Marchak and Chang 2011). Electron (Coburn and Winters 1979a) and photon (Shin et al. 2012) stimulated etching can also contribute. The contribution of each etching mechanism depends on the plasma conditions and the material to be etched. For a typical RIE process, the main contribution to the etching rate comes from the synergy between neutrals and ions given by Eq. (7.4). The mechanism that is described in Eq. (7.4) is sputtering of a chemically modified, weakened surface.

Analysis of the sensitivity of the etching rate ER to the ratio of the two fluxes, the ion flux J_i and neutral flux J_n, using Eq. (7.4) reveals that the slope is least steep when both fluxes are of similar magnitude (after considering the other parameters in Eq. (7.4) such as neutrals sticking coefficient and ion energy). Under this condition, the etching rate is the least sensitive to flux variations. Therefore, RIE processes with balanced ion and neutral fluxes are more repeatable. They also have the highest etching rate.

Figure 7.7 shows results from a mixed chlorine neutral and Ar^+/Cl^+ beam experiment by Chang and Sawin (1997) and Chang et al. (1997). An initial sharp rise in the etching yield is observed at low flux ratios where the reaction is limited by the supply of reactive neutrals. This etching regime is called neutral-limited regime. The etching yield then gradually saturates as the reaction becomes ion-flux limited at high flux ratios. Here, the etching yield is a function of the square root of ion energy (see Eq. (2.6)). These trends are observed regardless of whether chlorine or argon ions were used. Figure 7.8 illustrates schematically the neutral and ion-limited process regimes.

The ion-limited regime plays an important role in the etching of high aspect ratio features. The reason is that the angular distribution of neutrals is isotropic. They

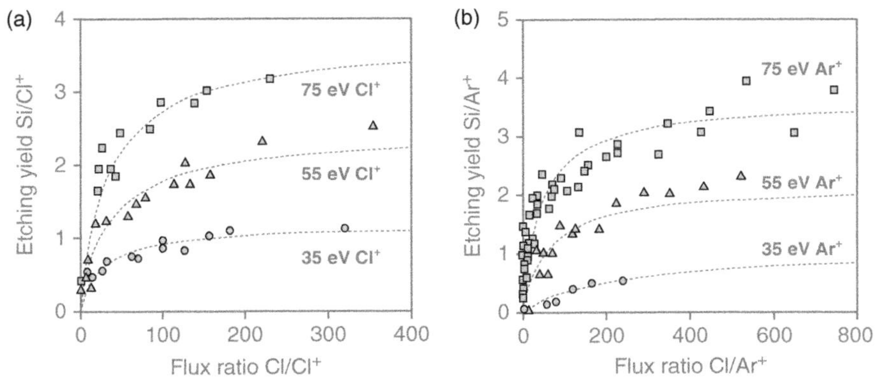

Figure 7.7 Ion-enhanced polysilicon etching by chlorine atoms and ions (a) and by chlorine atoms and argon ions (b) at three ion energy levels. Source: Chang and Sawin 1997; Chang et al. (1997). © 1997 AIP Publishing.

Figure 7.8 Illustration of neutral and ion-limited process regimes.

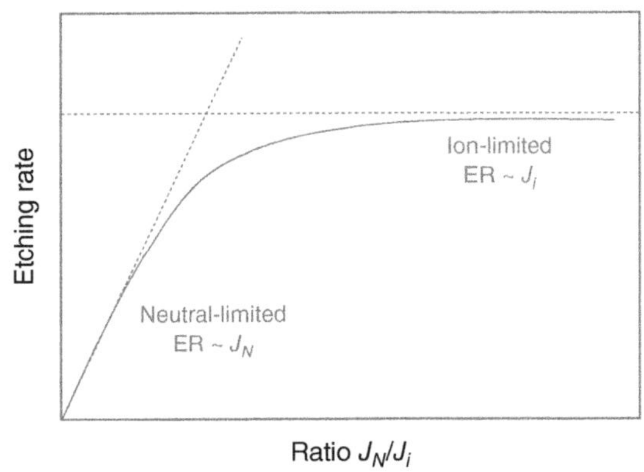

collide with the feature sidewall and reach the bottom by Knudsen transport (see Section 2.8.1). This leads to a fast attenuation of the neutral flux as the aspect ratios increase. In contrast, ions carry an electrical charge and are accelerated toward the etching surface by the plasma sheath (see Section 9.2). Therefore, their angular distribution is directional, and their flux is attenuated to a lesser degree compared to neutrals. The implication of neutrals and ions having different angular distributions is that the neutral and ion fluxes attenuate with different rates as the aspect ratio increases and according to Eq. (7.4), the etching rate changes. Because the ion flux attenuates less than the neutral flux as a function of aspect ratio, it is advantageous to operate in the ion-limited regime. The etching rate will be lower compared to the balanced regime but features with different top opening dimensions will be etched more equally.

This approach is frequently deployed in processes that do not use polymerizing passivation gases such as etching of silicon with a chlorine-based plasma (Huard et al. 2017). This regime is to a certain extend related to directional ALE, for instance, silicon ALE with Cl_2/Ar^+. Here, the removal step starts with a fully chlorinated

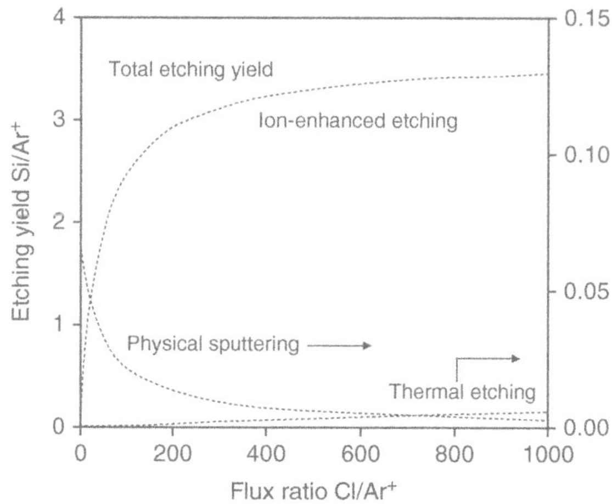

Figure 7.9 Graphical illustration of the ion-enhanced polysilicon etching with Cl and 100 eV Ar^+. The etching yield for both physical sputtering and thermal etching of polysilicon are shown in the right-hand *y*-axis. Source: Chang et al. (1997). © 1997 AIP Publishing.

surface. Each impinging argon ion will remove surface silicon as long as chlorine is in the vicinity of the impact site. As the removal step progresses, this becomes less probable and the removal step reaches self-limitation. In ion-limited etching of silicon with a Cl_2/Ar plasma, there is always chlorine at the ion impact site. The removal by ions is not self-limited but the chlorine adsorption step is saturated in the ion flux-limited regime. Of course, RIE will create a mixed layer even in the ion-limited regime.

There are additional etching mechanisms that are not considered in Eq. (7.4). Firstly, fast ions, either chemically reactive or nonreactive, can contribute to direct sputtering of bulk material if the energies are high enough and the neutral fluxes are too low to ensure continuous coverage. If the surface is completely covered by reactive neutrals, sputtering of bulk material has a low probability because of the shallow depth of origin of sputtered atoms (see Figure 2.8). Secondly, neutrals and radicals can etch with thermal energies spontaneously.

Figure 7.9 shows the contributions from synergistic ion-neutral etching, from pure physical sputtering, and from thermal etching for a chlorine radical/argon ion beam (Chang et al. 1997). Ii is clear that operating in the ion-limited regime suppresses physical sputtering and enhanced chemically enhanced sputtering, i.e. synergistic ion-neutral etching. The physical sputtering yield in this experiment is about 1 order of magnitude lower than for chemically assisted sputtering because the ion energy is only 100 eV. The yield will increase with the square root of the ion energy (see Eq. (2.6)). For this system, the thermal etching rate is very low and hence operating in the ion flux limited regime does not introduce significant thermal etching. Etching of a chemically modified surface can also be initiated by electrons (Coburn and Winters 1979a) and photons (Shin et al. 2012). For an RIE process with all these components, the etching rate is determined by the contributions from the various species fluxes. To guarantee wafer to wafer and tool to tool repeatability, these fluxes must be precisely controlled.

So far, we analyzed etching rates as a function of etching species fluxes. Of course, RIE rates are also dependent on the material properties. For a given plasma condition, different materials etch with different rates. The process chemistry is optimized to achieve high etching rates for each material. Materials properties are considered in Eq. (7.4) in the form of the parameters v, v_n, and s, which are the volume removed per unit bombardment energy, the volume removed per reacting neutral, and the neutral sticking probability, respectively. The volume that is removed per unit bombardment v is proportional to the sputtering yield and can be described by Eq. (2.8) where E_O is the binding energy for the chemically modified, mixed layer. This is analogous to directional ALE where the modification is achieved by chemisorption and in some cases diffusion into a thin surface layer.

The volume that is removed per reacting neutral v_n is a function of the neutrals in weakening the bulk material. The binding energy of the chemically mixed layer is a function of the binding energy of the bulk material, the chemical reactivity of the adsorbing neutral species, and their concentration.

In a simple framework, the modified layer can be envisioned as a solid-state lattice with a network of bonds that make up an effective binding energy. The reactive species used in RIE have a valency of one (hydrogen, halogens) or two (oxygen) with typical elements used in semiconductor manufacturing. Therefore, these reactive species form terminal bonds in the solid, which interrupt the lattice and reduce the effective binding energy. The case of etching with oxygen is special. Oxygen forms volatile CO and CO_2 but also very strongly bonded SiO_2 and metal oxides.

The bond termination effect is concentration dependent. The higher the concentration of reactive species, the lower the effective binding energy is. If the concentration is high enough, fully reacted species can be formed. Depending on the boiling point of these species, they may leave the surface at thermal energies or may require kinetic energy from an ion impact. It is a common method to compare the boiling point of the hydrates, oxides, and halogens to select the appropriate etching gas for a given element. The lower the boiling points of the compounds, the higher the etching rate. The binding energy of the unreacted bulk material is considered important for ion-driven processes such as high aspect ratio etches. Mask materials are optimized to achieve high binding energies for high selectivity. A proxy for the effective binding energy is the modulus of the material.

A special case of the influence of material properties on RIE rates is the etching of doped silicon. Heavily n-doped silicon etches faster, and p-doped silicon etches slower than undoped silicon in a fluorine plasma (Lee and Chen 1986; Winters and Haarer 1987). This doping effect is more pronounced in fluorine radical etching than in RIE. The effect is stronger in chlorine fluorine plasmas followed by fluorine and bromine as shown in Figure 7.10. The experiments in Figure 7.10 were conducted in an inductively coupled plasma reactor at a pressure of 4 mTorr, 475 W source power, and 70 W bias power.

The corresponding profiles tend to be more undercut for n-doped silicon and more tapered for p-doped silicon if the undoped silicon profile is vertical. Lee and Chen proposed a space charge model to explain the doping effect in silicon etching. This model assumes that the etching rate is limited by a field-enhanced diffusion

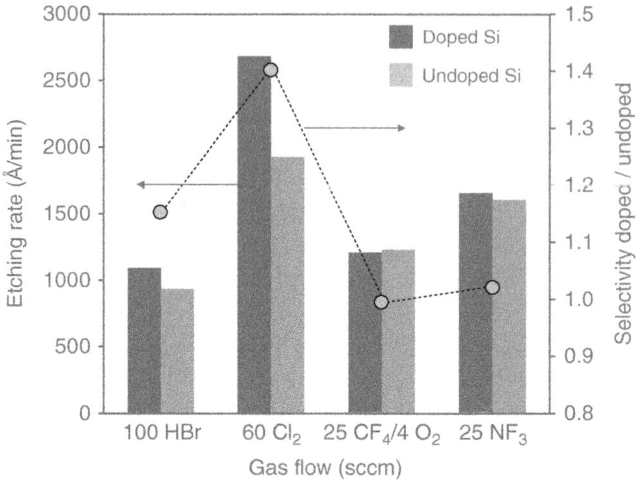

Figure 7.10 Etching rates of n-doped and undoped polysilicon for HBr, Cl_2, NF_3, and CF_4/O_2 plasmas. All other reactor parameters such as pressure and power settings were kept constant.

mechanism, i.e. Cabrera–Mott (CM) diffusion (Winters et al. 1983). Doping modifies the space charge at the surface. Field-enhanced diffusion in n-doped silicon drives the halogen ion into the bulk and enhances etching, while a repulsive field in p-doped silicon pushes the halogen ion to the surface and hence suppresses etching (Lee and Chen 1986). The same Cabrera–Mott diffusion mechanism was discussed in Section 6.1.2 in the context of the depth of modification in directional ALE.

The RIE mechanism for silicon oxide and silicon nitride is related to directional ALE with reactive layer deposition in the modification step. RIE of silicon oxide relies on the deposition of reactive fluorocarbon polymers for etching and ion bombardment to form a selvedge layer. Joubert et al. observed three different regimes as a function of RF bias for an oxide surface exposed to an electron cyclotron resonance (ECR) fluorocarbon plasma (Joubert et al. 1994). These regimes are a fluorocarbon film deposition regime for small RF bias voltages, a fluorocarbon suppression regime for intermediate bias voltages, and an oxide chemically enhanced sputtering regime for high RF voltages (see Figure 7.11). The experimental results were obtained in an ECR reactor with a microwave power of 1000 W at a pressure of 1 mTorr. These are not typical silicon oxide etching conditions, which operate at medium-density plasmas at pressures above 10 mTorr. The results are nevertheless insightful and important. Bias voltage thresholds for the transition from deposition to etching are 35 V for CHF_3 and 60 V for C_2F_4. The transition from polymer-suppressed regime to chemically enhanced sputtering regime occurs at 55 V for CHF_3 and 80 V for C_2F_4. C_2F_4 requires higher ion energies to etch because it has a higher polymer deposition rate than CHF_3 (Joubert et al. 1994).

The fluorocarbon polymer is not only an etch-inhibiting layer, but also the source of fluorine, which reacts with silicon and carbon, which reacts with oxygen. Schaepkens et al. found that SiO_2, Si_3N_4, and silicon are covered with a several

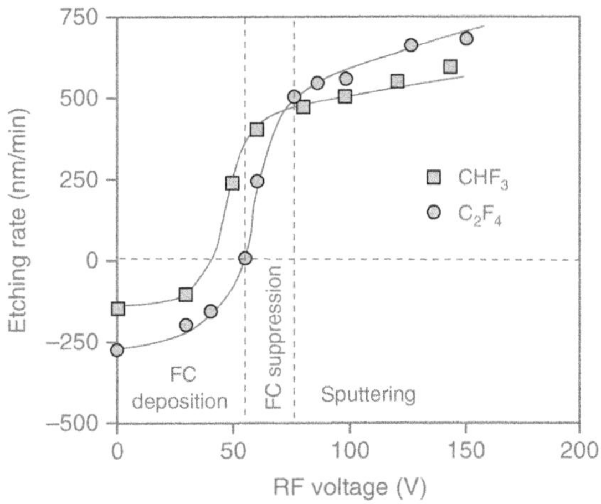

Figure 7.11 Silicon oxide etch rates in CHF_3 and C_2F_4 plasmas as a function of RF voltage. Source: Joubert et al. (1994). © 1994 AIP Publishing.

nanometer thin fluorocarbon film during steady-state etching and at sufficiently high bias voltages. A general trend is that the substrate etch rate is inversely proportional to the thickness of the fluorocarbon film (Schaepkens et al. 1999). Figure 7.12 shows this thickness dependence for SiO_2, Si_3N_4, and silicon. The variable parameters in Figure 7.12 are feed gas chemistry (CHF_3, C_2F_6, C_3F_6, and C_3F_6/H_2) and operating pressure (6 and 20 mTorr). The RF bias power level corresponds to a self-bias voltage of 100 V. The thickness dependence of the etching rate for all three materials and for all process conditions falls onto one trend curve. Independent of the process conditions, the SiO_2 surfaces are covered only with a 1.5 nm or thinner fluorocarbon film. The ion penetration depth in the investigated energy range is around 1 nm. The SiO_2 surfaces can therefore in principle be etched directly by a mechanism of chemically enhanced sputtering.

Silicon is covered with a relatively thick fluorocarbon film of roughly 2–7 nm. The film thickness for Si_3N_4 is between SiO_2 and silicon with about 1–4 nm. This difference in fluorocarbon film thickness and its correlation with etching rate allow etching SiO_2 selectively to Si_3N_4 and silicon with fluorocarbon or hydrofluorocarbon feed gases. The underlying mechanism causing the difference in film thickness

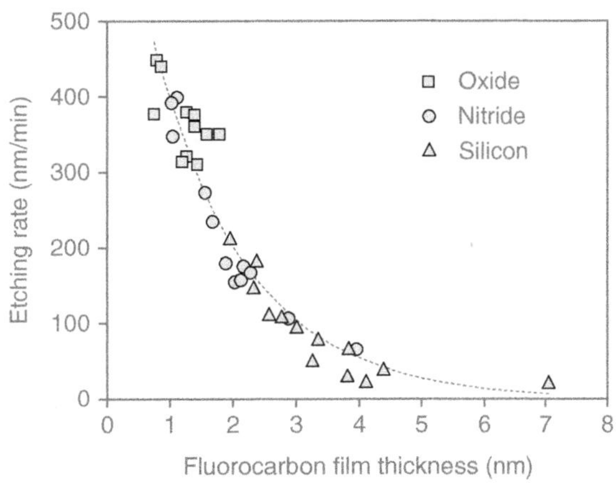

Figure 7.12 Etch rates of SiO_2, Si_3N_4, and silicon samples plotted vs the thickness of the fluorocarbon film present on the surface during steady-state etching conditions. Source: Schaepkens et al. (1999). © 1999 AIP Publishing.

on the various substrate materials is a difference in the ability of a substrate material to consume carbon species from the fluorocarbon film in chemical reactions during substrate etching (Schaepkens et al. 1999). Schaepkens et al. proposed a mathematical description of the etching yields, which includes the fluorocarbon deposition rate, net fluorocarbon etch rate, and a carbon consumption factor.

The thickness dependence of RIE with fluorocarbon feed gases in Figure 7.12 is different from what we observed for directional ALE with polymer film activation (see Figure 6.15, Metzler et al. 2014). In RIE, the source of reactive fluorine and carbon species is the continuous flux of these species. The process is limited by the ion-assisted diffusion of the reactants to and the reaction products from the selvedge layer. If this layer is too thick, etching will cease. In ALE, etch per cycle (EPC) is limited by the amount of carbon and fluorine in the polymer layer. If the layer is thicker than the projected range of the incoming ions, i.e. if these ions cannot reach the selvedge layer to stimulate reactions with the etching surface, it will be sputtered until it reaches the critical thickness. This will increase the step time to reach saturation but not the EPC.

7.2.2 ERNU

The complexity of the RIE etching mechanism creates severe engineering challenges to achieve etching rate uniformities that meet the requirements of advanced semiconductor devices. ALE etching technologies leverage self-saturation of the steps to achieve EPC uniformity across the wafer. Thermal and radical etching use in most implementations just one flux of reactive species, and control of the flux uniformity and surface temperature are sufficient.

RIE, however, is characterized by continuous ion and neutral fluxes, which act synergistically according to Eq. (7.4). Thermal etching as well as physical and chemical sputtering can also contribute to the local etching rate non-uniformity (ERNU). Passivating species are deposited onto the surface. In addition, dissociation and redeposition of reaction products contribute to local etching rates. These processes are affected by complicated neutral transport patterns (Kiehlbauch and Graves 2003). To achieve good etching uniformity across the wafer, all neutral and ion fluxes must be uniform across the wafer. Because the plasma is the source of all etching and passivating species, its uniformity across the wafer is paramount. Because of the complex behavior neutral fluxes inside the etching reactor, partially due to the fact that the wafer itself blocks the gas flow, it is nearly impossible to achieve perfect uniformity with a non-tunable, static plasma generation and gas injection system. Therefore, solutions to tune the uniformity of RF power, gas injection, and wafer temperature are mandatory in advanced RIE etching tools.

7.2.3 ARDE

RIE is prone to ARDE as can be easily seen from Eq. (7.4). For a given point at the etching front inside a feature, the net ion and neutral fluxes entering Eq. (7.4) are dependent on the solid angle that has a line of sight to the plasma. There is also an energy dependence because the scattering probability of ions and neutrals off the sidewalls is energy and impact angle dependent (see Sections 2.8.2 and 7.3.3). Ions, radicals, and neutral reactive species have very different angular and ion distributions, which gives rise to ARDE in RIE.

A comprehensive analysis of ARDE reveals that there are several contributing factors. The most important ones are, according to Gottscho et al. (1992), (i) Knudsen transport of neutrals (Coburn and Winters 1989), (ii) ion shadowing (Shaqfeh and Jurgensen 1989), (iii) neutral shadowing (Giapis et al. 1990), and (iv) differential charging (Arnold and Sawin 1991; Ingram 1990). For etching of silicon oxide, two mechanisms specific to the deposition of fluorocarbon polymers must also be considered (Joubert et al. 1994): (v) charging of polymer covered sidewalls (Hayashi et al. 1996), and (vi) shading of polymer precursors (Joubert et al. 1994). With the emergence of etching of dielectric features with aspect ratios above 50 : 1, ion scattering is becoming an important contributing factor (Huang et al. 2019).

Knudsen transport: The fundamentals of this neutral transport mechanism can be found in Section 2.8.1. The probability that a neutral atom or molecule reaches the bottom of a feature depends on the sticking coefficient s and the transmission probability K. The attenuation of the neutral flux can be calculated using Eq. (2.18). Gottscho et al. pointed out that this expression is true only in the regime of low surface coverage where the neutral sticking coefficient is not a function of depth or flux. This is the case for a neutral limited regime. They derived a modified expression for the more realistic case of flux-dependent Langmuir adsorption (Gottscho et al. 1992).

Ion shadowing: Because the etching rate depends on the ion energy flux, the ion angular distribution contributes to ARDE. If the width of the ion angular distribution would be zero degree and the orientation would be vertical to the surface normal, ion shading would not exist. However, this ideal scenario cannot be realized for fundamental reasons. The width of the ion angular distributions can be narrow, but it is finite. It is determined by plasma properties and will be discussed in Section 9.2.

Neutral shadowing: Under typical RIE process conditions with pressures between 0.1 and 100 mTorr, the mean free paths for collisions are much longer than characteristic dimensions of the etching structures. Therefore, neutral transport is in the molecular flow regime and gas-phase collisions can be neglected relative to collisions with the feature sidewalls. If one excludes fast neutrals created by charge exchange reactions with ions or sidewall surfaces, the neutral angular distribution is nearly isotropic. If the etching rate is limited by the neutral flux in Eq. (7.4), the effect of neutral shading in the center of a hole can be expresses as (Gottscho et al. 1992)

$$\mathrm{ER}_n = v_n S J_n \sin^2\left[\arctan\left(\frac{\mathrm{AR}}{2}\right)\right] \tag{7.5}$$

Here, AR is the aspect ratio of the hole defined as the ratio of the width divided by the depth of the feature ($\mathrm{AR} = w/d$). The expression for a trench is

$$\mathrm{ER}_n = v_n S J_n \sin\left[\arctan\left(\frac{\mathrm{AR}}{2}\right)\right] \tag{7.6}$$

For other portions of the profile besides the center of the bottom surface or for structures other than trenches or vias and for situations where sidewall etching is important, the solid angle expressions become more complicated (Gottscho et al. 1992). The implication from Eqs. (7.5) and (7.6) for practical application is that ARDE is more severe for holes compared to trenches for the same top opening dimension.

Differential charging: Charging of the mask and feature sidewalls will impact ion trajectories and attenuate the ion flux to the bottom of the feature. For a given charge, the electrical field will be stronger for a small distance between the opposing charged surfaces. Thus, feature charging is certainly CD dependent. Since it attenuates the ion flux, features with smaller CD and stronger electrical field should etch slower and result in a shallower depth and hence cause ARDE. Differential charging can occur as a result of the different angular distributions for ions and electrons. As will be shown in Section 9.1, ions are more directional than electrons. This leads to a more negative charge of a nonconductive mask compared to the bottom of the feature, which is bombarded by positive ions.

Simulations by Ingram revealed that the trajectories of normally incident positive ions can be perturbed by the local field structure during trench etching. This leads to an enhanced flux at the trench sidewalls (Ingram 1990). The surface potential evolves until the electron and ion fluxes are equal along all points of the surface. The result is a dispersion of the ion trajectories and considerable ion bombardment of the feature sidewall (see Figure 7.13).

Arnold and Sawin modeled surface charging and ion bombardment distortions for a rectangular trench in a planar surface. The electron flux was assumed to be isotropic with a thermal energy of 5 eV. The ion flux was modeled as monodirectional and normal to the perfectly isolating substrate surface. The ion energy was 100 eV. Under these conditions, they find that 70% of all ions impact the sidewall for an aspect ratio of four (Arnold and Sawin 1991). The results are shown in Figure 7.14. The ion energy in these simulations is much lower than that used in advanced high aspect ratio etching where ion energies of several thousand electron volts are generated. Higher ion energies lead to higher potentials on the mask

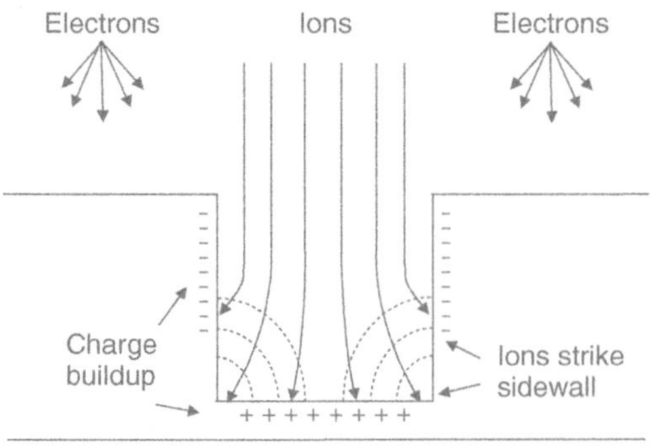

Figure 7.13 Schematic illustration of the origin and effect of local surface charging. Source: Arnold and Sawin 1991). © 1991 AIP Publishing.

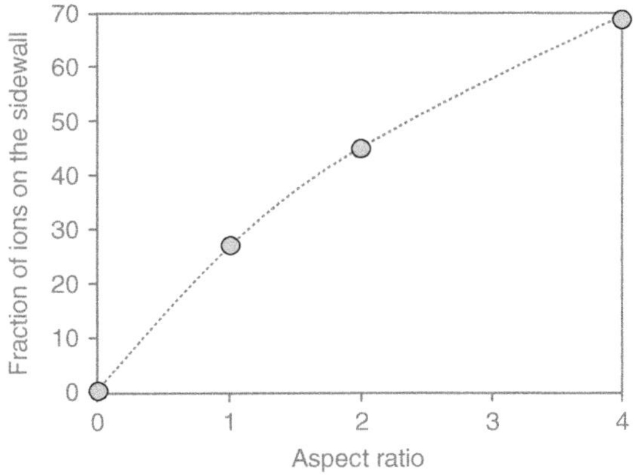

Figure 7.14 Effect of trench aspect ratio on percentage of ions that impact the sidewall. Source: Arnold and Sawin 1991).

and sidewall if the surface is perfectly isolating. Also, ions travel longer in deeper features and are longer under the influence of the charged sidewall. Therefore, this is a very important effect in high aspect ratio dielectric etching, which will be discussed in Section 7.3.3.

Charging of polymer-covered sidewalls: In etching of SiO_2 and Si_3N_4, the reactive species are fluorocarbon polymers. This leads to a special case of differential charging where the sidewall is a fluorocarbon polymer layer. Hayashi et al. conducted elegant experiments using capillary plates with aspect ratios of above 10 : 1. The reactor was a magnetically enhanced RIE chamber and the gas mixture was comprised of C_4F_8 and CO at a pressure of 40 mTorr. They found that the ion flux was attenuated by the high aspect ratio capillary plate while the ion energies and the ion mass spectra were unchanged. The ion flux was partially restored when oxygen was added to the plasma. Oxygen addition has the effect of less sidewall polymer deposition. Therefore, the authors concluded that the fluorocarbon polymers at the feature sidewall induce charging of the sidewall (Hayashi et al. 1996).

In principle, a higher carbon concentration in the fluorocarbon layer should increase the conductance of the film. This effect will therefore depend on the process conditions, which can result in different polymer thicknesses and compositions. The passivation layers for etching of silicon and metals are oxides or fluorides, which are most likely nonconductive. Thinner passivation layers will reduce charging but will also be less effective in protecting the feature from lateral attack and bowing caused by impact of charge-deflected ions. This can create a trade-off between bowing and charging. Plasma pulsing is an effective method to allow charge dissipation.

Shading of polymer precursors: A commonly deployed technique to reduce or eliminate ARDE does not address the root cause but leverages the competing aspect ratio effects of etching vs deposition. RIE processes that use polymerizing gases to protect the sidewall or as reactive species are characterized by simultaneous or sequential (in the case of MMP) etching and deposition processes. Figure 7.12 shows that the etching rate in a fluorocarbon-based plasma is strongly influenced by the deposition of fluorocarbon film on the oxide surface during the etching process. This deposition is achieved primarily by neutrals for aspect ratios up to 10 : 1 and by neutralized ionic species for even higher aspect ratios (Huang et al. 2019).

The flux of polymerizing species attenuates as the aspect ratio increases. The degree of attenuation depends among others on the angular distribution and sticking coefficients of the deposition species and will be different from the etching species. If they attenuate slower than the energy flux of the etching species, ARDE will be amplified (Joubert et al. 1994). If they attenuate faster than the etching species, this will reduce ARDE. In the extreme, this will lead to what is called reverse ARDE. In some cases, deposition in the open areas of a chip can be observed while small features are still etching. This effect can be obtained by flux balancing or by using separate deposition steps that deposit more in the larger structures or open areas. Using polymerizing reactions to reduce ARDE has the negative effect of reduced etching rates. The required degree of polymerization is also driven by profile performance requirements, for instance, in high aspect ratio dielectric etching. Here, higher RF power is used to increase the energy flux of the etching species to the bottom of the feature and to balance profile and ARDE performance.

7.2.4 Selectivity

RIE without significant deposition exhibits "intrinsic" selectivity. Intrinsic selectivity is a function of "excess" removal energy, which can be several hundred or even thousand electron volts for RIE. Obviously, these energies exceed the energy needed to break any chemical bond, and hence the sputtering or etching rate difference between materials will be diminished. If the RIE has a dominant sputtering component, the yields can be estimated according to Eq. (2.8). For a sputtering-driven process, selectivity is determined by the difference in mass, binding energy, and density.

Infinite intrinsic selectivity is not attainable for RIE without deposition. Generally, the intrinsic selectivity of RIE is much worse than for thermal and radical etching or directional and thermal isotropic ALE. Therefore, depositing species must be added to RIE processes to deposit a protective layer where etching is not desired while etching in locations where it is desired.

Figure 7.15 illustrates this conceptual framework for etching selectivity. Thermal etching and thermal isotropic ALE have the highest intrinsic selectivity because the excess energy is close to zero. Infinite intrinsic selectivity is possible. Radicals are less selective because they have unpaired electrons. They can overcome activation barriers, which may otherwise provide selectivity in Thermal Etching.

Lower yet is the intrinsic selectivity of directional ALE. Here, selectivity is achieved by carefully selecting the ion energy such that the sputtering thresholds of the activated etching material are higher than the sputtering threshold of the materials, which must not be etched. The sputtering thresholds for most materials are several tens of electron volts (see Table 2.1). Energy is lost in inelastic collision inside the solid material. Only a fraction of the kinetic energy of the impinging

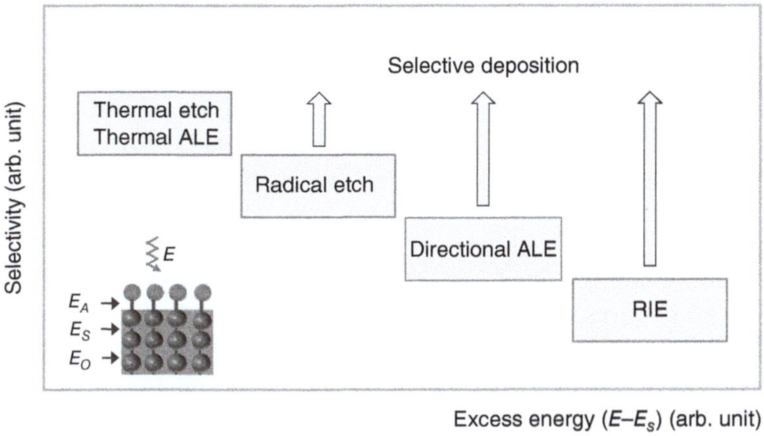

Figure 7.15 Conceptual etching selectivity framework.

ion is used to eject surface atoms. The remaining energy can cause damage to the material, which can negatively impact the performance of the final device. Vacancies and other defects can, for instance, impact the performance of FinFET gate devices where very narrow gate channels are etched (Eriguchi et al. 2014). Damage of the material can also lower the sputtering threshold of compound materials. An example is silicon ALE with Cl_2/Ar^+ selectively to SiO_2 where a thin gate oxide was eventually damaged and sputtered away (see Section 6.2). The proposed mechanism is a sub-threshold preferential sputtering of oxygen atoms, which makes the remaining silicon available to chlorination and sputtering. As mentioned above, typical RIE processes operate with ion energies far above the sputtering threshold. The excess energy is the highest among the various etching technologies and hence intrinsic selectivity is typically low.

Deposition reactions are frequently used to boost selectivity. We described this approach for directional ALE with reactive layer deposition in Section 6.1.3 where we discussed the C_4F_6/Ar^+ of SiO_2. In continuous RIE, the deposition and removal reactions proceed simultaneously. The origin of selectivity is enhanced polymer buildup on Si_3N_4 and silicon compared to SiO_2 due to the presence of oxygen in the latter. This creates a pathway for the removal of carbon via the formation of CO or CO_2, which does not exist in Si_3N_4 and silicon. The etching rate is controlled by the thickness of the fluorocarbon layer on the surface. It is consistent with a diffusion-limited supply of fluorine across the fluorocarbon film to the interface and out-diffusion of the etching product (Oehrlein and Lee 1987).

Polymer-enhanced selectivity enables patterning and high aspect ratio etching of SiO_2, Si_3N_4, and low-k materials with carbon-based masks. Low-k materials can be fluorine-doped SiO_2 or SiOCH-based materials. Selectivity is achieved via polymer deposition on carbon. Applications include contact etching, etching of trenches and via holes for metallization structures, high aspect ratio SiO_2 holes for DRAM capacitors, and high aspect ratio holes and trenches in multilayer stacks of SiO_2 and Si_3N_4 or polysilicon for 3D NAND flash memories. This makes RIE with C_xF_y

chemistries one of the most important etching applications in the semiconductor industry.

Another example of deposition-enhanced selectivity is etching of silicon selectively to silicon oxide. A typical application is patterning of polysilicon gates with a few nanometer thin gate oxide stop layer. Toward the end of the 1990s, Moore's Law drove the thickness of gate oxides to less than 50 nm. At that time, the so-called soft-landing steps where introduced. The idea is that the majority of the profile is defined by a chlorine- or fluorine-containing main etch; the soft-landing step clears the majority of the silicon with reasonably vertical profile while an over-etch removes silicon residues and straightens out the bottom profile. A preferred soft-landing contains HBr and oxygen. The etching product in this process is SiO_xB_y, which is redeposited onto the gate oxide and increases gate oxide selectivity. The latter cannot be precisely measured on blanket silicon and SiO_2 wafer because the amount of removed SiO_2 is very small and because of loading effects on patterned wafers.

Reflectometry can be used to measure the lifetime as a proxy for gate oxide selectivity in situ as it is exposed and as it breaks through (Lill et al. 2001). Figure 7.16 depicts experimental results that were obtained in an inductively coupled plasma at 10 mTorr. The HBr to oxygen ratio was about 10 : 1. The lifetimes of 20 and 14 Å thick gate oxides were measured on coupons mounted onto wafers with different amounts of exposed silicon and SiO_2 to produce different degrees of silicon loading. Gate oxide lifetime is shorter when silicon loading increases. This is the result of oxygen consumption by silicon and a reduction of protective polymer deposition onto the exposed gate oxide.

7.2.5 Profile Control

So far, we discussed performance metrics that painted a rather mixed picture for RIE compared to the other etching technologies. Because etching in RIE is the result of synergistic effects of simultaneous species fluxes, RIE exhibits poor intrinsic uniformity on wafer scale (ERNU), on feature scale (ARDE), and on atomic level scale (surface roughness). Owing to large excess energy, intrinsic selectivity is

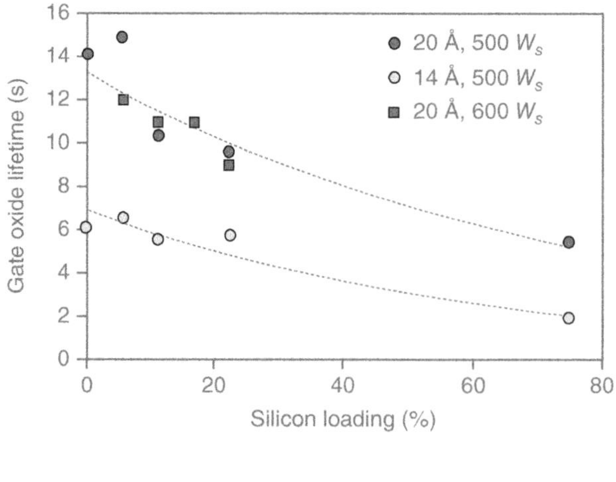

Figure 7.16 Lifetime of gate oxide in an ICP plasma as a function of silicon loading. Source: Lill et al. (2001).

worse compared to thermal etching, thermal ALE, radical etching, and directional ALE. On the positive side, RIE etching rates are generally higher than any of the ALE technologies because the latter use time in each step to reach saturation and frequently require purging between steps.

What makes RIE the patterning workhorse of the semiconductor industry is its profile tuning versatility. RIE can create a wide range of different shapes and sidewall angles, some of which are shown in Figure 2.2. Pure RIE according to Eq. (7.4) is a chemically assisted sputtering process. Sputtering cannot produce perfectly vertical profiles because the sputtering yield is a function of impact angle. Nonreactive ions have angular distributions of the sputtering yield with a maximum between 40° and 60° (see Figure 2.9). For chlorine ion sputtering of silicon, such a maximum was not observed. The yield is flat from normal incidence to about 40° and then drops gradually to zero for glancing incidence as shown, for instance, in Figure 7.3 (Jin et al. 2002, see also Chang and Sawin (1997), Chang et al. (1997), Mayer et al. (1981), and Jin et al. (2002).

The profile of a sputtered feature is determined by the angular distribution of the sputtering yield. The lower the sputtering yield is for large incidence angles, the more tapered the profile will be. Hence, tapered profiles can be achieved using chemical and chemically enhanced sputtering. Most RIE processes, however, have also radical and thermal etching components. For instance, when etching silicon with a fluorine plasma, the default profile will be strongly re-entrant. This means that in the absence of deposition processes, vertical profiles can be achieved by balancing tapering from sputtering and lateral etching from radical and thermal etching.

John Coburn lists three approaches for controlling sidewall angles with RIE (Coburn 1994a): Sidewall protection or passivation, selection of etching species, and temperature. Let us analyze these approaches in more detail.

7.2.5.1 Sidewall Passivation

Figure 7.17 depicts sidewall passivation mechanisms in RIE. The source of the passivation layer can be the plasma via CVD or line-of-sight deposition of sputtered species. To achieve vertical profiles, the rates of the simultaneous deposition and etching reactions must be balanced. Too much deposition will lead to tapered profiles (<90° profile angle) while insufficient passivation will cause re-entrant profiles (>90° profile angle).

In order to passivate sidewalls from the plasma, gases are added to the gas mixture, which can form passivation species directly or in combination with etching by-products. For instance, oxygen mixes with chlorine to etch silicon with vertical profiles. The oxygen reacts with silicon and chlorine to form nonvolatile silicon oxychloride at the surface. When this material is removed by ion bombardment, it may redeposit onto the sidewall of the feature. The latter effect is highlighted as "line-of-sight deposition" in Figure 7.17.

The same process chemistry can passivate the sidewall also by CVD-like deposition from the plasma. This mechanism is depicted as "plasma deposition" in Figure 7.17. It relies on material from the wafer to form passivation species in the plasma. In the case of silicon RIE with chlorine and oxygen, the silicon can

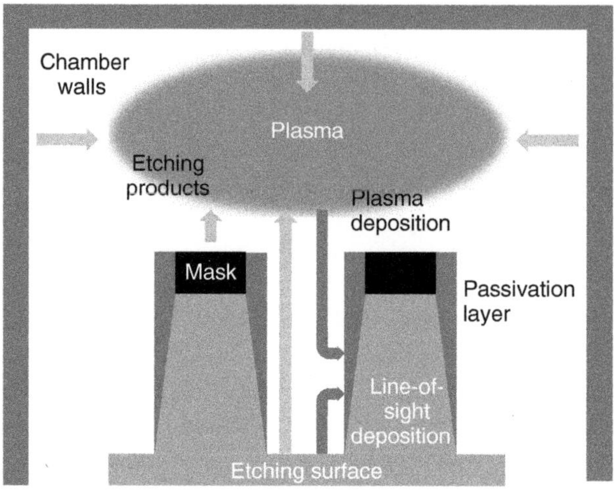

Figure 7.17 Sidewall passivation mechanisms in RIE.

form nonvolatile as well as volatile etching products. When these species reach the plasma excitation region above the wafer, they can dissociate and ionize via electron impact. Dissociation of $SiCl_4$, for instance, will form nonvolatile $SiCl_x$ neutrals and ions. The chlorine that is formed in the process is pumped away. Nonvolatile $SiCl_x$ species can deposit onto the wafer. They can also coat the reactor walls and from there be sputtered onto the wafer. Thus, reaction products have a complex life cycle, which is important for etching performance (Kiehlbauch and Graves 2003).

The passivation layer thickness can be tuned by changing the feed gas mixture. Figure 7.18 shows that for the same process, total flow, and power, HBr/O_2 forms a thicker sidewall passivation layer than Cl_2/O_2. The passivation layer thickness is strongly dependent on the oxygen concentration in the gas phase. Higher oxygen flows lead to thicker passivation layers. The results in Figure 7.18 were obtained using angle-dependent x-ray photoelectron spectroscopy (XPS) measurements of etched silicon lines and spaces with a silicon oxide mask. The wafers were etched in an inductively coupled RIE reactor (Desvoivres et al. 2001).

Etching of features involves a mask, and the etching selectivity for most mask materials is finite. Hence, etched mask material will also enter the plasma region and participate in the deposition process. Depending on the kind of mask material and the selectivity, this may influence the sidewall thickness and composition. Figure 7.19 shows sidewall thickness and composition for etching of silicon with $HBr/Cl_2/O_2$ with photoresist and silicon oxide masks (Bell and Joubert 1997). The thickness of the sidewall can be deduced from the metallic silicon signal, which is attenuated when the photoelectrons travel through the passivation layer. The passivation thickness is very similar for both masks. With the resist mask, 8% carbon is incorporated into the sidewall. This carbon was originally removed from the top of the photoresist and subsequently redeposited from the gas phase.

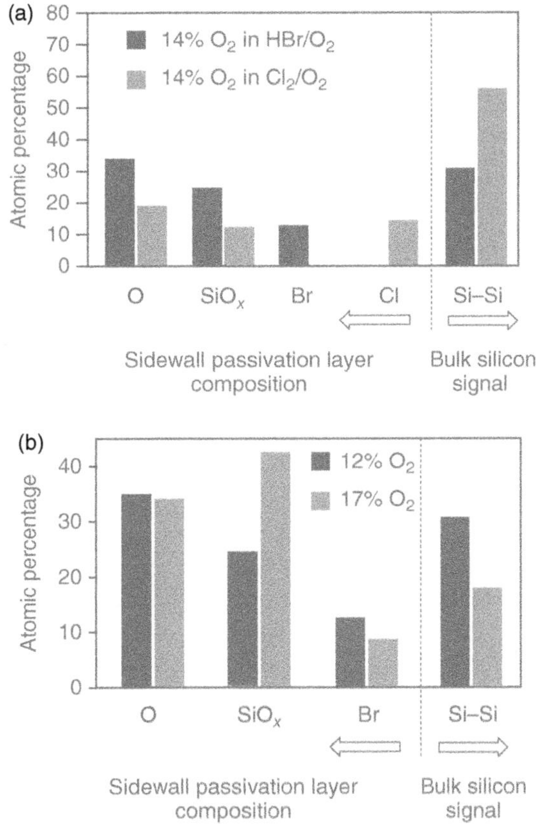

Figure 7.18 Sidewall passivation for silicon etching with Cl_2/O_2 and HBr/O_2. Source: Desvoivres et al. (2001). © 2001 AIP Publishing.

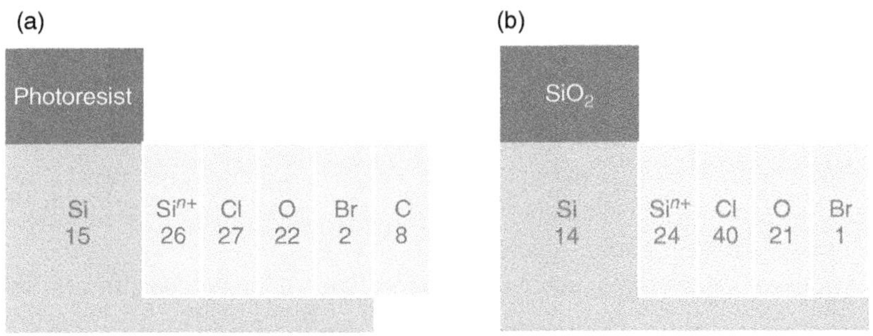

Figure 7.19 Sidewall passivation for silicon etching with $HBr/Cl_2/O_2$ comparing a resist and a silicon oxide mask. Source: Data from Bell and Joubert (1997).

When CF_4 is added to the $HBr/Cl_2/O_2$ etching process, the formation of non-volatile silicon oxychlorides and oxybromides is suppressed, and volatile silicon fluorides are formed instead. The oxychloride and oxybromide passivation film is largely replaced by a CF_xCl_y layer (see Figure 7.20). Adding of CF_4 shifted the passivation mechanism from a mixed plasma/line-of sight deposition to a plasma deposition mechanism. The silicon signal from the bulk material beneath

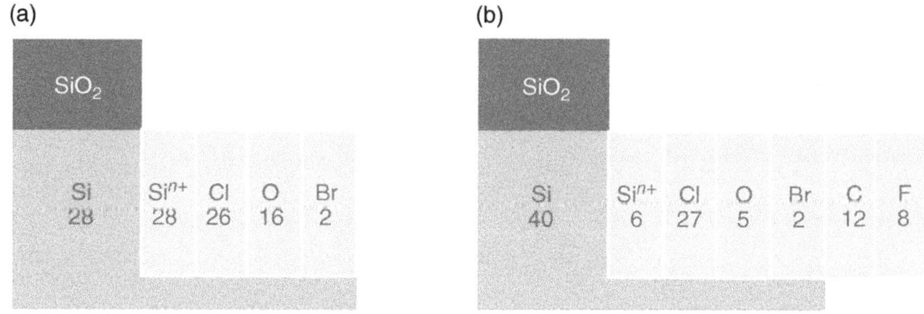

Figure 7.20 Sidewall passivation for silicon etching with $HBr/Cl_2/O_2$ and $HBr/Cl_2/O_2/CF_4$. Source: Vallier et al. (2003).

the passivation layer is increased from 28 to 40 at.%, which indicates a thinner passivation layer.

Next, we will discuss plasma passivation in more detail. In particular, we will analyze why it is possible to protect a vertical surface via plasma passivation while etching at the same time a horizontal surface at the bottom of a feature. On blanket wafers, the addition of deposition-inducing precursors reduces the etching rate. Etching will stop if the flow of passivating gases is too large.

In 3D structures, ion and neutral fluxes of etching species to vertical surfaces are lower than to horizontal surfaces. This allows etching at the bottom of the feature faster than laterally. However, we must also consider that plasma deposition reactions are CVD processes, which have higher deposition rates on horizontal surfaces than on vertical surfaces.

Preferred etching of horizontal surfaces is possible for two reasons. Firstly, CVD deposition is driven by neutral species. Their flux will attenuate faster than the flux of high energy ions due to the wider angular distribution. Thus, ions can clear the bottom of the feature via sputtering. Secondly, the sputtering yield rapidly drops for impact angles larger than 60° as is the case for sidewalls while the ion scattering yield increases (see Figure 2.16). The parameter ν in Eq. (7.4), which represents the volume of material removed per unit bombardment energy, goes to zero and synergistic RIE is suppressed on near vertical sidewalls.

Another sidewall passivation mechanism is direct line of sight redeposition of nonvolatile etching products. We discussed this mechanism in the context of profile performance of directional ALE (see Section 6.2). Figure 6.24 illustrates that the profile becomes less bowed and eventually tapered when redeposition of sputtered products is increased. Here, redeposition was modeled by increasing the sticking coefficient (Berry et al. 2018), while in practical applications, this is frequently accomplished by increasing the ion flux or ion energy to enhance sputtering.

Tuning of the ion energy by changing the RF bias power is a profile tuning knob, which is commonly used in RIE process development. It is based on sidewall passivation direct line-of-sight redeposition of nonvolatile etching products. The flux of the passivation species to the sidewall depends among others on the surface area of the etching front. A narrow feature will see a lower flux of passivating species compared to a feature with large adjacent open areas such as an isolated line or a line at the end of an array. This leads to more tapered isolated features. The effect is called profile microloading and can be observed for plasma passivation and line-of-sight passivation. It is an undesired effect because it introduces differences in the taper angle and the CD of the features.

Besides synergistic etching as expressed by Eq. (7.4), RIE mechanisms can include simultaneous thermal and radical etching as well as direct physical and chemical sputtering. Sidewall passivation must prevent lateral etching by radicals and reactive neutrals as well as from ion bombardment. Etching of aluminum in a chlorine plasma is a good example for protection from radical etching. $AlCl_3$ has a boiling point of 180 °C, and hence RIE of aluminum in chlorine plasmas has a large radical etching component. Boron- and carbon-containing gases are used to passivate the sidewall while the bottom of the feature is kept clean of passivating species by ion bombardment.

The role of sidewall passivation in high aspect ratio SiO_2 etching is the prevention of etching from synergistic etching and also direct physical and chemical sputtering. The ion energies in this process are so high that non-synergistic sputtering plays an important role (see Section 7.3.3). Ion impact will cause sputtering of the sidewall passivation layer, but as long as the passivation layer is replenished, the bulk material underneath is protected. An equilibrium between deposition and etching must be achieved via balancing of the species fluxes. If the passivation layer is too thick, the aspect ratio of the etching feature is increased and ARDE slows the etching rate. This is for instance a challenge in contact etching for advanced logic devices. Directional ALE with C_4F_8/Ar^+ is increasingly used instead of C_4F_8-based RIE because the polymer is removed in the Ar ion sputtering step, which prevents clogging of the feature.

7.2.5.2 Selection of Etching Species

In this approach to achieve vertical profiles, isotropic etching caused by radical or thermal etching is suppressed by selecting etching gases, which form less volatile reaction products (Coburn and Winters 1979b). Coburn gives the example of etching of silicon with fluorine-, chlorine-, and bromine-containing gases (Coburn 1994a). Silicon etches spontaneously when exposed to fluorine radicals or neutrals such as F_2 and XeF_2 and hence forms re-entrant profiles. The isotropic etching component of chlorine plasmas is much less and absent for bromine. This etching behavior of silicon when exposed to different halogen plasmas can be explained by the volatility of the reaction products. The boiling points for SiF_4, $SiCl_4$, and $SiBr_4$ are −86, 57.6, and 153 °C, respectively. Hence, SiF_4 is much more likely to desorb at thermal energies than $SiBr_4$, which requires ion bombardment to stimulate desorption. Based on this observation, Coburn suggested to use mixtures of fluorine, chlorine, and bromine gases to achieve vertical profiles as schematically shown in Figure 7.21. This approach is widely used in the development of RIE processes that use halogen gases.

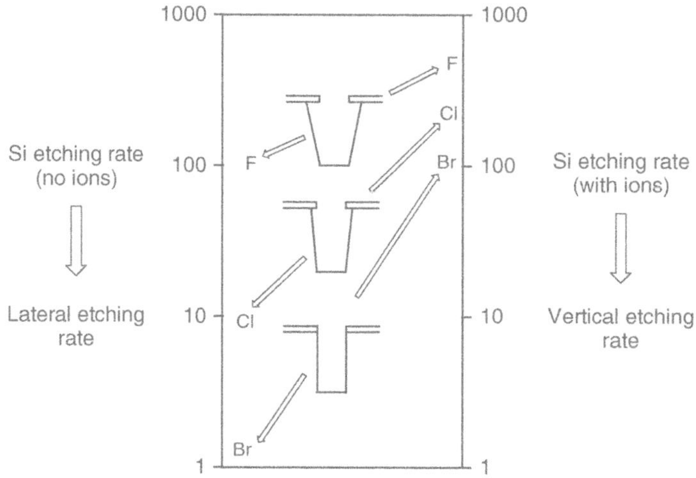

Figure 7.21 Illustration of the role of different halogens on RIE profiles. Source: Coburn (1994a). © 1994 Springer Nature.

7.2.5.3 Temperature

Spontaneous etching can also be reduced by lowering the wafer temperature. Thus, vertical silicon profiles can be achieved even with SF_6 plasmas at surface at −110 °C and below (Tachi et al. 1988). This approach does not require additional passivation species. Tachi et al. proposed multilayer physisorption of SF_6 as a possible sidewall passivation mechanism. Figure 7.22 shows that the silicon etching rate increases and the photoresist etching rate decreases when the temperature is reduced below −100 °C. At the same time, the profile becomes vertical without undercut. These trends are being used in deep silicon etching, for instance, for through silicon via (TSV) and micron-electromechanical (MEMS) applications. In some cases, lateral

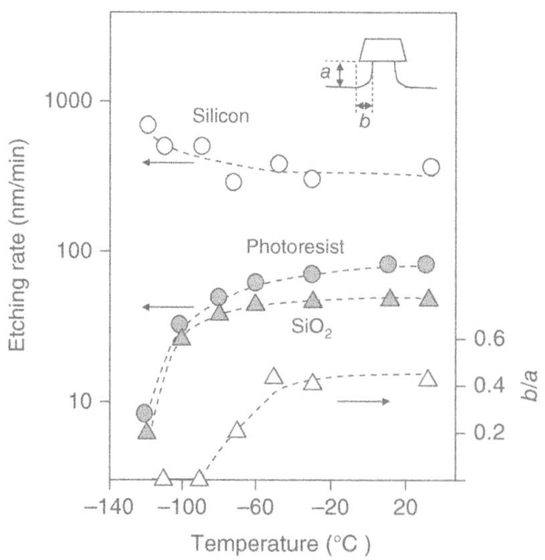

Figure 7.22 (a) Temperature dependence of the etching rate of silicon and photoresist in an SF_6 plasma. (b) The profile performance for etching of silicon lines with resist mask expressed by the ratio of undercut to depth. Source: Tachi et al. (1988). © 1988 AIP Publishing.

etching cannot be prevented even at temperatures below −100 °C and roughly 10% of oxygen was added to SF_6 to provide better sidewall passivation (Bartha et al. 1995). Bartha et al. suggested that the results by Tachi and coworkers were influenced by small amounts of oxygen from the reactor walls. Cryogenic etching is recently becoming the focus of renewed interest in the plasma etching community.

At the other end of the temperature spectrum is high-temperature etching of the so-called nonvolatile materials. These are materials that form halides and hydrates with boiling points of several hundred or thousand centigrade. In manufacturing of ferroelectric random access memory (FeRAM) structures, the challenge of patterning the platinum and iridium contact layers as well as lead zirconate titanate (PZT) and barium strontium titanate (BST) ferroelectric layers drove the introduction of high-temperature RIE (Yokohama et al. 1995). Still, vertical profiles are nearly impossible to obtain even when high etching temperatures are used. The reason is the absence of meaningful lateral radical or neutral etching. Even at elevated temperatures in reactive plasmas, the main RIE etching mechanism is sputtering. Vertical profiles cannot be obtained under these conditions.

These examples illustrate that temperature is an important parameter for profile control in RIE. The surface temperature influences the rates of etching and deposition reactions that take place on the sidewall of a feature, and hence the resulting profile. The etching processes on the sidewall are the same that drive RIE etching rates: Synergistic chemically enhanced sputtering, thermal and radical etching, and physical sputtering. The deposition processes include CVD-type gas-phase deposition and line-of-sight sputtering redeposition. All of these processes have their distinct temperature sensitivities.

Let us first discuss physical sputtering. The ions in RIE carry energies between several ten to several thousand electron volts and can cause physical sputtering in a neutral limited regime. Upon impact onto the surface, these ions create a collision cascade in the solid material. The temperature of the atoms in the collision cascade can be estimated using the relationship between energy and temperature for the ideal gas:

$$E = \frac{3}{2} N k_{\mathrm{B}} T \tag{7.7}$$

Assuming the collisions are elastic, the energy of the impinging ion is 100 eV and the collision cascade is comprised of $N = 10$ atoms, and the temperature of this ensemble of atoms is over 77 000 K according to Eq. (7.7). Raising the wafer temperature by even several hundred Kelvin will not make a difference. Hence, physical sputtering is not temperature sensitive. The profile will not change with temperature if the process is primarily a physical sputtering process. The same arguments are valid for chemical sputtering (sputtering with chemically active ions).

RIE with ion-neutral synergy according to Eq. (7.4) is a chemically enhanced sputtering process where the surface bonds are weakened by adsorption of reactive species, and the removal of the weakened material is the result of ion bombardment. The sputtering process itself is not temperature sensitive. The adsorption process will be more sensitive for molecular and less sensitive for radical adsorption (see Figure 6.3). RIE creates large fluxes of radicals to the wafer and it is safe to assume that radical adsorption dominates most RIE processes, not neutral adsorption. Therefore, the etching rate of pure synergistic RIE with dominant radical adsorption (without parallel thermal or radical etching or deposition processes) is most likely not temperature sensitive.

Based on these arguments, we can exclude physical, chemical, and chemically enhanced sputtering as the primary root cause for the temperature sensitivity of the profile shape in RIE.

Let us now discuss etching processes in RIE, which require only thermal energies for etching. The process window for thermal etching is schematically shown in Figure 3.1. It depicts an optimum temperature below and above which the etching rate drops. The lower rates at low temperatures are driven by the energy needed to activate dissociative chemisorption and desorption. The rate drop at higher temperatures is caused by physisorption becoming rate limiting. Molecules may just scatter off the surface at higher temperatures rather than form weak van der Waals bonds. Without physisorption, the subsequent step, chemisorption, cannot occur. In conclusion, raising the temperature can influence the etching rate in both ways for thermal etching.

Chemisorption of radicals does not require intermittent physisorption (see Figure 5.1). Therefore, the activation barrier for chemisorption of radicals is low and chemisorption rates are largely temperature independent. At the same time, desorption rates increase exponentially with temperature (see Eq. (2.3)). Hence, the rates of radical etching typically increase when temperatures are raised. Accelerated isotropic radical etching is one of the main reasons why profiles become more vertical or undercut when the temperature is increased.

Most RIE processes deploy deposition processes to passivate the sidewall of the etching structure. The deposition processes can be classified into CVD-type plasma deposition and line-of-sight sputtering. What is the temperature sensitivity of these deposition processes? In the case of line-of-sight sputtering, the reaction products are nonvolatile or else they would have desorbed thermally. When they impact a nearby sidewall, they will redeposit. This situation does not change much with changing temperature. Since we concluded previously that sputtering-based etching is temperature independent, the flux of the sputtered and redeposited species is temperature independent.

CVD deposition rates are a function of neutral, radical, and ion fluxes similar to RIE. In the case of neutrals and radicals, the sticking coefficients of neutrals drop as the temperature is increased due to the need for physisorption. The sticking coefficients of radicals and ions are less temperature sensitive. Generally, thinner sidewall passivation is observed in RIE when the temperature is increased. This is a very important effect for profile tuning in RIE.

In conclusion, RIE profiles can be very effectively tuned using wafer temperature. This is the result of accelerated lateral thermal and radical etching and suppression of CVD deposition rates. This has profound implications for RIE reactor design. Because the etching profile is determined by the flux of passivating species, gas composition uniformity of the reactive and passivating species including reaction products must be extremely uniform across the wafer to achieve the same profiles everywhere on the wafer. At the same time, the temperature must be extremely uniform across the wafer. Because sidewall passivation is the result of flux and sticking coefficient of passivating species and temperature dependent radical etching, it is possible to compensate flux nonuniformities with wafer temperature to a meaningful extend. This method is widely used in the industry.

7.2.6 CD Control

During etching of semiconductor devices, the critical dimensions from the mask are transferred to the underlying layer. The fidelity of this transfer is determined by the local etching profiles. The change in CD between the mask and the etched feature is zero if the sidewall is perfectly vertical without isotropic recessing or tapering. Therefore, the discussion in section 7.2.5. on profile tuning is directly applicable to CD control in RIE. CD control is essentially sidewall passivation thickness control across the chip, across wafer, and wafer to wafer.

The difference between the mask CD and the feature CD is called CD bias or ΔCD. From the discussion of the profile formation mechanisms, it is obvious that ΔCD depends on the aspect ratio of the opening that is being etched. The processes involved in the formation of a sidewall profile such as chemically enhanced and physical sputtering, radical and thermal etching, line-of sight redeposition as well as plasma deposition are governed by their own within-feature transport mechanisms, which result in aspect ratio dependent profile angles and ΔCD. This effect is called CD microloading. It is to lateral etching in RIE what ARDE is to vertical etching. It is expressed as the difference of the ΔCD for the features of interest, for instance, isolated and dense lines:

$$\text{CD microloading} = \Delta CD_{\text{iso}} - \Delta CD_{\text{dense}} \tag{7.8}$$

To characterize a complex layout, CD microloading can be characterized by the difference of the largest and smallest ΔCD:

$$\text{CD microloading} = \Delta CD_{\text{max}} - \Delta CD_{\text{min}} \tag{7.9}$$

The method to measure ΔCD depends on the application and what is important for overall chip performance. For instance, it makes sense to define ΔCD in a hardmask open process as the difference of the bottom CDs of the photoresist and the etched mask because the bottom CD of the mask defines the CD of the next etch. The ΔCD of a metal line may be defined as the difference between the bottom CD of the mask and the CD at half-height of the metal line because it describes the cross section and hence the resistivity of the metal line.

The preferred method of measuring CDs in the manufacturing of integrated circuits (ICs) is top down scanning electron microscopy (SEM) or CD-SEM, which produces a 2D image. The mask areas and the etched features are represented by lighter or darker colors or gray shades. The edge of a line or hole is defined by the brightness transition. Sophisticated algorithms are used to extract the dimensions at the desired feature height. CD-SEMs produce values that provide a good proxy but not the real physical dimension at a given height of the sidewall. Calibration of the

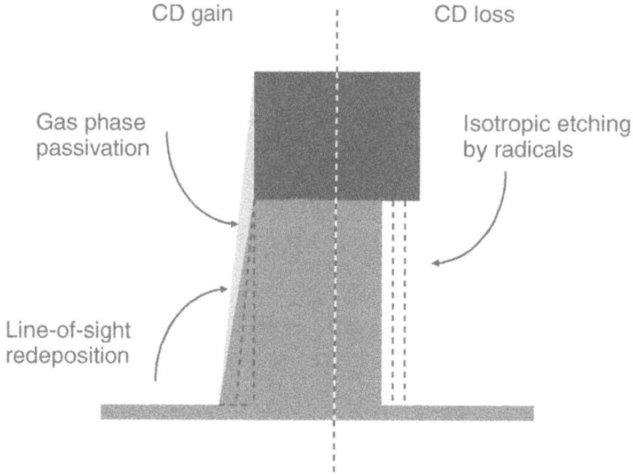

Figure 7.23 Schematic illustration of the origin of CD bias with and without sidewall passivation.

CD-SEM with cross section SEM or transmission electron microscopy (TEM) is used to address this shortcoming.

The benefits of top-down SEM are that it is nondestructive, has a high throughput, and allows measurement of complex 2D shapes such as L- or T-shaped structures, holes, elliptical openings, ends of lines in close proximity to each other, and other features. A metrology recipe for a chip may include a whole range of dimensions of various features. To measure ΔCD, these features must be measured before and after etching in the same locations. The result is a table of ΔCD for the different locations. The difference between the chip layout, which is the intent of the designer, and the final etch CD is also called edge placement error (EPE). The factors that contribute to EPE include systematic errors from lithography and etching. These systematic errors must be considered in the design of the mask. In the case of systematic lithography errors, this is called optical proximity correction and computational models exist to predict optical EPE. Etch proximity correction is still largely based on iterative experiments. While significant progress has been made in feature scale modeling of RIE, etch proximity effects cannot be predicted with computational models yet due to the complexity of the underlying processes. The discussion of RIE sidewall passivation mechanisms in section 7.2.5 underscores this point.

A schematic illustration of the origin of CD bias for etching processes with and without sidewall passivation is depicted in Figure 7.23. The mask is shown with its dimensions before etching to provide a point of reference. The case of process with sidewall passivation is shown on the left side. A sidewall with net deposition (light gray color) gives a tapered profile (dark gray color). The CD bias is the difference between the CD at the bottom of the feature and the bottom of the mask.

Passivation can originate from the plasma via CVD deposition or from line-of-sight sputtering. In the case of CVD deposition from the gas phase, isolated features exhibit a larger CD bias than dense features because the transport of deposition precursors in dense areas is attenuated as the aspect ratio of the etching feature increases. Line-of-sight sputtering does not require transport from the top of the feature. The source of the deposition is the etch front at the bottom of the feature.

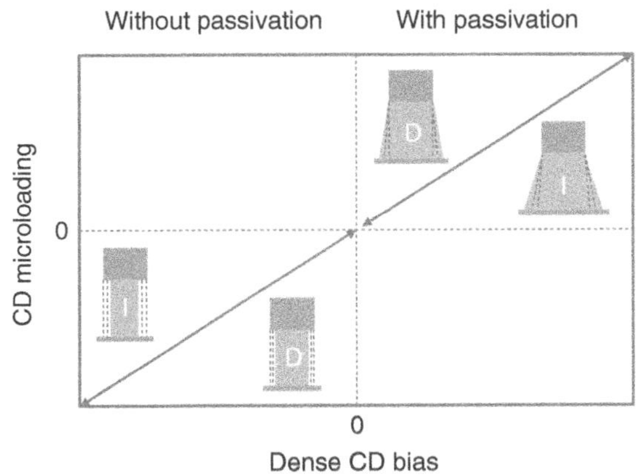

Figure 7.24 CD bias and microloading trends with and without sidewall passivation.

Still, the isolated feature will have a larger CD because the area from which nonvolatile sputtered deposition species originate is larger for isolated features. A larger surface area is exposed to etching next to an isolated feature compared to a dense feature.

In the case of etching without sidewall passivation, the isolated feature will shrink faster than the dense line because the flux of etching species to the sidewall is larger compared to a dense feature. This leads to the insight that etching processes with and without sidewall passivation generally exhibit opposite CD microloading trends. This point is schematically illustrated in Figure 7.24.

The magnitude of CD microloading increases with the CD bias. This is important for processes that have the goal to grow or shrink the CDs. For instance, the so-called photoresist trimming has the goal to shrink the CD of a line or increase the CD of a trench by isotropically etching the photoresist. Resist trimming is frequently used to reach small CDs that cannot be printed directly. It is also used as a tuning parameter in multiple patterning. We will discuss this in more detail in Section 7.3.1.

Figure 7.25 shows the trade-off between CD bias of the dense line and the CD microloading for a 130 nm polysilicon gate process with a dielectric hardmask. For the traditional $HBr/Cl_2/O_2$ process, it is impossible to reach the required 0 nm dense bias and 0 nm CD microloading. If the passivation is thick enough to achieve a vertical profile in the dense array, for instance, by reducing the flow of oxygen, the isolated line is too tapered. The CD bias of the isolated line is around 40 nm in this case.

When CF_4 is added to the process, the sidewall passivation changes from silicon oxychloride and oxybromide to carbon based (see Figure 7.20). The carbon-based passivation layer is thinner while still providing protection from lateral etching. This thinner and stronger passivation layer allows achieving zero CD microloading without CD bias of the dense and isolated lines as shown in Figure 7.25. This illustrates the importance of the properties of the passivation layer material for the design of RIE processes with excellent CD control.

In summary, RIE is unique among etching technologies in that it enables etching of high aspect ratio features with vertical sidewalls. This is the result of the interplay between chemically enhanced sputtering, which leads to tapered profiles, and

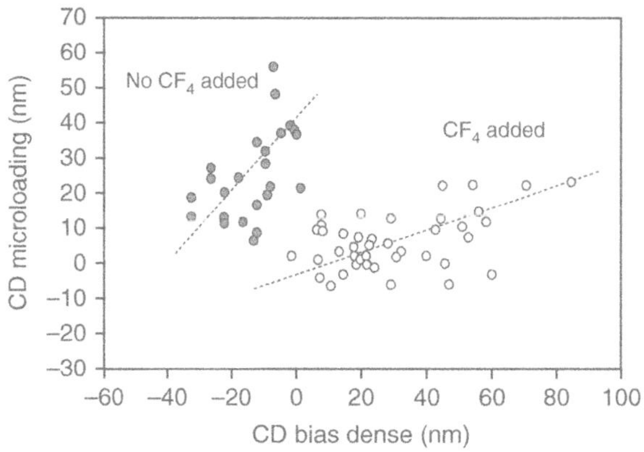

Figure 7.25 CD microloading for polysilicon gate etch with $HBr/Cl_2/O_2$ and $HBr/Cl_2/O_2/CF_4$.

radical etching, which drives lateral etching and undercutting. These processes have different aspect-dependent etching behavior, which can lead to profile microloading where the profiles of the dense and isolated features are not the same. CD bias and CD microloading are the consequence of profile microloading. It is the goal of the process engineer to select etching and passivation chemistries that are best suited for a given application and to find the balance of all species fluxes, which gives the desired CD performance.

7.2.7 Surface Smoothness

RIE processes are characterized by the presence of a mixed or selvage layer. MD simulations by Graves and Humbird for etching of silicon with fluorine (Graves and Humbird 2002) and Feil et al. for etching of silicon with chlorine (Feil et al. 1993) led to the conclusion that the selvage layer is caused by the implantation of reactive ions. Reactive ion impacts increase the silicon surface area by creating crevices and cracks into which reactive neutral can diffuse and terminate the bonds. This leads to local microroughness and under certain conditions to surface roughness on the nanometer scale as shown in Figure 7.26. RIE does not allow atomic precision control of the surface smoothness. Directional ALE, however, smoothens the surface or at least maintains it due to inert ion bombardment (Kanarik et al. 2018).

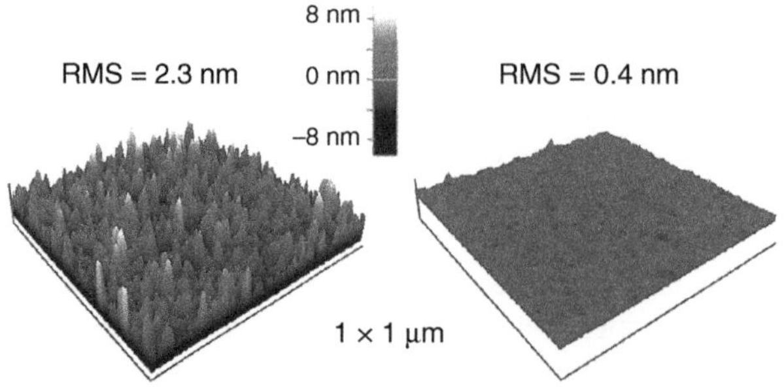

Figure 7.26 Silicon surface roughness measured AFM for directional ALE with alternating chlorine plasma and argon ion bombardment with 50 eV Cl_2/Ar RIE at 50 eV. Source: Kanarik et al. (2018). © 2018 American Chemical Society.

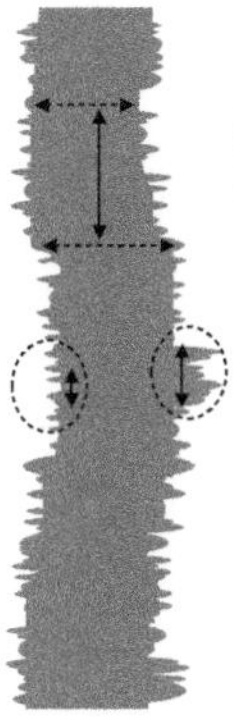

Figure 7.27 LER/LWR features and frequency of patterned photoresist. Source: Lill et al. (2014).

7.2.8 LWR/LER

When variations in the width occur over a certain length of a feature, this variation is called linewidth roughness (LWR). When these variations are measured along just one edge it is called line edge roughness (LER). LWR and LER create local CD bias and EPE, which impact device performance. As such, they are part of the overall EPE budget and an increasingly important component of CD control. The roughness of a feature edge has not only a magnitude but a frequency component as well. Lithography engineers measure and express WLR and LER as the power spectral density (PSD) vs the spatial frequency of the edge roughness. Etch process engineers have adopted this methodology and nomenclature.

The initial mask for any stack is the photoresist pattern, which is characterized by typical features as shown in Figure 7.27. They are crevices with high spatial frequencies of larger than 1/50 nm^{-1}, resist feet for medium spatial frequencies between 1/50 and 1/200 nm^{-1}, and bulk features with low spatial frequencies <1/200 nm^{-1}.

RIE can improve the high and medium spatial frequencies by filling in crevices via deposition and lateral etching and by removing feet via directional etching. RIE cannot address low-frequency features because the length scales are too large to be filled in or removed.

Lateral etching and deposition reactions, which are used to improve high-frequency LWR/LER, are also influencing CD microloading. To achieve optimum performance, lateral etching (trimming) and deposition can be separated into individual so-called resist treatment steps. Figure 7.28 illustrates a CD microloading vs LWR trade-off curve and the improved performance with photoresist treatment.

RIE can also *introduce* high- and medium-frequency LER and LWR. Because feature sidewalls are exposed to reactive ion bombardment, the surface roughening mechanism we discussed in section 7.2.7 applies in principle to LER as well. Also, when etching structures are comprised of several layers, it is important to identify the roughening or smoothening effects for each step. Atomic force microscopy (AFM) techniques have been developed to measure the roughness of sidewalls directly on photomasks (Reynolds et al. 1999) and etched features on the wafer (Goldfarb et al. 2004).

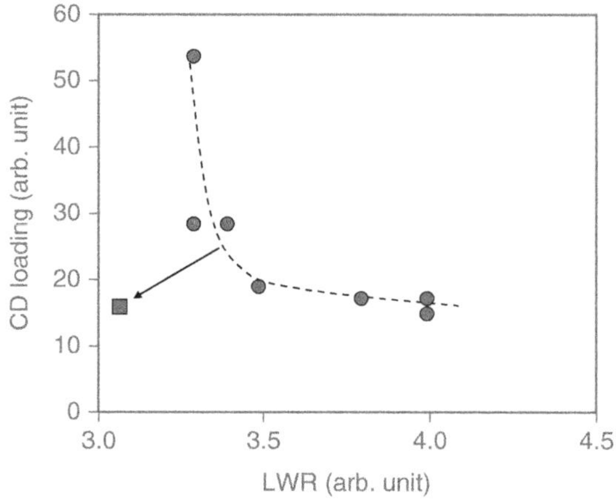

Figure 7.28 CD microloading/LWR trade-off can be overcome by separating photoresist etching and deposition reactions. Source: Lill et al. (2014).

Goldfarb et al. demonstrated that isotropic resist roughness found on the resist walls after the lithographic step becomes anisotropic during the etching of dielectric underlayer with fluorine-based plasmas. These anisotropic resist roughness features are called striations. Striations that are generated on the resist layer act as templates for the propagation of such topography into the etched underlayers (Goldfarb et al. 2004). Roughness propagation is reduced in case of more remaining resist, either by starting with taller resist or by increasing selectivity. The general trend in the industry, however, is toward thinner and less etch-resistant photoresists. This places a high value on highly selective mask open etch processes.

A typical gate stack contains the following films from the top: photoresist/bottom antireflective coating (BARC)/SiO_2 hardmask/polysilicon. Atomic Force Microscopy (AFM) measurements revealed that the LWR after the SiO_2 or carbon hardmask open step was transferred into the polysilicon gate during the gate etching step with $HBr/Cl_2/O_2$ (Pargon et al. 2008). This implies that the control of the final polysilicon LWR/LER strongly depends on the lithography and plasma etching steps preceding the gate etching step and that these steps must be optimized to succeed in minimizing the final polysilicon gate LWR/LER.

A proven approach to reduced roughening of the photoresist includes plasma treatment steps (Pargon et al. 2009; Azarnouche et al. 2013). Pargon et al. studied the effect of HBr and Ar plasmas on the composition of 193 nm photoresist. They were able to distinguish the effects of vacuum ultraviolet (VUV, 110–210 nm) light, radicals, and ions using windows with different cutoff wavelengths patched onto the photoresist film. Both HBr and Ar plasma cure treatments induce surface and bulk chemical modifications in the resist films. Synergistic effects of low energetic ion bombardment and VUV plasma light lead to roughly 10 nm deep surface graphitization or cross-linking, while the plasma VUV light is responsible for ester and lactone group removal from the resist bulk.

Since HBr plasma emits VUV light near 160–170 nm and can chemically modify the photoresist deeper, Pargon et al. found that it can modify 193 nm photoresist at least 240 nm deep. Argon plasmas emit VUV light near 100 nm and the effect is more shallow (Pargon et al. 2009). Chemical effects during the HBr treatment lead to the formation sp^2 bonds most likely due to the diffusion of hydrogen into the resist film. As the result of cross-linking and removal of hydrogen- and oxygen-containing ester and lactone groups, HBr plasma is established as a preferred method to increase resist etch resistance and also to significantly reduce the LWR/LER of 193 nm photoresist patterns.

Azarnouche et al. compared HBr plasma and VUV treatments of 193 nm resists. They found that both treatments reinforce the etch resistance of the photoresist exposed to fluorocarbon plasma etching process used to open dielectric antireflective coating (DARC) layers (Azarnouche et al. 2013). The study shows that the LWR is degraded at the top of the resist pattern and propagates along the pattern sidewalls. The high- and medium-frequency components of the roughness are not completely transferred during the gate patterning, allowing an LWR decrease at each plasma step. An overview on resist mask durability during plasma etching can be found in a review paper by Oehrlein et al. (2011).

RIE cannot improve low-frequency LER and LWR because the dimensions are too large to be removed by etching or deposition. However, RIE can cause resist and hardmask bending or toppling which worsens low frequency LER and LWR. The effect of line bending and toppling is a function of the aspect ratio of the line, as well as the stress and Young's modulus of the material. It is also called line buckling or wiggling. Stan et al. compared experimental and modeling results for thin lines made of a low-*k* dielectric (interlayer dielectric, ILD) and etched with a TiN hardmask (Stan et al. 2015). Such structures are used in patterning of trenches for copper metallization in the back end of line (BEOL) module of logic and memory devices.

The etched structures were measured by means of AFM. The etching depth was found to be one of the main control parameters for buckling. For the same width, buckling is more pronounced for the lines that are patterned deeper and have larger aspect ratios. The height of the TiN mask and the Young's modulus of the ILD, and the Young's modulus and compressive stress of the TiN mask are other parameters influencing buckling behavior. Line bending is induced by the compressive stress in the TiN mask and the work of the compressive forces is converted into bending energies of the structure. The stress of the ILD fin was 2 orders of magnitude lower than for TiN in the experiment and could be ignored. The line bending mechanism is illustrated in Figure 7.29.

Figure 7.30 shows the contour lines that separate the buckled and unbuckled states for the structure in Figure 7.22 as a function of the height of the structure, h_{ILD} and h_{TiN}, and the stress of the TiN cap or mask, Σ_{TiN}. The lines represent modeling results and the two datapoints are experimental results where sample A buckled, and sample B did not buckle. Σ_{TiN} was 1.4 GPa in the experiments.

This specific case of ILD line buckling illustrates the general trends that can be observed in etching of lines. To avoid buckling, the films must be stress free. Compressive stress is particularly decremental for buckling performance. This is a challenge for ion-assisted deposition technologies because ion bombardment tends to generate films with compressive stress. Ions are also implanted into the film during RIE and make the outer layers compressive. Generally, the occurrence of line

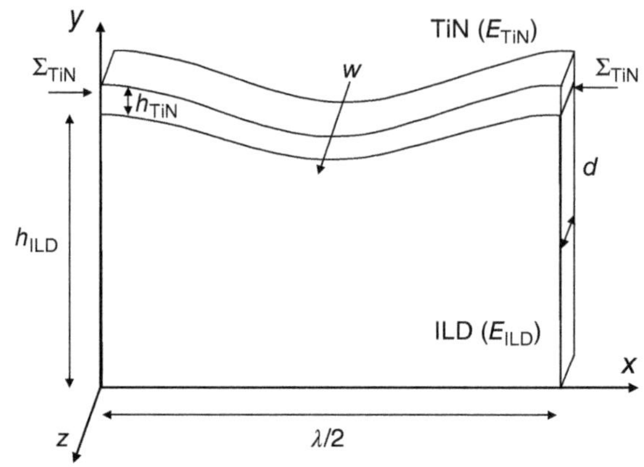

Figure 7.29 Out-of-plane deflection w of a buckled TiN-capped ILD fin over half-wavelength $\lambda/2$. Under the residual compressive stress Σ_{TiN}, the buckling of the TiN beam is limited by the deformation of the ILD plate. E_{TiN} and E_{ILD} refer to the corresponding Young's moduli. Source: Stan et al. (2015). © 2015 American Chemical Society.

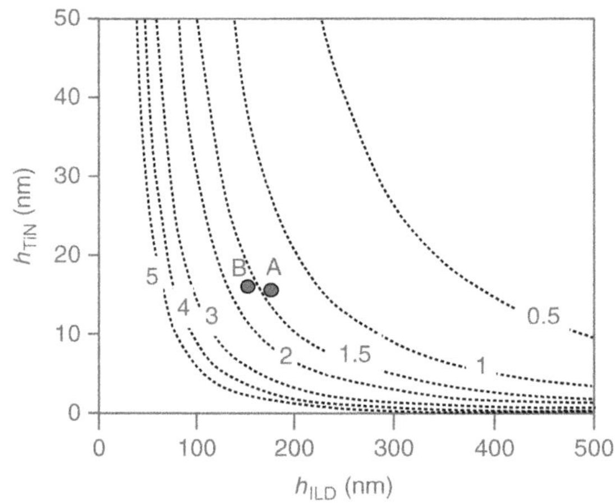

Figure 7.30 Contours of the minimum critical buckling stress Σ_{TiN} from 0.5 to 5.0 GPa in the (h_{ILD}, h_{TiN}) plane (fin width 32 nm). Each contour is labeled with the critical stress value in GPa and separates the unbuckled (left) from buckled (right) states of fins of various dimensions. Source: Stan et al. (2015). © 2015 American Chemical Society.

buckling is suppressed when the RF power in the RIE process is reduced. Weak mask materials with low modulus can be stabilized by encapsulating it with a low stress film in the RIE reactor before or during etching. If the film has tensile stress, line buckling can even be improved.

On the other hand, deposition designed to generate vertical profiles can have the unintentional side effect buckling if the film is compressively stressed. If line buckling is observed, changing the sidewall passivation chemistry can lead to success.

Figure 7.30 shows that for a given feature width (CD) and film stress, there is a maximum stack height beyond which lines buckle. In other words, the aspect ratio of a line is limited by the film stress. However, this does not mean that for films without any stress, the aspect ratio can be unlimited. Based on experience, the aspect ratio limit of etched lines is about 10–15.

LWR and LER of holes and via features are dominated by high- and medium-frequency roughness such as striations. The mechanism for striation formation in mask open is similar to lines. Striations that are created in the photoresist are transferred to the underlayers. Buckling is not important because round features are self-supporting. However, etching of high aspect ratio holes has unique challenges. They are the potential loss of circularity and twisting. As the name suggests, loss of circularity means that the shape of the bottom cross section of the high aspect ratio hole is not a circle anymore. Twisting denotes an effect where the center of the hole is offset from the center of the hole on the top. We will discuss the mechanism for loss of circularity and twisting in Section 7.3.3.

The LER/LWR challenges we discussed so far are even more pronounced for extreme ultraviolet (EUV) resists. High-frequency LER and LWR of EUV resists is a function of exposure dose due to stochastic effects. Exposure time in EUV is extremely expensive and hence etching and materials solutions are needed to repair high-frequency LWR/LER in EUV patterning. Also, EUV resists are mechanically not as stable as UV resists, which causes resist buckling challenges.

7.3 Applications Examples

There are hundreds of etching applications used in the manufacturing of advanced integrated circuits when considering various IC device types, functional elements of these devices, and IC manufacturer-specific integration solutions. There are several approaches to classifying etching applications.

Materials: RIE applications can be classified by the materials to be etched. In the 1990s, etching tools were segmented into the three materials that needed to be patterned by RIE: silicon oxide, silicon, and metals, specifically aluminum. As aluminum was replaced by copper to create the interconnects of IC devices, aluminum etching was replaced by etching of low-k dielectrics to form trenches and vias, which were filled with copper. At the same time, tungsten appeared as a common metallization material. This resulted in the classification of RIE into conductor and dielectric applications. Different types of reactors are used for these applications: high-density transformer coupled plasma (TCP), inductively coupled plasma (ICP), or electron cyclotron plasma (ECR) reactors for conductor etch applications and capacitively coupled plasma (CCP) reactors for dielectric applications. The decision about the category of an etch application is primarily a decision about what type of reactor to use and hence is very important.

Classification into conductor and dielectric etch applications is still used today; however, the resistivity of the material is not a definitive property for classification. For instance, silicon nitride is frequently etched in both conductor and dielectric etch tools. High aspect ratio silicon vias are etched primarily in dielectric etch CCP tools with fluorocarbon gases. Nonvolatile metals such as cobalt and iron were patterned in ICP or CCP reactors before ion beam etch (IBE) tools were introduced to the industry.

What drives the choice of the etch tool are the plasma conditions and chemistry needed to etch the devices successfully. If an application requires high density of radicals to achieve a strong isotropic etching component, a high-density plasma such as TCP, ICP, or ECR with deep dissociation is needed. Ion energy is a secondary requirement even though advanced TCP and ICP reactors can generate ion energies of several thousand electron volts. Applications that rely on the formation of chemically reactive polymer layers at the wafer surface for etching need a medium-density plasma where the feed gases are not completely dissociated. To be reactive at the surface the polymer layer typically requires high-energy ions to be activated and removed. For medium-density plasmas and high ion energies, CCP reactors are the right choice. We will discuss the plasma physics that drives these properties in Chapter 9.

Aspect ratio: It turns out that plasma requirements for patterning of 4 to 10 : 1 aspect ratio features with extreme CD control and conductor etch applications overlap to a large extend. At the same time, CCP tools emerged as the tools of choice to etch high aspect ratio (50 : 1 and higher) trenches and holes. This classification by aspect ratio is today the more important one because it drives the reactor development roadmap.

Device type: Etch applications can also be grouped by device type: logic, Flash, dynamic random access memory (DRAM), and emerging memories. This classification is more important for market analysis and has limited technical implications. One can say that, in general, high aspect ratio etch applications dominate the fabrication of memory ICs. Dual damascene low-k etching is unique to logic ICs. The latter also drive the roadmap for defectivity because logic devices are extremely defect sensitive due to the lack of redundancy. Etching reactor materials development is pushed forward by etching of logic devices.

IC manufacturing modules: Manufacturing of advanced semiconductor devices may include several hundred processing steps. The entire flow is typically grouped into front end of line (FEOL), mid end of line (MEOL), and back end of line (BEOL) modules. Typically, transistors are created in the FEOL module, the contacts and memory devices are formed in the MEOL, and the metallization is realized in the BEOL module. This classification varies between IC manufacturers. The classification by IC manufacturing module is important to prevent metal cross-contamination.

In the following, we will discuss important etching applications with unique requirements.

7.3.1 Patterning

7.3.1.1 Self-aligned Patterning

The goal of most etching processes is to transfer a pattern generated by lithography to form a structure that stays in the integrated circuit. In the early days of IC manufacturing, lithography created the pattern and RIE transferred it into silicon, silicon oxide, or aluminum to form transistor gates, isolation features, and metal lines to conduct the current. At that time, special "patterning etches" did not exist.

Device shrinking leads to higher aspect ratios and the need for better CD control. This trend started with silicon etches because silicon RIE requires oxygen addition to the chlorine- and bromine-based etching chemistry and resists did not provide sufficient selectivity. This necessitated the introduction of hardmasks for etching of

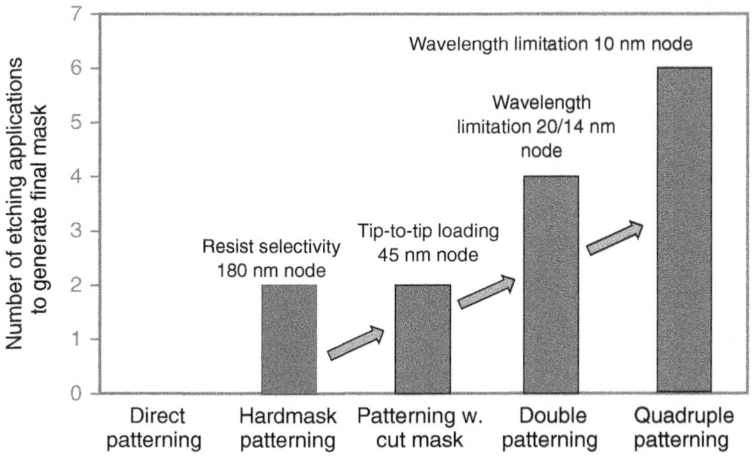

Figure 7.31 Number of etching steps needed to generate the final mask for device patterning in FEOL logic patterning. Source: Lill et al. (2014).

silicon gates and silicon isolation trenches and introduced a new class of applications, mask open or hardmask patterning, at the 180 nm node. Later, carbon layers were introduced to open thick dielectric hardmasks and to etch dielectric features. Because carbon cannot be etched selectively with resist, this required a thin dielectric layer on top of the carbon, which also acted as a DARC for lithography. Because DARC layers contained nitrogen, which interacts with the resist to form a foot at the bottom of the profile, organic antireflective coating where introduced. These films were also called BARC. As a result, simple hardmasks evolved into complex stacks designed to address lithography and etching challenges.

In most cases, these multiple mask layers are etched in situ in the same RIE chamber. This triggered the introduction of complex, multilayer hardmask or patterning etches. The number of etching applications that were needed to generate the final mask increased steadily from there as shown in Figure 7.31. At the 45 nm node, cut masks were introduced to reduce the tip-to-tip space between lines. Direct lithography and etching would create too large a gap between the line ends due to optical and etch proximity effects. In the cut mask approach, the top DARC layer is patterned first as lines and spaces. Then, the resist is stripped, and a new resist layer is spun on. Lithography then prints just small windows, which allow cutting the lines via RIE. Subsequently, the rest of the mask step is etched. Cut masks utilize the idea to divide the pattern into two or more parts and to process each conventionally. The complete pattern then emerges when all lithography and etching processes are completed. This is a brute force method to generate features beyond the resolution of lithography and is called Litho-etch-Litho-etch (LELE). It is obviously expensive and also suffers from the challenge to align the different parts of the overall pattern precisely. This alignment is called overlay. Overlay errors contribute to EPE just line CD bias, CD microloading, LWR, and LER.

In this section, we will discuss self-aligned double and quadruple patterning, which increases the number of etch applications further as shown in Figure 7.32.

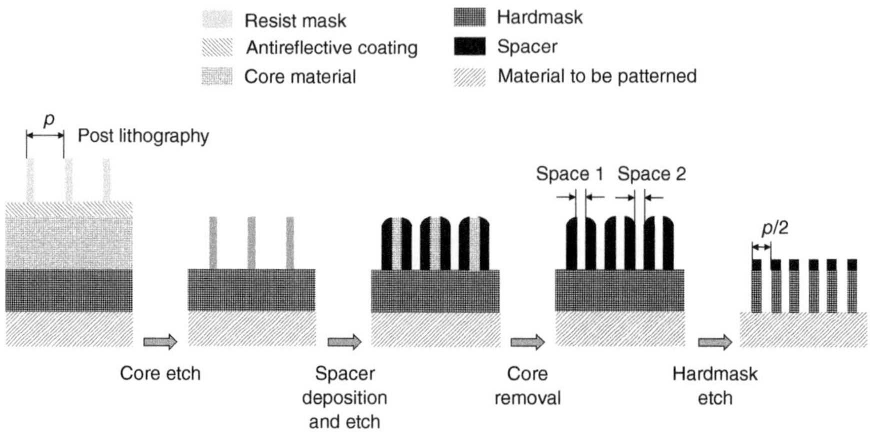

Figure 7.32 Schematic illustration of self-aligned double patterning with positive spacer tone. Source: Lee et al. (2014). © 2014 IOP Publishing.

At the core of self-aligned multi-patterning is the sidewall image transfer (SIT) or spacer patterning technique. Here, a first line is patterned with 193 nm immersion lithography. The pitch, which is the sum of the line width and space between the lines, is twice the desired final pitch. This first line is called mandrel or core. Depending on the performance and cost requirements, the core can be made of different materials. In the performance of critical logic FEOL applications, the core is typically made of polysilicon. In memory applications, CVD or spin-on carbon materials are used. In the next step, a liner is deposited over the mandrel. To provide selectivity, silicon oxide is typically used as the liner material. After the spacer etch, the mandrel is etched with high selectivity to the spacer. Finally, the oxide spacer is used to etch the next layer, which can be the final hardmask in the case of low aspect ratio etches. For high aspect ratio etches yet another transfer layer is needed to pattern the final hardmask. We use the term high and low aspect ratio in the absence of quantitative cutoff criteria. The complete self-aligned double patterning (SADP) flow is shown in Figure 7.32. The technique shown in Figure 7.32 is called SADP with positive spacer tone because the spacer determines the location and width of a line.

Let us now discuss the requirements for each step in more detail from an etching technology point of view. Depending on whether the mandrel is made of carbon or polysilicon, oxygen-, or halogen-based processes are used. The aspect ratio of the final mandrel is only around 3 to 1. Hence, high-density plasmas are typically used for mandrel etch. Current state-of-the-art photolithography scanners using 193 nm wavelength lasers and numerical apertures of 1.35 have fundamental printing limits of 40–45 nm half-pitch. To form equal width lines and spaces, the ratio of the mandrel to space width must be around 1 : 3. Lithography prints 45 nm equal lines and spaces. Subsequently, the line is trimmed in an oxygen-based plasma to about 23 nm, which increases the space to 67 nm. This trim process also has the benefit that high- and medium-frequency LER/LWR is reduced.

The mask to etch mandrel is a dielectric layer, in most cases DARC, which is a silicon oxynitride (SiON). If the mandrel is carbon, the selectivity is intrinsic. A deposition process is not needed to obtain selectivity. Oxygen etches carbon spontaneously and sidewall passivation is obtained by adding HBr- or sulfur-containing compounds such as COS or SO_2 to the oxygen plasma. The passivating effect of bromine is based on the fact that CBr_4 is not very volatile – it has a boiling point of 190 °C. Bromine can also combine with the silicon, which is sputtered from the hardmask, and oxygen to form silicon oxybromides. The passivating effect of sulfur is thought to be caused by the ability of sulfur to form polymers with carbon. The passivation is deposited from the plasma.

In the case of a polysilicon mandrel, typical RIE processes are halogen based. If chlorine or hydrogen bromide gases are used, selectivity to the mask is achieved via plasma deposition of silicon oxychloride and oxybromide reaction products. These compounds also deliver sidewall passivation. For fluorine-based plasma chemistry, fluorocarbon polymer gases are added to provide mask and sidewall passivation by carbon compounds.

The mandrel profiles must be absolutely vertical and without bottom foot. Otherwise, the spacer profiled will be asymmetric and in extreme cases the spacer can topple. Thin yet effective passivation layers such as sulfur or carbon are preferred.

The next step in the SADP flow is spacer deposition, typically by conformal CVD or ALD of SiO_2 or Si_3N_4. TiO_x spacers are used in BEOL SADP. Because the outer side of the spacer is formed by deposition, LWR/LER performance is excellent at the outer edge. The inner edge is determined by the mandrel lithography, trim, and etch process. The trim step smoothens the sidewall. Therefore, one of the most outstanding properties of SADP is the excellent high- and medium-frequency LWR/LER performance.

Low-frequency LWR/LER is impacted by potential bending or buckling of the mandrel or the spacer. Both affect the space between the spacers, but not the width of the spacers, which is determined solely by deposition. To leverage this effect, positive and negative spacer schemes have been introduced. The positive tone spacer scheme is used to pattern structures where the width of the line is the critical dimension. Examples are logic fins and FinFET gates.

Negative tone spacer double patterning is preferred for patterning of structures where the space between the lines is more important. Examples are trenches. SADP with negative spacer tone is schematically depicted in Figure 7.33. The bifurcation of SADP into positive and negative spacer tone schemes is implemented after spacer deposition.

In the positive spacer tone approach, the spacer is etched with a fluorine-based process. Both ICP/TCP and CCP reactors are used for this etch. The mandrel is usually etched in situ after the spacer etch. The key requirements for spacer etch/mandrel removal are no spacer loss, no spacer foot, no recess into the underlayer, and minimum spacer top profile tapering. Fluorocarbon-based plasma chemistries are used for spacer etching. Carbon mandrels are removed with oxygen and polysilicon mandrels with fluorine plasmas. To mitigate spacer width loss, in situ deposition steps can be used to protect the sidewall of the spacer during etching.

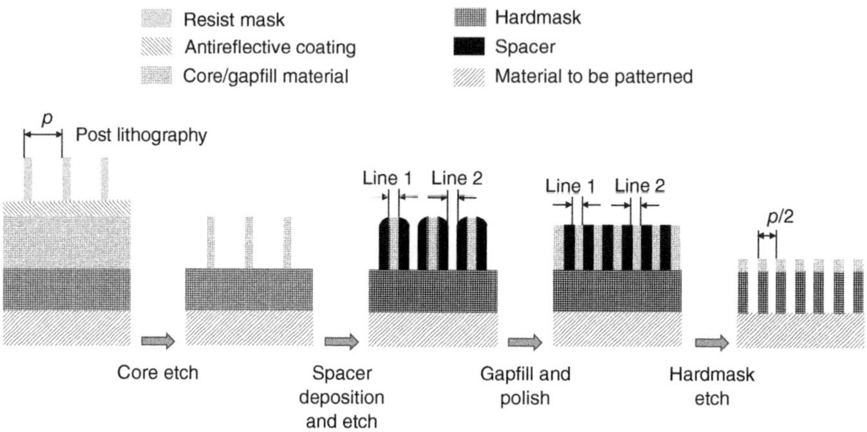

Figure 7.33 Schematic illustration of self-aligned double patterning with negative spacer tone.

The mandrel height is an important parameter for a successful SADP process as it determines the height of the spacer. The height of the mandrel must be at least two to three times the spacer width so that the spacer has a vertical sidewall before etching. Because the mandrel and spacer widths are the same, the aspect ratio of the spacer after mandrel etch is also about 2–3. The mandrel film must not have any intrinsic stress to prevent buckling. Taller mandrels are implemented if the device is patterned with the spacer directly.

In the negative tone spacer approach, the gap between the spacers is filled and polished back. Typically, the fill material is the same as the mandrel material. If the fill material is any material other than the mandrel, subsequent cuts can be selective to either the mandrel or the gap fill material. This allows a self-aligned cut mask and relaxes the overlay requirements. The next step after polishing the gap fill material is the selective removal of the spacer material. The process must not be too isotropic or the fill material can be undercut and mechanical integrity can be compromised.

The mandrel and the filled gap serve as the mask materials for etching the final hardmask or yet another transfer layer. Therefore, the mandrel height is primarily driven by selectivity requirements.

SADP can exhibit the so-called odd–even effects as shown in Figures 7.32 and 7.33. In the case of positive tone spacers, the effect manifests itself in alternating larger and smaller space CDs (space 1 and 2 in Figure 7.32). In negative tone spacer SADP, the odd–even effect impacts the line CDs (line 1 and 2 in Figure 7.33). The solution to odd–even effects is precise CD control of the mandrel patterning, spacer deposition, and spacer etching. CD measurements after mandrel lithography and spacer deposition is frequently used to re-target the mandrel trim and spacer etch processes for optimal performance.

SADP can also be used to pattern hole masks. Figure 7.34 shows that square and hexagonal hole arrangements can be obtained by repeating the SADP process twice. In the case of hexagonal layout, the two SADP patterns are rotated by 60° with respect

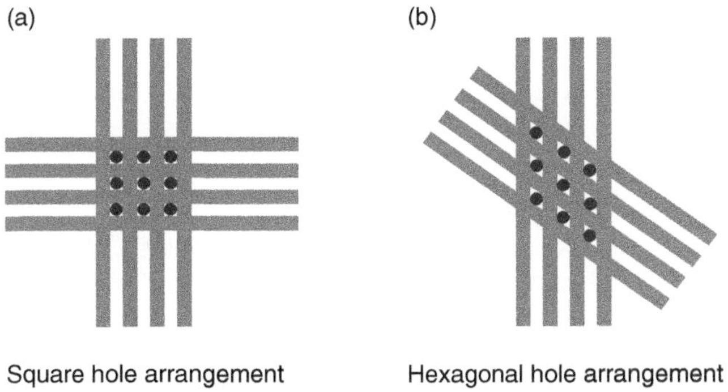

Figure 7.34 Schematic illustration of cross-SADP for hole structures. (a) Square hole arrangement and (b) hexagonal hole arrangement.

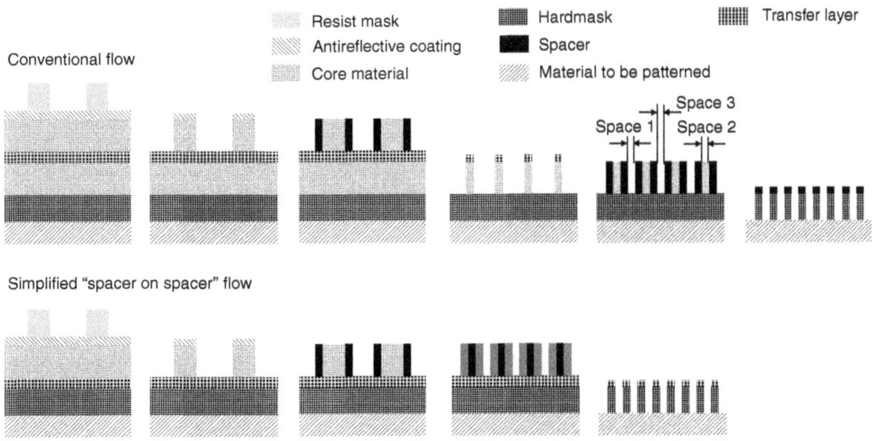

Figure 7.35 Schematic illustration of positive tone self-aligned quadruple patterning.

to each other. This gives rhombus-shaped holes. The mask open etching rounds them and turns them into circular holes. Double SADP is used for instance to pattern the mask for advanced DRAM cells. Half pitches of about 20 nm can be produced in high volume manufacturing using cross-SADP.

To generate half pitches of about 10 nm, self-aligned quadruple patterning (SAQP) is used. The process flow is depicted in Figure 7.35. As the name suggests, the SIT is applied two times in sequence to reduce the pitch by a factor of 4. The spacers are used in the positive tone approach. The etching processes are the same as for SADP except that the final spacers are only about 10 nm wide and line buckling is a formidable challenge, which requires film adhesion and stress engineering.

Figure 7.35 also shows a simplified SAQP approach utilizing the so-called spacer-on-spacer (SOS) technique. In this implementation, the second spacer is deposited directly onto the first spacer and the deposition and etching steps to form a second mandrel are saved. This requires a square profile of the first spacer to minimize asymmetries. Spacers can be etched with a square profile by optimizing deposition and etching processes. SOS-SAQP also requires a second spacer material. Positive tone SAQP creates three types of spaces as illustrated in Figure 7.35. Precise control of the mandrel and first spacer CD is needed to ensure that all three spaces are equal.

7.3.1.2 Extreme Ultraviolet (EUV) Lithography

Extreme ultraviolet (EUV) scanners use a wavelength of 13.5 nm, which is generated in a laser-driven tin plasma light source. The short wavelength of EUV lithography enables patterning of lines and spaces with sub-60 nm pitch without SADP. Sub-20 nm lines and spaces can be patterned with SADP instead of SAQP (Raley et al. 2018). However, EUV lithography faces trade-offs between line edge roughness and exposure dose. While EUV offers a 2.5 times resolution improvement over leading edge 193 nm exposures, LER is the limiting factor for overall resolution improvement. The latest EUV illumination sources produce over 10 times fewer photons than 193 nm sources. The random distribution of these photons, which is called shot noise, can lead to line edge roughness due to underexposure in some areas. While increasing exposure dose can reduce underexposure-related defects, it also can make overexposure-related defects worse. Increasing the dose for a given source also decreases the throughput and increases cost tremendously.

EUV resists are thinner than 193 nm resists because the line CDs are only about 20–30 nm wide, which can lead to buckling if the resist is too tall. To increase etching resistance of EUV resists, metal oxide nanoparticles are being tested as additives. Despite these efforts, patterned EUV resist is thin and has high LER. New smoothening techniques are being introduced, which combine deposition and etching processes similar to the LWR/LER treatments for conventional photoresists. These processes are being run in ICP/TCP reactors either standalone or in situ with subsequent mask etches. Alternating selective carbon deposition and smoothening with directional ALE show promising performance (Wise and Shamma 2018). This is an example of how the smoothening effect of ALE is used to aid with the implementation of EUV lithography.

So far in this chapter, we have seen that there are many different patterning applications: multilayer mask open, LELE, SADP, SAQP, and EUV patterning. The etching processes that are deployed in these patterning techniques have the common feature that they are driven primarily by chemistry, not ion bombardment. The most important performance metric in patterning is CD control. CD bias is driven by the balance of lateral, chemical etching, and sidewall passivation deposition, primarily from the plasma. This means that RIE reactors for patterning must run a wide range of etching chemistries and must have hardware features that allow tuning the reactant and etching product flows to each point of the wafer. The temperature must be tunable across the wafer with high resolution.

The fundamental challenge is that RIE reactors are far from being ideal chemical reactors, for instance, continuous flow stirred tank reactors (CFSTRs), which are used in chemical engineering to model chemical reactions. CFSTR models assume perfect mixing and work for liquids, gases, and slurries. Etching of wafers involves not only gases but also a solid wafer, which is placed right in the center of the gas

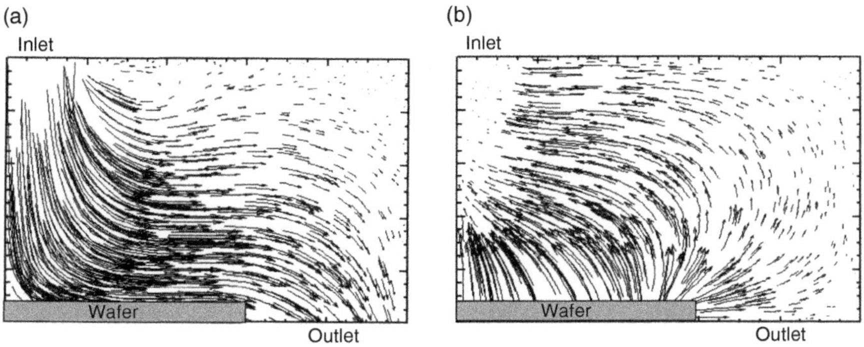

Figure 7.36 Modeling of silicon etching in a chlorine plasma: (a) convective chlorine flux and (b) diffusive $SiCl_2$ product flux. The picture depicts one half of the symmetric reactor. The gas inlet is in the center on the top of the reactor. Source: Kiehlbauch and Graves (2003). © 2003 AIP Publishing.

flow from the top of the reactor. Figure 7.36 illustrates the complexity of the reactant and reaction product fluxes in the reactor because of the blockage by the wafer (Kiehlbauch and Graves 2003). The model in Figure 7.36 assumes a pressure of 50 mTorr, 50 W of inductive RF power, and 500 sccm of chlorine flow. The resulting etching rate is 600 nm/min and the dominant etching product is neutral $SiCl_2$.

The incoming chlorine gas is shown in Figure 7.36a. At any point within the reactor, the length and direction of the arrow indicate the magnitude and direction of the flow. The chlorine flux hits the center of the wafer and then sweeps across the wafer surface toward the edge where it is pumped away. Because the wafer is etching, the concentration of the reaction products increases from the center to the edge of the wafer. The reaction products diffuse toward the point of their lowest concentration, which is at the chlorine gas inlet (see Figure 7.36b). This results in more tapered profiles and hence larger CDs at the edge of the wafer.

To suppress this effect, the center gas injection system can be equipped with nozzles that face sideways instead of down. In addition, gas nozzles can be installed at the reactor walls to inject reactants closer to the edge of the wafer. When the reactant flux is diffusive, the etching products tend to accumulate in the center of the wafer, which results in larger CDs in the center of the wafer (Kiehlbauch and Graves 2003). Diffusive reactant fluxes are observed when the gas injection is distributed across the entire top lid of the reactor in the form of a showerhead (Kiehlbauch and Graves 2003). They also dominate mass transport at pressures below 10 mTorr.

Based on these discussions we can see that the CD bias uniformity pattern across the wafer can be flipped just by changing the pressure within a range that is entirely covered, for instance, by polysilicon gate main and overetches. The main etch may be conducted at 4–5 mTorr and the overetch at 50 mTorr or thereabout. This means that the different gas injection points must be addressable independently and in combination to create a reactant and reaction product uniformity across the wafer. Also, process engineers must tune the center and edge gas flows to find the optimum setting.

Gas tuning may not be sufficient to achieve CD uniformities across the wafer of less than 5 Å, which is a common requirement for advanced devices. The CD at the very edge of the wafer needs more local tuning than gas flows can provide. Also, the incoming lithography pattern may have a CD nonradial nonuniformity pattern. For cases of such residual CD nonuniformity, temperature is used to change the sticking coefficient of the passivation species locally. For this purpose, electrostatic chucks (ESCs) with a multitude of heating zones and local heaters have be developed for advanced patterning. Needless to say, the overall geometry of the reactor must be as symmetric as possible.

In summary, patterning etching processes have a strong chemical etching component. They are typically conducted in RIE tools with high-density ICP or TCP sources. The main parameters to tune CD are gas flow and wafer surface temperature.

7.3.2 Logic Devices

7.3.2.1 Fin Etch

The evolution from planar transistor to fin field effect (FinFET) and gate-all-around (GAA) transistors is illustrated in Figure 7.37. The formation of a FinFET transistor includes two important RIE applications, Fin etch and FinFET gate etch. First, we will discuss Fin etch.

The most important part of the Fin is the top 30–50% of the fin profile. After the trenches next to the fins are filled with isolating silicon oxide which is then selectively recessed. This exposes the top of the fin, which will serve as the channel for the transistor. The gate is placed in vertical direction to the fin and wraps around it. The resulting structure is depicted in the middle panel of Figure 7.37. In a planar transistor, the channel is horizontal to the wafer surface. Silicon oxide filled trenches next to the channel serve as shallow trench isolation (STI, see Figure 7.37a).

Modeling results of a silicon Fin etch are shown in Figure 6.24 and germanium fins etched with Directional Cl_2/Ar ALE are depicted in Figure 6.25. These examples illustrate some of the main challenges in Fin etch: ARDE, profile shape, and profile loading. Since the etching materials are silicon and germanium, the etching

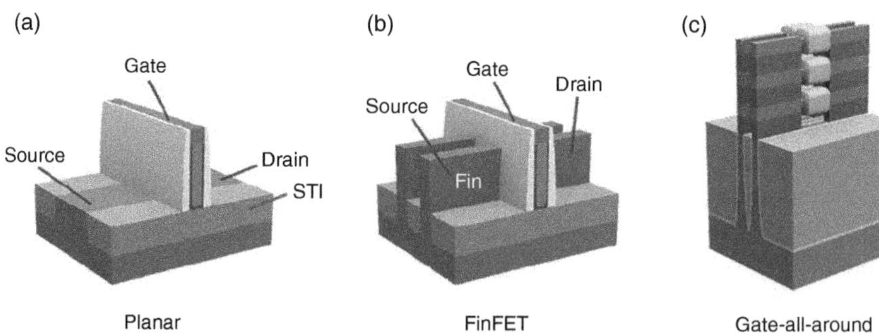

Figure 7.37 Schematic illustrations of planar, FinFET, and GAA transistors. Source: Draeger (2016).

process has a strong chemical component. Hence, Fin etches are typically conducted in ICP/TCP RIE reactors.

The etch does not land on a stopping layer and therefore any ARDE will translate directly into depth variations. ARDE can be minimized with bias pulsing, MMP, and directional ALE. The reason why these cycling techniques reduce ARDE have been discussed in Sections 6.1 and 7.1. The profile performance of the top 30–50% of the fin is extremely important. Typically, vertical profiles without any bow are desired. Since fins have a relatively high aspect ratio of 10–20, the critical section of the profile is exposed for a long time to ions and radicals. Profile bowing in the dense areas is a challenge that can be mitigated by intermittent sidewall passivation via in situ CVD. Advanced RIE reactors offer in situ dielectric atomic layer deposition (ALD) capability. When periodic passivation is used to protect the top fin profile, it is important to tune the passivation breakthrough step carefully to avoid profile discontinuities.

Even though Fin etch is a high aspect ratio etch, ion energy must be used prudently to avoid ion implantation or chemical damage. Such damage can negatively impact the performance of the Fin as the channel of the FinFET such as operating speed. Eriguchi et al. conducted MD simulations of plasma-induced physical damage of FinFET devices caused by straggling of incident ions (Eriguchi et al. 2014). We encountered the term "straggle" previously in Section 2.6 (see also Table 7.2). It is due to the nature of the collision cascade. Impinging ions penetrate the solid surface not only in the vertical direction but also in the lateral direction. During the etching of a fin structure, an impinging ion penetrates into the crystalline also in the lateral direction, resulting in lateral damage in the sidewall region.

Modeling results of the number of implanted argon and chlorine atoms for an ion energy of 200 eV are depicted in Figure 7.38. The numbers exclude a 0.25 nm thick amorphous surface layer, which would be removed in a wet clean after etching. The defects are argon and chlorine interstitials and dumbbell silicon. The depth

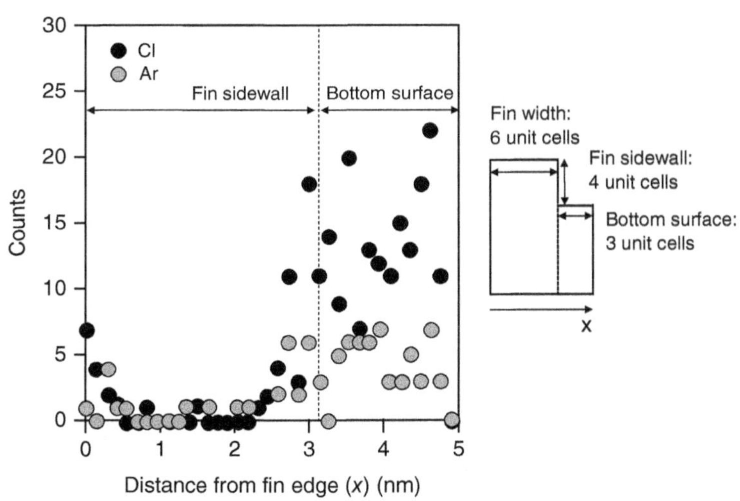

Figure 7.38 Number of argon and chlorine atoms in the fin after fin etching with 200 eV. Source: Eriguchi et al. (2014).

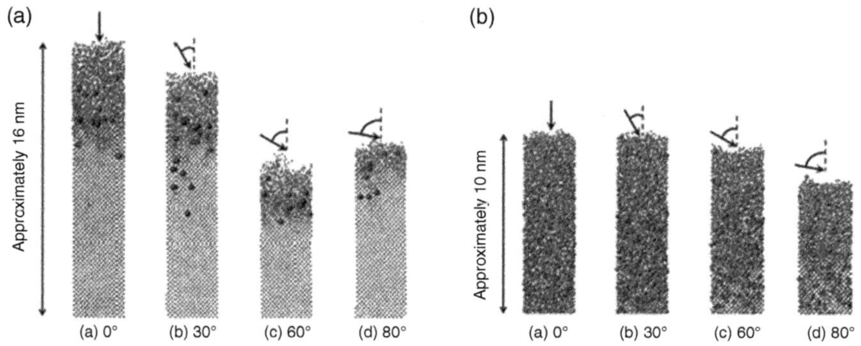

Figure 7.39 Implantation of silicon with 500 eV argon (A) and 300 eV hydrogen ions (B) for various impact angles. Source: Mizotani et al. (2015). © 2015 American Vacuum Society.

of the defects is about 1 nm. This means that for a 10 nm wide Fin, the model predicts that the damaged region constitutes about 20% of the entire fin width even at a fairly low ion energy of only 200 eV. As the fin width is decreased and the ion energy is increased to etch fins with higher aspect ratios, the ratio of damaged to undamaged fin material increases and repair by annealing becomes a challenge. If defects remain, they degrade the electrical performance of the device.

Mizotani et al. studied damage formation mechanisms at a vertical silicon wall by energetic incidence of hydrogen ions using MD simulation (Mizotani et al. 2015). They found that the penetration depth of hydrogen ions into a silicon substrate depends only weakly on the incident angle and therefore hydrogen ions at grazing incidence can form deep damage. Hydrogen can penetrate deeper than larger and heavier ions into a silicon substrate and can cause significant amorphization of the silicon substrate (see Table 2.2). This behavior has important implications for the use of HBr gas for etching of fins and silicon gates.

Figure 7.39 depicts modeling results comparing 500 eV argon and 300 eV hydrogen implantation for various impact angles. The argon ion dose was 7.3×10^{15} cm^{-2} and the hydrogen ion dose was 9.4×10^{16} cm^{-2} in each case. The argon and hydrogen atoms are depicted disproportionally large to make them more visible. In the cases of normal and 30° incidence of 300 eV hydrogen ions, the damage caused by the ion bombardment is so deep that almost 10 nm of the substrate is amorphized. Even at 80° incidence, the amorphized Si layer is much deeper than that by more energetic 500 eV argon ions at the same angle of incidence (Mizotani et al. 2015). The depth over which implanted hydrogen is observed extends over a similar distance as the damage profile.

7.3.2.2 Gate Etch

Etching of transistor gates has changed tremendously with the evolution from planar to FinFET gates as illustrated in Figure 7.37. One of the most important performance requirements of the gate etch application is CD control because the length of the gate is determined by the CD and LWR/LER of the etched line. In fact, the gate CD was so important that the technology nodes of logic devices were named after the CD of the transistor.

Integrated circuits are manufactured using complementary metal–oxide–semiconductor (CMOS) technology, which uses complementary and symmetrical pairs of p-type (PMOS) and n-type (NMOS) MOSFETs for logic functions. NMOS and PMOS silicon transistor gates are doped with phosphorus and boron, respectively. To reduce cost, the NMOS and PMOS gates are etched at the same time. This requires solving doping effects, for instance, by using fluorine-containing etching chemistries (see Section 7.2). Fluorine plasma and plasma passivation with carbon polymers are also proven solutions to minimize CD microloading (see Section 7.2).

Selectivity to gate oxide is very important. In fact, the requirement is not just to stop on only a few Angstrom of gate oxide, but to avoid damage of the crystalline silicon underneath. Plasma oxidation of the silicon substrate occurs during overetching with HBr/O_2 RIE. This silicon oxide is removed in subsequent wet cleans. This results in recessed regions next to the gate, which increase the resistance between the source and drain regions and the channel. Plasma oxidation and silicon loss are reduced by using a shorter polysilicon overetch time, lower source and bias power, lower substrate temperature, and lower O_2 flow (Vitale and Smith 2003). The effect can be eliminated using directional ALE with chlorine modification and xenon ion removal (Tan et al. 2015). Figure 6.21 shows a comparison of the gate oxide loss for Cl_2/Xe ALE (Lill et al. 2018) and conventional HBr/O_2 RIE polysilicon gate overetches (Vitale and Smith 2003).

A big change in transistor gate technology was implemented at the 45 nm technology node when metal gate materials and high-k dielectric materials such as hafnium oxide were introduced. The replacement of silicon oxide with hafnium oxide reduced the leakage currents by 2–3 orders of magnitude. High-k material can be physically thicker without appearing electrically thicker. The electrically relevant gate oxide thickness is called effective gate oxide thickness.

The replacement of silicon with metals eliminates the so-called depletion region in the gate just above the gate oxide. This depleted region in a semiconducting gate acts like a thicker gate oxide when the gate voltage is applied. Thus, the implementation of gate metals reduces the effective gate oxide thickness.

High-k/metal gates were implemented at the 45 nm technology node using two competing integration techniques. In the so-called "gate first" approach, the gate stack with the high-k gate oxide, metal gate, and mask layers were deposited onto a planar surface and etched. In the "gate last" approach, a polysilicon gate is etched to define the dimensions of the gate. After spacer deposition and etch, the space between gates is filled with dielectric material, typically silicon oxide. This material is polished back to expose the polysilicon gate to remove it. This removal step is important because the gate channel is exposed just below a thin protective silicon oxide layer. Combinations of RIE, radical etching, and wet etching are used to remove the polysilicon gate. After that, the high-k and metal films are deposited into the exposed trench. Thus, the gate is formed as the last step, which explains the terminology "gate last."

The challenge of the gate first approach was that hafnium oxide is a nonvolatile material. It is difficult to etch the thin hafnium oxide layers without leaving any residues at the bottom of the gate line to achieve high selectivity to the channel at

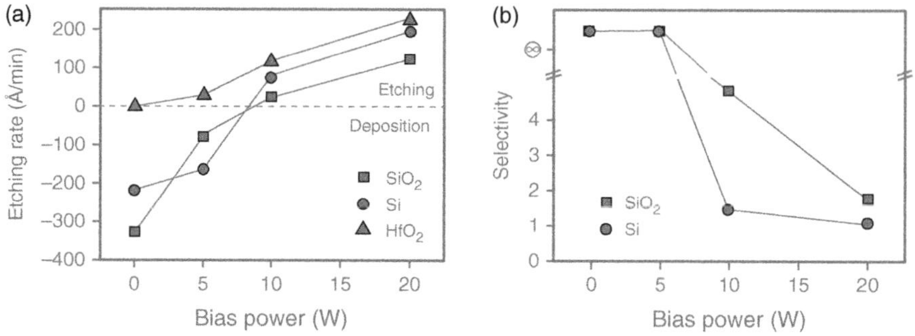

Figure 7.40 Etching of HfO_2 at room temperature with a BCl_3 plasma. Source: Joubert et al. (2006). © 2006 IOP Publishing.

the same time. Plasma using BCl_3 can etch HfO_2 selectively to silicon and silicon oxide, but the bias power must be controlled carefully to achieve selectivity as shown in Figure 7.40. In this experiment, a bias power of only 5 W was needed to achieve an etching rate of 30 Å/min and infinite selectivity to silicon and silicon oxide. The experiments were conducted in a 200 mm ICP reactor at 5 mTorr, a source power of 800 W, and 100 sccm BCl_3. To improve the volatility of the etching products such as hafnium chloride, high temperature etching was introduced to gate etching (Helot et al. 2006). In another approach, hafnium oxide is etched at room temperature and the residues are removed in a wet etch process. The exposure to halogen radicals in combination with ion bombardment eases the removal by liquid-phase chemistries.

The mechanism for ion-enhanced chemical etching of hafnium aluminate thin films in Cl_2/BCl_3 plasmas at room temperature was investigated by Martin et al. (2009). Several compositions of $Hf_{1-x}Al_xO_y$ thin films ranging from pure HfO_2 to pure Al_2O_3 were etched in BCl_3/Cl_2 plasmas and their etch rates were found to scale with $\sqrt{E_i}$ in both Cl_2 and BCl_3 plasmas. This indicates that the process is a sputtering process. In BCl_3 plasma, deposition dominates at ion energies below 50 eV, while etching occurs above that energy with an etch rate three to seven times that in Cl_2. This can be explained by the formation of more volatile boron-containing etching products such as $(BOCl)_3$. This reaction pathway facilitates the abstraction of oxygen and etching of metallic aluminum and hafnium by chlorine species (Martin et al. 2009).

Because of etching challenges and limited materials integration flexibility in the "gate first" approach, the "gate last" integration dominated the industry beyond the 45 nm node. This choice turned out to be the right one to enable the next change in gate integration, the change from planar to FinFET gates (see Figure 7.37). Here, the polysilicon layer for sacrificial gate is deposited over tall fins. The etching challenge is to etch vertical gates inside the tall fins without residues in the corners. This is a challenge directional ALE of silicon with Cl_2/Ar^+ can solve. Figure 6.31 shows feature scale modeling results comparing continuous RIE and directional ALE. Better corner removal performance of ALE is clearly visible. These modeling results have been reproduced in experiments. To improve the throughput and reduce the cost of the process, directional ALE is used only as an overetch step to clear the corners while the main etch is RIE. Selectivity to the fin is similarly important as in planar gate etching. The gate oxide is wrapped around a vertical fin, which can cause corner erosion challenges.

7.3.2.3 Spacer Etch

Traditionally, silicon nitride is the material of choice for gate spacers. Nitride acts as a better barrier than silicon oxide to prevent diffusion during downstream processing. The purpose of the spacer is to isolate the source and drain contacts from the gate. As transistors continue to shrink, capacitive coupling is becoming a challenge that impacts device performance, and low-k spacer materials are being investigated.

Spacer etch was one of the first applications where Si_3N_4 was etched (Regis et al. 1997). It is an archetypical Si_3N_4 RIE application. Si_3N_4 is also etched in fluorocarbon plasma such as silicon oxide. However, it does not contain oxygen to consume carbon. Therefore, Si_3N_4 etching requires hydrogen addition to the plasma. Gas mixtures containing CHF_3, CH_2F_2, and CH_3F are commonly used.

The main challenges of spacer etching are CD control due to isotropic loss of spacer width, profile performance due to footing, and selectivity. Lack of selectivity leads to silicon loss, which is a challenge we encountered in section 7.3.2.2 on gate etch. There is a trade-off between footing at the bottom of the profile and selectivity. To remove a bottom foot, higher ion energies are needed. However, higher ion energies remove the thickness of the fluorocarbon layer, which boosts selectivity between Si_3N_4 and silicon.

The underlying reason for the profile vs selectivity challenge in Si_3N_4 RIE is the lack of intrinsic selectivity due to high ion energies. This makes directional ALE a potential alternative etching solution. Posseme et al. demonstrated vertical bottom profiles using directional ALE as shown in Figure 6.30 (Posseme et al. 2014). In their approach, Si_3N_4 was directionally damaged by hydrogen or ion bombardment in a first step. This modified layer was removed with diluted HF acid, which in an isotropic etching process. Posseme et al. demonstrated less than 6 Å recess of the SiGe channel layer without Si_3N_4 foot (Posseme et al. 2014). Sherpa et al. demonstrated an all-dry directional ALE process with removal by a capacitively coupled fluorine plasma with SF_6 or NF_3 feed gas instead of diluted HF (Sherpa and Ranjan 2017).

7.3.2.4 Contact Etch

Contact etching is a very critical application. Every single contact must be open for the logic chip to function and there can be billions of contacts on an advanced logic chip. Source/drain contacts are the electrodes of the transistor, injecting and removing carriers to and from the transistor's channel. The gate contact connects the device circuitry to the transistor gate. Source/drain contact etch is more complicated because the aspect ratios are higher.

The evolution of source/drain contact integration schemes is schematically illustrated in Figure 7.41. Originally, contact patterning comprised photolithography of holes and transferring these holes into silicon oxide via RIE with high selectivity to the silicon substrate. Owing to scaling, the aspect ratios of the contact

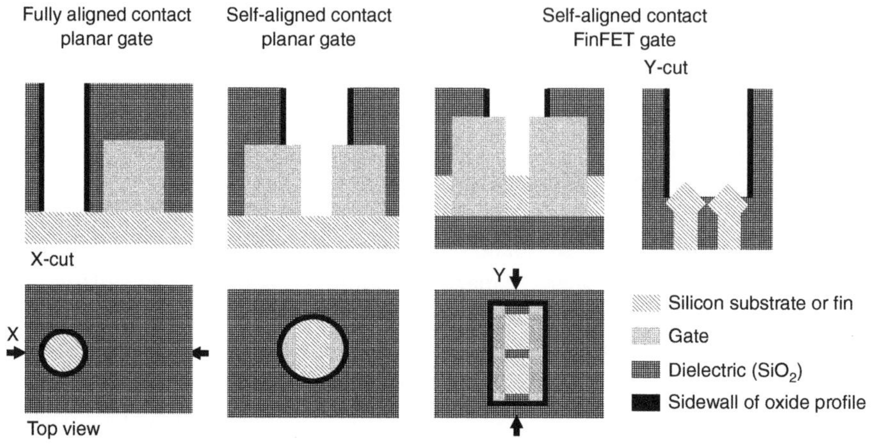

Figure 7.41 Schematic illustration of the evolution of transistor contacts.

holes increased, which led to increased ion energies. This triggered research of the selectivity mechanism for etching of silicon oxide and silicon with CF_4/H_2 plasmas and the discovery of fluorocarbon-based layers as selectivity boosters (Oehrlein and Lee 1987). We discussed deposition-enhanced selectivity mechanisms in Section 7.2. Carbon-rich gases such as C_4F_8 were soon discovered to offer great selectivity to silicon, and ARDE studies were conducted due to the realization that the aspect ratios for contacts were reaching 6 : 1 and higher (Hayashi et al. 1996).

In the late 1990s, further scaling of logic and also DRAM devices reduced the space between gate lines to such a degree that the contact hole started to overlap with the gate, either by design or by misalignment. The resulting contact was called a self-aligned contact. In such a contact the etch exposes the corner of the gate structure before it reaches the silicon substrate layer (see Figure 7.41). This reduces the diameter of the contact hole and increases the aspect ratio. Depending on the degree of misalignment, the aspect ratios of a self-aligned contact can be 10 : 1 and higher. This gave etching technologists a first preview of the challenging high aspect ratios to come because high aspect ratio structures could be generated by simply changing the alignment between the contact and the gate.

Besides ARDE, the main challenge of self-aligned contact etching is selectivity to the exposed corner. This corner is made of Si_3N_4 from the gate hardmask and a liner that encapsulates the gate and electrically isolates gate and contact. If the selectivity is too low, the Si_3N_4 corner can break through and create contact-to-gate shorts. The solution to this challenge is again passivation-enhanced selectivity, for instance, by using a C_4F_8 plasma RIE process. As discussed in Section 6.1.3, fluorocarbon polymers tend to deposit on silicon nitride but not on silicon oxide. The reason is the consumption of fluorine and carbon when intermixed upon ion bombardment, which leads to the formation of SiF_4, CO, and CO_2. Silicon nitride does not contain oxygen, which leads to accumulation of carbon-rich layers.

The selectivity requirements increased further with the advent of FinFET gates. When landing on fins, their corners are exposed. To decrease contact resistance, epitaxial silicon or SiGe is grown around the fins as illustrated in the right panel in Figure 7.41. Selectivity to SiGe is lower than to silicon. Epitaxy forms rhombus-shaped fin extensions, which increase the contact area and reduce contact resistance. Removal of the surface peaks due to lack of selectivity would increase contact resistance.

To increase selectivity, the thickness of the passivating polymers has to be increased even more. This can lead to excess polymer deposition at the top of the feature and increase the risk of not-opened contacts. Directional ALE with C_4F_8/Ar^+ removes the polymer in each argon bombardment step as discussed in Section 6.1.3. At the same time, the ion energy is kept as low as possible. These benefits propelled directional ALE with C_4F_8/Ar^+ to become the first industrially used ALE process (Hudson et al. 2014). The etching mechanism of this process was studied in detail (Metzler et al. 2014; Metzler et al. 2016; Huard et al. 2018).

7.3.2.5 BEOL Etch

Among all the logic etch applications, BEOL etching underwent one of the most dramatic changes. Before the turn of the century, logic device metallization was realized with aluminum, which was etched directly with a resist mask or a hardmask. Aluminum etches readily with chlorine. The isotropic etching component is strong. To achieve vertical profiles, boron (via addition of BCl_3) or carbon (via addition of CH_4) polymer deposition processes were used.

Toward the end of the last century, it became clear that aluminum had to be replaced with copper to meet resistance requirements. RIE of copper was attempted and initial feasibility was demonstrated (Schwartz and Schaible 1983; Choi and Han 1998). However, copper does not easily form volatile etching products. Therefore, higher etching temperatures and ultraviolet (UV) radiation were investigated with the goal to enhance the etching rate (Choi and Han 1998). Still, many challenges such as corrosion from chlorine residues remained.

A so-called copper damascene interconnect approach was developed as an alternative and transformed into high-volume production. It is the integration approach that is used in all logic devices today. There are two embodiments of damascene interconnect integration: "trench first" and "via first." Simplified process flows for both approaches are depicted in Figure 7.42. For the "via-first" integration scheme, the via is patterned first followed by etching the trench using a second mask, typically carbon. Following the etch step, this carbon mask must be removed using oxygen-based radical etching. Oxygen radicals can damage the porous low-k dielectric. Therefore, the industry adopted the "trench-first" metal hardmask scheme that is the dominant BEOL integration strategy now. Here, the trench pattern is defined in a TiN hardmask. The via is partially etched using a photoresist mask. The resist is removed by plasma and the bottom of the via and the trench are etched together using the metallic mask. The metallic mask is finally removed by chemical mechanical polishing (CMP) during the metallization, which avoids using damaging plasma processes (Lionti et al. 2015).

Besides the etching steps, BEOL integration involves other processing steps such as post etch cleaning, low-k sealing, barrier deposition, copper fill, and CMP.

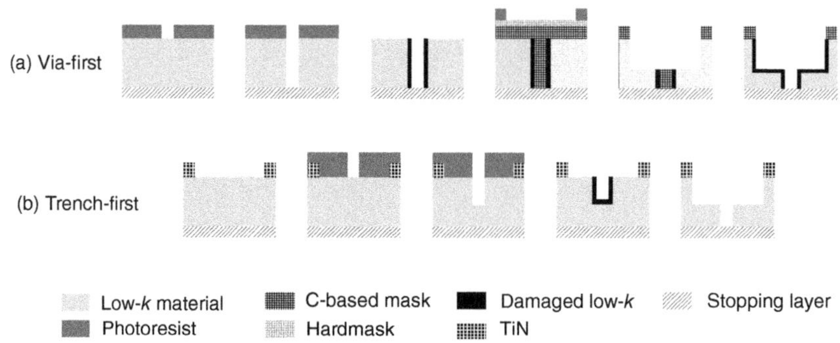

Figure 7.42 Simplified process flow illustrating (a) "via-first" and (b) "trench first" metal hardmask integration flow for interconnect fabrication. Source: Lionti et al. (2015). © 2015 IOP Publishing.

A detailed description of the different integration schemes and challenges goes beyond the scope of this book. The interested reader is referred to publications that focus on the subject of BEOL metallization (Lionti et al. 2015).

The BEOL dielectric materials have evolved over time. Initially, the material was SiO_2 with a k value of about 4.2. Fluorine-doped SiO_2 was introduced to decrease the dielectric constant of the material to around 3. Carbon-containing silicon oxides with ultra-low k values below 3 were introduced at the 90 nm node. The next steps of evolution were porous low-k materials at the 45 nm node. By introducing pores, k values below 3 were obtained. Intel implemented air gaps at several interconnect levels from the 14 nm node. Air has a k value of ~1.0.

RIE of low-k materials can be considered as the archetypical application for studying reactive plasma damage of dielectric materials, specifically porous dielectric materials. Etching of porous materials has very unique challenges. The sidewall that is created during the etching process is not continuous, which means that reactive species can diffuse into the sidewalls. A porosity above the percolation threshold of about 20% leads to interconnected pores, and etching species can penetrate deep into the bulk of the material. Chemical damage of carbon-containing silicon oxides typically means a loss of carbon groups and the formation of Si–OH and Si–F groups. As a result, the insulator electrical properties, such as dielectric constant, leakage current, and breakdown voltage are compromised. This chemical damage mechanism is present for porous and nonporous carbon-containing silicon oxides.

Therefore, the approach to porous low-k etching is to use heavy sidewall passivation to close the pores. Porous low-k materials are typically etched in a C_4F_8-based process in dual-frequency CCP reactors. C_4F_8 is a carbon-rich gas that effectively generates fluorocarbon polymers in plasmas as discussed in section 7.3.2.4 about contact etching. CCP reactors generate plasma densities where the gases are not completely dissociated and larger radicals and ions are present. Plasma pulsing has been demonstrated to reduce plasma damage presumably due to further increasing the concentration of carbon-rich radicals (Jang et al. 2019).

RIE at cryogenic temperatures was recently demonstrated to significantly reduce low-k material damage (Iacopi et al. 2011). The idea behind this approach was to slow down the diffusion of damaging reactive species into the low-k material at lower temperatures (see Eq. (2.15)). Iacopi et al. found in their studies, however, that the effect is largely unrelated to changes in the volume diffusivity. Rather, it is determined by the increase of the sticking coefficient and the radical recombination and reaction coefficients, favoring an early irreversible surface adsorption of the plasma radical species (Iacopi et al. 2011). This effect reduces the penetration depth into the porous matrix and limits the damage.

Atomic layer etching has been used to extend the BEOL RIE process (Lutker-Lee et al. 2019). Lutker-Lee et al. demonstrated that directional ALE successfully suppressed RIE while keeping low-k damage to a minimum. In addition, the ALE showed improved hardmask selectivity and resulted in lower line edge pattern roughness.

Recently, etching of copper and alternative low-resistance materials such as cobalt, molybdenum, and ruthenium is being investigated (Toyoda and Ogawa 2017; Altieri et al. 2017). The reason for this renewed interest in metal etching is that diffusion barriers for Electro Copper Plating (ECP) fill too much volume of the metallization lines with higher resistance material. If the low-resistance material could be etched, a dielectric diffusion barrier could be used. Graphene films are looked at for that purpose. However, etching of metals is also inflicted by physical and chemical damage during plasma etching. Etching species that diffuse into copper, for instance, form compounds with much higher resistance. If the thickness of this damaged layer is similar to the thickness of the diffusion barrier, there is no benefit to be gained by copper etching.

7.3.3 DRAM and 3D NAND Memory

7.3.3.1 DRAM Capacitor Cell Etch

Until about 2010, there were two competing DRAM technologies. Deep trench and stack capacitor cells were used to manufacture DRAM chips. In the case of deep trench cells, the capacitor in the shape of a cylinder was etched into the single crystalline silicon substrate, which drove the development of high aspect ratio silicon etching. Alternatively, stack capacitors are etched into a SiO_2 layer that is deposited onto the silicon substrate. Today, DRAM devices are produced with the stack capacitor method. One of the reasons is high aspect ratio etching performance.

Advanced deep trench etch processes were able to achieve aspect ratios of 80 : 1 (see Figure 2.13). The challenge for deep silicon etching is the risk of lateral etching by fluorine radicals. High aspect ratio silicon etching of DRAM capacitors was conducted in CCP reactors in most cases with enhancement of the plasma density by magnetic fields (Rudolph et al. 2004). The process gases were HBr, NF_3, and O_2 and the hardmask was made of silicon oxide. During etching, silicon-containing etching products exit the trenches and are re-introduced and dissociated on the plasma to serve as precursors for CVD type of passivation from the gas phase. At the same time, direct line-of-sight sputtering from the etching front also occurs. The resulting sidewall passivation is comprised of fluorine- and bromine-containing silicon oxide.

This process generates sufficient fluorine radicals so that isotropic etching of silicon can occur in locations where the passivation is too thin. Such isotropic etching must be avoided because it forms side pockets, the so-called "mouse bites." If such a pocket is deep enough to reach a neighboring trench, the result is an electrical short and device failure. Increasing the passivation layer thickness can lead to partial closure of the trench top, so-called necking. This decreases the cross section through which ions can enter the trench and slows down the etching rate. Thus, one of the biggest challenges in silicon deep trench etching is the balance between necking and isotropic silicon etching.

The trade-off between lateral etching and too much passivation also exists in high aspect ratio etching of silicon oxide to form stack capacitors. While sidewall passivation breakthrough in the case of silicon holes has catastrophic consequences, the impact on oxide trenches is more gradual in the form of bowing. The reason for this is that silicon oxide is not easily etched by fluorine radicals. The dominant etching mechanism is chemically enhanced or chemical sputtering.

In summary, the fact that deep trench silicon etching has a chemical or radical etching component makes it easier to etch with high selectivity to a silicon oxide hardmask. Aspect ratios of over 80 : 1 have been achieved. However, the chemical etching component is also the biggest challenge of high aspect ratio silicon etching. Breakthrough of the sidewall passivation layer must be avoided because it leads to catastrophic failures. The challenge of controlling random isotropic etching ultimately contributed to the demise of Deep Trench DRAMs. This is also the reason why cyclic C_4F_8/SF_6 processes, which are successfully used in MEMS device manufacturing and can routinely achieve aspect ratios of 100 : 1 and higher, are not used in the manufacturing of DRAM deep trenches. The isotropic component of the SF_6 step cannot be controlled to maintain sidewall integrity inside features with sub 40 nm diameter.

High aspect ratio silicon oxide capacitor etching is the processing technology that enables the DRAM roadmap today. Silicon oxide capacitors are etched with C_4F_8- and C_4F_6-based chemistries in double- or triple-frequency CCP reactors. SiO_2 capacitor cell etching is the archetypical high aspect ratio dielectric etching application. Aspect ratios of advanced capacitor cells are around 60 : 1 and are reaching 100 : 1. Several horizontal Si_3N_4 layers are part of the stack as braces for the isotropic wet process removal of the silicon oxide after the capacitor is formed. The capacitor etch may stop on these layers or etch through them nonselectively.

The mask is a hexagonal arrangement of identical holes. The pitch is typically 40 nm and the hole CD about 20 nm. Device shrinking is accomplished by reducing both the pitch and the holes size. This introduces ARDE and mask integrity challenges. The mask is typically made of amorphous silicon and has itself an aspect ratio of 40 : 1. It is etched with another SiO_2 hardmask in dual- or triple-frequency CCP reactors with processes similar to the silicon deep trench process. One can say that the legacy of silicon deep trench lives on as the mask formation step for stack capacitors. The mask for the oxide hardmask is generated by dual SADP as illustrated in Figures 7.25 and 7.27. Both SADP steps must maintain CD extremely precisely because even small deviations in the mask CD will impact ARDE and mask integrity.

It is predicted that this patterning approach will be one of the first steps that will be replaced by EUV lithography.

The critical technical challenges of high aspect ratio dielectric etching are etching rate and ARDE, profile performance including necking, bowing, hole distortion, and twisting, as well as mask selectivity. These challenges are being solved with a good understanding of the etching mechanism.

As discussed in Section 7.1.3, RIE of SiO_2 (and Si_3N_4) is predominantly synergistic, chemically enhanced sputtering. RIE of silicon oxide relies on the deposition of reactive fluorocarbon polymers for etching and ion bombardment to form a selvedge layer. The ion energy has to be high enough for the ions to penetrate the selvedge layer and to activate the formation of volatile etching products at the interface.

On blanket wafers and in low aspect ratio features, a reactive fluorocarbon layer is generated by neutral species from the plasma because the neutral fluxes are larger than the ion fluxes. This is due to lower threshold energies for dissociation compared with ionization (see Section 9.1). In C_4F_8 and C_4F_6 plasmas, CF_x neutrals are produced by electron impact dissociation. These neutrals typically arrive at the wafer surface with isotropic angular distributions. They are also called thermal radicals.

Etching of high aspect ratio features into dielectrics such as SiO_2 and Si_3N_4 faces the challenge to deliver thermal radicals to the etching front. The transport mechanism of these radicals is Knudsen transport (see Section 2.8.1). Owing to the limits of vacuum conductance, the probability of neutral radicals reaching the etch front at the bottom of the feature decreases with the aspect ratio (Coburn and Winters 1989). The reason is diffusive reflection from sidewalls, which causes some fraction of the incident flux to be reflected back out of the feature to the plasma. The same transport challenge applies to the etching products that are produced at the bottom and must be transported out of the feature without redeposition on the sidewalls.

Attenuation of thermal neutrals would quickly lead to etch stop if the only etching mechanism is chemically assisted sputtering. Therefore, another etching mechanism must be considered. This alternative mechanism is based on ion scattering and neutralization, which was discussed in Section 2.8.2 and chemical sputtering.

Scattering ions interact with the electronic structure of the surface atoms and can exchange charges. The probability of charge exchange depends on the effective interaction time of the projectile with the surface. Grazing impact angles in high aspect ratio etching extend the duration of the interaction. Ions are neutralized and become fast neutrals that maintain an anisotropic angular distribution. They can deliver kinetic energy to the etch front and are a source of chemical reactivity, which allows etching to proceed (Huang et al. 2019).

The majority of the reactive ions generated in C_4F_6 and C_4F_8 plasmas are CF_2^+ and CF_3^+. Etching of SiO_2 has been studied with energy-controlled and mass-selected beams of CF_2^+ and CF_3^+ ions (Toyoda et al. 2004) and simulated by a quantum chemical MD model of fast radical interactions (Ito et al. 2013, 2014).

According to the simulations, fast CF_2 is the main etchant to break the Si—O bond due to its higher chemical reactivity at low energies of 10 eV. At 150 eV, however, CF_3 becomes the main etchant due to the production of more reactive F atoms, resulting in the formation of more Si—F bonds (Ito et al. 2013, 2014). The conclusion from these calculations is that fast ions with energies as low as several tens of electron volts can etch SiO_2. Chemical sputtering via fast CF_2 and CF_3 is a possible etching mechanism for high aspect ratio dielectric etching.

Experimental beam studies by Toyoda revealed that CF_3^+ etches SiO_2 for energies above 50 eV (Toyoda et al. 2004). The etching yield is 0.25 at 100 eV, 0.6 at 200 eV, and 1.0 at 400 eV. Here, the yield is defined as the number of silicon atoms removed per incident CF_3^+ ion. Reactive fluorocarbon polymer layers were detected on the SiO_2 surface upon CF_3^+ ion bombardment. This means a selvedge layer is formed by ion and fast neutral impact onto the SiO_2 surface. Hence, ions and fast neutrals are the source of reactive species and at the same time provide the energy to intermix the fluorocarbon layer with the SiO_2 surface. Based on these results, it appears that high aspect ratio dielectric etching is the result of chemical and chemically assisted sputtering by ions and fast neutrals, which are the result of ion neutralization at the feature sidewall.

Huang et al. conducted computational investigations that confirm this mechanism. They simulated the plasma and used the resulting fluxes and energies to investigate the profile evolution for 80 : 1 AR features in SiO_2. The feed gases were argon, C_4F_8, and oxygen. The plasma was energized by three CCP-coupled RF frequencies. The specific operating conditions were gas flow ratios $Ar/C_4F_8/O_2 = 75/15/10$, pressure 25 mTorr, total flow 500 sccm, RF power 80/10/5 MHz = 0.4/2.5/5 kW. Under these conditions, the incident ions have energies of up to several kilo electronvolts and incident angles of less than 4°.

Figure 7.43 shows the calculated fluxes and power delivered to the etch front as a function of aspect ratio. The first observation is that a relatively small flux of ions reaches the bottom of the feature. Ions striking the etch front did not undergo collisions with the sidewalls since ions neutralize upon striking a surface and become hot neutrals in this model. The flux of ions to the etch front is limited by the decreasing view angle from the bottom of the feature subtending the ion angular distribution at that aspect ratio. Hence, the ion flux to the etch front decreases strongly with AR.

The flux of fast neutral to the etching front is about two to three times higher than the ion flux for aspect ratios above 5. This is because ions neutralize and thus contribute to the neutral flux. The flux of fast neutrals more than doubles with increasing aspect ratios from 1 to 5. This means that the fast neutral to ion flux ratio at the very beginning of the etching process is largely controlled by the thickness of the mask, which can easily exceed 20 : 1.

The power density that is delivered to the bottom of the feature is similar for fast neutrals and ions. The reason is the energy loss when ions collide with the sidewall. In addition, they can lose even more energy in subsequent collisions. Ions, on the other hand, possess the full initial energy because they did not collide with the sidewalls.

The thermal neutral flux attenuates faster than the other fluxes till the SiO_2 aspect ratios of about 10, beyond which fast neutrals are the dominant species reaching the etching front. As the initially anisotropic fast neutrals collide with the sidewalls,

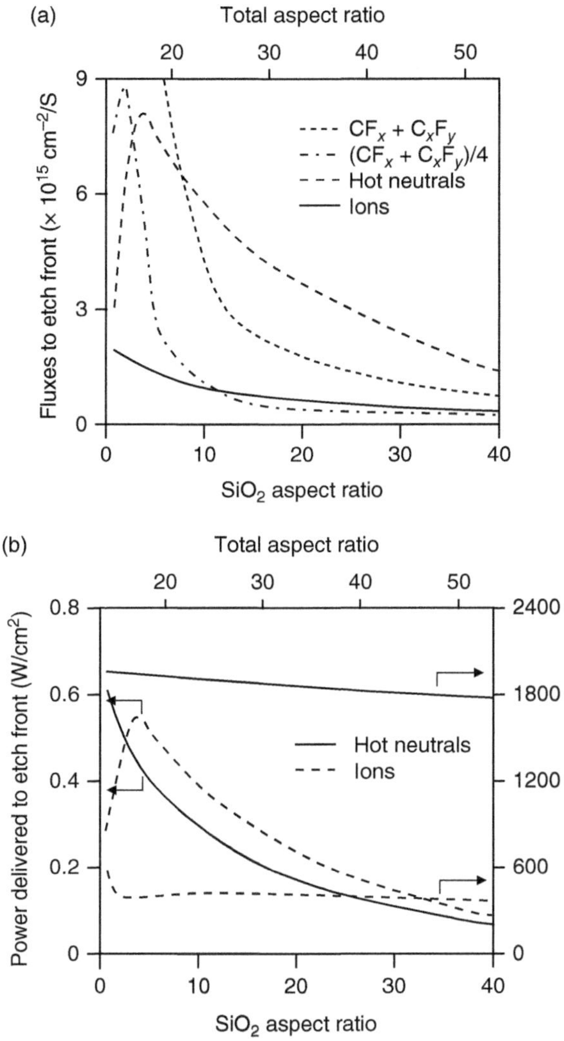

Figure 7.43 Fluxes and powers to the etch front as a function of aspect ratio. (a) Fluxes of ions, hot neutrals, and CF_x and C_xF_y radicals to the etch front and (b) power delivered to the etch front by ions and hot neutrals and the average energy of ions and hot neutrals to the etch front. Source: Huang et al. (2019). © 2019 American Vacuum Society.

their angular distribution becomes more isotropic and conduction limits begin to reduce their fluxes to the bottom.

The ion and angular distributions of the scattered ions govern how fast this "thermalization" of fast neutrals proceeds. The more the scattering angle deviates from specular to larger outgoing angles, the faster neutrals will lose energy. Huard et al. use the following simplified relationship for the energy and angular distribution of scattered ions and fast neutrals in their three-dimensional Monte Carlo feature profile model (Huang et al. 2019):

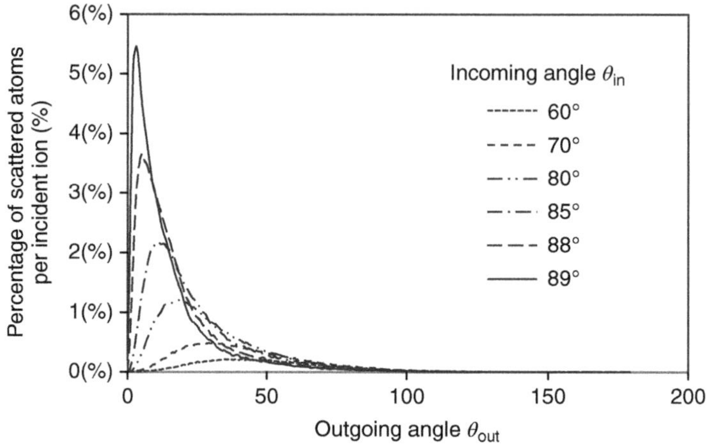

Figure 7.44 Angular distribution of scattered ions. Source: Berry et al. (2020).

$$E(\theta)_{out} = E_{in}\left(\frac{E_{in} - E_1}{E_2 - E_1}\right)\left(\frac{\theta - \theta_1}{90° - \theta_1}\right); \theta > \theta_1; E_1 < E_i < E_2 \tag{7.10}$$

Here, E_{in} is the incoming ion energy. E_1 is the cutoff energy for diffusive scattering, which Huard et al. define as 10 eV. Ions with energy of less than 10 eV will scatter like thermal atoms (see Section 2.8.1). E_2 is the threshold for complete specular scattering, which is set to 100 eV. Incident particles with $E_i > E_2$ are assumed to retain all of their energy. θ_1 is the lower cutoff angle for specular reflection, which is 70°. The origin of specular scattering was discussed in Section 2.8.2. Incident particles with $\theta < \theta_1$ or $E_i < E_1$ are assumed to diffusively scatter (Huang et al. 2019).

Berry calculated energy and angular distributions of scattered ions using a cellular, 2 1/2D Monte Carlo simulation (Berry et al. 2017, 2020). The angular distributions of the scattered ions as a function of impact angle for 1000 eV argon ion bombardment of SiO_2 are depicted in Figure 7.44. Specular reflection is given for an outgoing angle of $\theta_{out} = 90° - \theta_{in}$. The model predicts that the fraction of ions that undergo specular reflection is actually quite small. The peak fraction is 5.5% even for a very shallow impact angle of 89° and the peak fraction decreases by over a factor of 2 for an impact angle of 85°. The distributions have a long tail toward larger outgoing angles θ_{out}.

Figure 7.45 shows the ion energy distributions of the scattered ions. Near zero energy loss is observed for large incoming angles θ_{in} of 88° and 89°. Ions that have the same impact angles but are scattered with a larger outgoing angle lose energy as manifested by the long tail. For incoming angles of 80°, all ions lose more than 25% energy.

These Monte Carlo simulations predict that the main contributor to high aspect ratio etching are line-of-sight ions while Huang's model gives more equal weight to line-of-sight ions and scattered neutrals. Both models illustrate the importance of sidewall scattering in high aspect ratio dielectric etching. A mechanism, which is sometimes brought forward, where ions undergo multiple near specular sidewall collisions and get focused, appears highly unlikely.

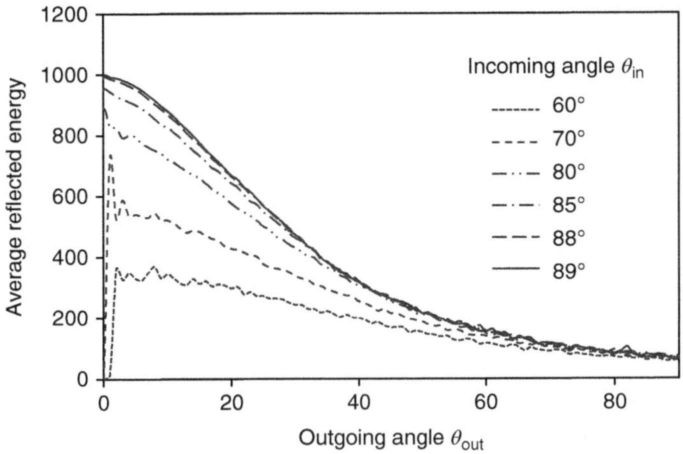

Figure 7.45 Energy distribution of scattered ions. Source: Berry et al. (2020).

This insight has important consequences for the interpretation of profile distortions and twisting. They are not the result of asymmetric ion focusing effects. Rather, fast thermalization of ions that are scattered from the sidewall ensure that direct, line-of-sight ion impacts drive the direction of the etch front. This provides high aspect ratio dielectric etching directional stability and robustness. Profile anomalies such as hole distortions and twisting have other root causes, which will be discussed below.

Based on these simulation results, a mechanism for high aspect ratio SiO_2 etching with two distinct regimes has been proposed by Huang et al. (2019). For low aspect SiO_2 ratios, abundant fluxes of CF_x and C_xF_y radicals passivate the oxide surface to form complexes, which are then removed by energetic species (hot neutrals and ions) through chemically enhanced sputtering with the formation of gas-phase SiF_x, CO_x, and COF (Huang et al. 2019). Hence, up to an aspect ratio of about 5 (excluding the contribution of the mask), etching proceeds according to this conventional SiO_2 etching mechanism as described in Section 7.1.1.

As the AR increases, conductance limits of the isotropic thermal radicals reduce their fluxes into the feature and the etching mechanism changes. The contribution of physical and chemical sputtering increases because the fluxes of energetic species to the etch front surpass those of the conduction constrained CF_x and C_xF_y radicals. For a sufficiently large AR, the neutral radicals reaching the bottom of the feature originate almost exclusively from neutralized ions that can overcome conductance limits by virtue of their initially anisotropic trajectories. The dominant oxide removal process transitions from chemically enhanced to chemical and physical sputtering as the fluxes of energetic species to the etch front surpass those of radicals. (For a discussion of chemically enhanced, chemical and physical sputtering see Section 7.1.2). The etching rate decreases with increasing aspect ratio due to decreased fluxes of energetic species and decreased power delivered by these species to the etching front as shown in Figure 7.46.

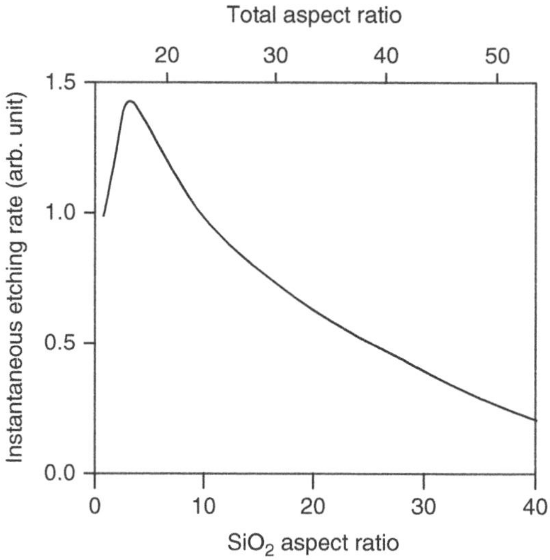

Figure 7.46 Decreasing etch rate with increasing aspect ratio for high aspect ratio SiO_2 etching. Source: Huang et al. (2019). © 2019 AIP Publishing.

This transition from chemically enhanced sputtering to chemical and physical sputtering drives the need for multistep processes. The starting chemistry deposits more CF_x polymers because these polymers contribute to the etching process and get consumed. As the aspect ratio increases, the process must be leaner because thermal neutrals will not reach the etching front anymore. They deposit on the top of the feature where they play the important role of protecting the feature sidewall from ion bombardment and where they prevent bowing. However, if their flux is too high, they will cause clogging and necking of the feature. This reduces the opening of the hole and decreases species fluxes. The effect is enhanced by the etching products, which originate at the bottom of the feature and get redeposited as they get transported out of the feature.

Kim et al. developed a semiempirical profile simulator and showed that the net deposition rate of polymer on the sidewall defines the amount of necking. Upon exposure to plasma, ion impact causes erosion of the mask and a primary facet forms as a result as shown in Figure 7.47. Mask erosion is strongly influenced by the angular dependence of the physical and chemical sputtering yields $\Gamma(\theta)$. Mask facet erosion is controlled by the maximum value of $\Gamma(\theta)$, which, as discussed before, is a function of the sputtered material and the nature of the incoming ions. The sputtering yield of high atomic mass materials peaks around normal incidence. Materials with lower atomic mass typically exhibit a maximum around 50–70° for chemically inert ions such as argon. Chemically reactive ions such as chlorine ions for silicon and CF_x for silicon oxide, however, peak around normal incidence even for low molecular materials (see Figures 7.3 and 7.4). Thus, the nature of the mask material and the plasma chemistry play an important role in the formation of the primary facet.

Kim et al. find that erosion rates and primary facet angle of photoresist (PR) show only a small influence on oxide etch profiles. This is the result of ion deflection and inelastic energy loss associated with scattering of the primary facet (Kim

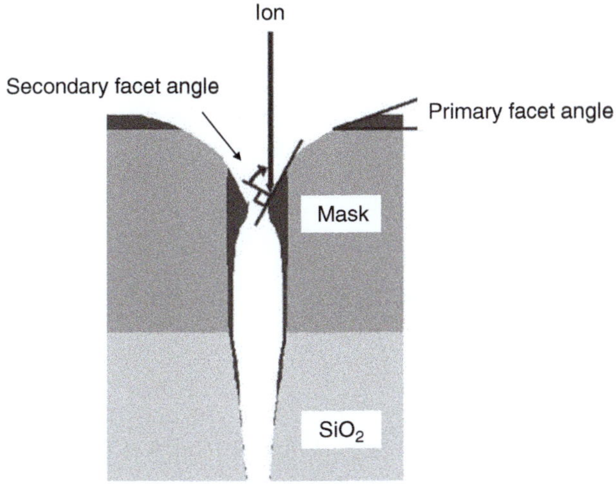

Figure 7.47 Simulated etch profiles on the top of high aspect ratio SiO_2 features. The mask material is photoresist. Source: Kim et al. (2007). © 2007 Elsevier.

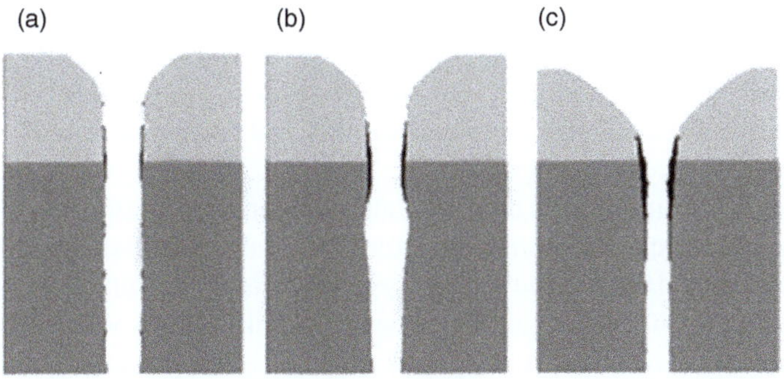

Figure 7.48 Effect of neutral depositor flux upon simulated etch profiles. Comparison of three different normalized values of neutral deposition precursor fluxes (a) 1, (b) 3, and (c) 6 using simulation times (a) 1, (b) 1.1, and (c) 1.9 adjusted to provide constant etch depth. Source: Kim et al. (2007). © 2007 Elsevier.

et al. 2007). The impact angle is too small with respect to the surface normal to allow specular reflection. Bowing is rather caused by scattering off the secondary facet, which is a facet of redeposited material. In other words, bowing is caused by ion reflections for the top of the polymer "neck" or secondary facet according to the model by Kim et al. (2007). The complex coupling between the deposition precursor flux and profile can be seen in Figure 7.48. Simulation time was adjusted to the same etch depth in each case, allowing direct comparisons (Kim et al. 2007).

The necking and bowing, which are derived from the simulated profiles, are shown in Figure 7.49 (Kim et al. 2007). Necking is defined as (top CD − neck CD)/top CD and bowing as (bow CD − top CD)/top CD. Top CD is the width of the opening at the top of the etched oxide feature. Necking increases continuously

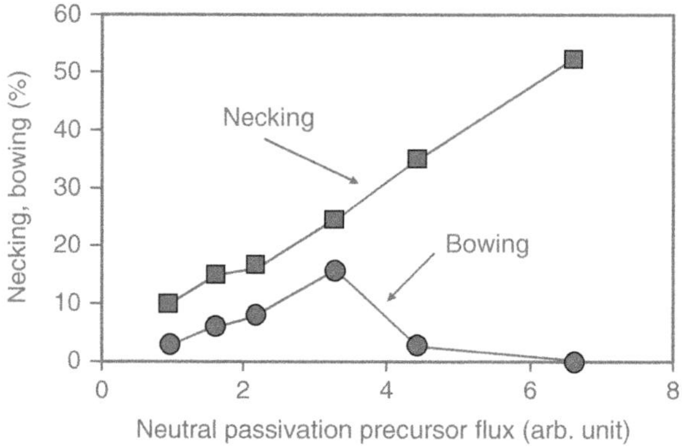

Figure 7.49 Necking and bowing of the simulated etch profiles vs variations in the neutral depositor flux. Source: Kim et al. (2007). © 2007 Elsevier.

with the neutral deposition precursor flux whereas bowing shows a maximum. When the neutral depositor flux is very low, the secondary facet angle is close to 90° and only a small number of incoming ions can be reflected from the facet. The bowing can be minimized as a result. Mask selectivity is, however, low under these conditions (Kim et al. 2007).

As the neutral depositor flux increases, necking increases. More ions are reflected and cause bowing. As neutral depositor flux increases further, the bowing starts to decrease. The sidewall below the neck becomes well passivated with polymer. This nonlinear relationship between deposition precursor flux and bowing is an important finding for practical process development.

The mask material in the simulations by Kim et al. is photoresist and the sputtering rates were derived from experiments using wafers with 220 nm holes patterned in 193 nm photoresist (Kim et al. 2007). The selectivity of photoresist is insufficient for advanced high aspect ratio SiO_2 etching, which is why hardmasks have been introduced. This raises the question of which hardmask material is the best choice and why. Are there other material properties besides etching selectivity that are important for the choice of the hardmask?

The mask material of choice for DRAM capacitor cell etching is amorphous silicon. In 3D NAND high aspect ratio memory hole etching, the mask materials are carbon and carbon compounds. Carbon and silicon are interesting materials from a sputtering point of view. Carbon has one of the highest binding energies of all materials in the form of sp^3 bonded diamond. Carbon mask materials that are used in semiconductor manufacturing are typically amorphous and have a high concentration of sp^2 bonds. A typical number for a carbon binding energy is around 6.7 eV. The sputtering rate is directly proportional to the binding energy according to Eq. (2.8). In comparison, amorphous silicon is a "weak" material with a binding energy of 4.7 eV.

Figure 7.50 shows Monte Carlo simulation results of freestanding lines of carbon and silicon after 1 keV argon ion bombardment. The model is a 2½ D model (Berry et al. 2017). It shows that carbon erodes slower than silicon but with steeper facet angle. The sputtering yield for carbon peaks at larger incoming angles. In the case of 1 keV argon, it is more than two times higher at 50° than at 0°. From this

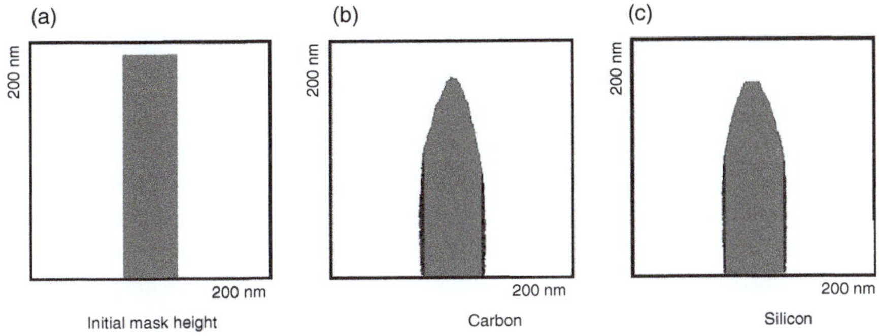

Figure 7.50 Erosion of carbon and silicon mask for 1 keV argon ion bombardment: isolated features. Source: Berry et al. (2020).

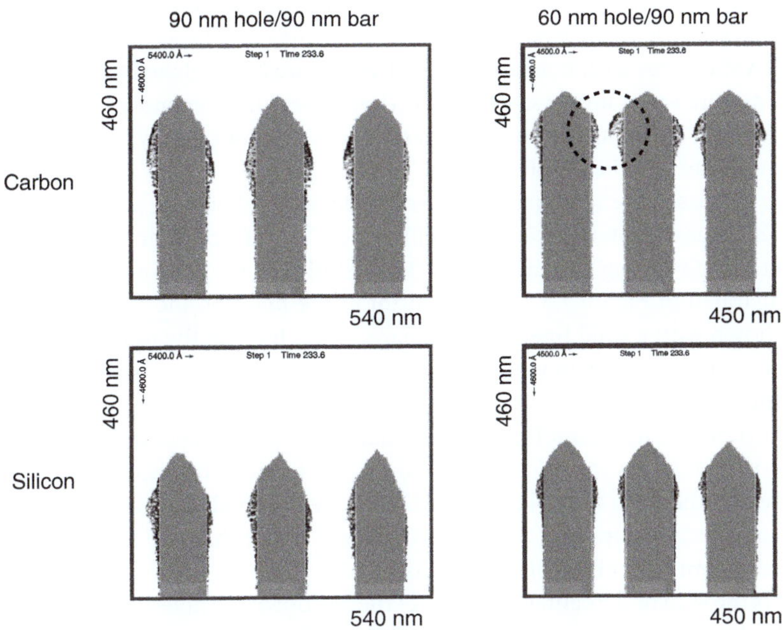

Figure 7.51 Erosion of carbon and silicon mask for 1 keV argon ion bombardment: dense features. Source: Berry et al. (2020).

simulation of the isolate mask one would conclude that carbon is the better mask material because it has higher selectivity and the steeper primary facet keeps preventing necking.

It turns out that these conclusions change when the mask forms a dense array of lines (see Figure 7.51). In the model, the sticking coefficient for the redeposition was selected to be one. Deposition from the plasma was not considered in these simulations. Necking is caused by redeposition of mask material. Figure 7.51 illustrates that the facet angle and the degree of necking are material and also spacing dependent.

For a hole diameter of 90 nm, the primary facets for carbon and silicon are similar due to mask sputtering and redeposition. The reason why the fact angle of carbon is changed in comparison to the case of the isolated mask in Figure 7.50 is that the flux toward the opposing sidewall is larger for the steeper facet angle. For a hole diameter of 60 nm, the necking of the carbon mask is larger than for silicon. The holes are almost closed. This is because the redeposited material is carbon, which has a lower sputtering rate than silicon. Necking decreases the effective diameter and increases the apparent aspect ratio of the feature. Owing to ARDE, the etching rate slows down when necking occurs. The implication is a lower selectivity for carbon despite having lower sputtering and etching rates.

The argon sputtering results ignore aspects of a chemically enhanced RIE etching process. The mechanism is chemically enhanced and chemical sputtering. Fluorine and oxygen radicals contribute to mask etching. Simultaneous polymer deposition from the gas phase contributed to the neck formation and is also sputtered (Kim et al. 2007). Nevertheless, the results correctly reflect aspects of experimental observations. DRAM devices with hole diameters of 20 nm are being etched with silicon hardmasks despite lower intrinsic selectivity. High aspect ratio memory holes with diameters of about 90 nm use carbon hardmasks because the selectivity can be leveraged while necking is still controllable.

So far, we discussed the relationship between necking and bowing as well as the role of the hardmask material. Next, we will explore the critical high aspect ratio etching metrics of twisting and distortions. The definitions for twisting and distortion are illustrated in Figure 7.52. The footprint of the holes on the bottom of the feature is shown with a solid line, and the shape of the hole on the top of the hole as a dotted line. Figure 7.52 shows the top down overlay of the shapes of the top and bottom of the hole.

Hole distortion is a deviation from the circular shape of the hole. If the hole on the bottom of the high aspect ratio hole has an elliptical shape, the distortion can be characterized by the ratio of the large axis divided by the small axis. An ellipticity of 1.1, for instance, denotes that the long axis is 10% larger than the short axis. A typical requirement is an ellipticity of less than 1.03.

Twisting is the shift of the center of gravity of the bottom of the hole with respect to the top hole. Asymmetric distortions will result in a shift of the center of gravity and

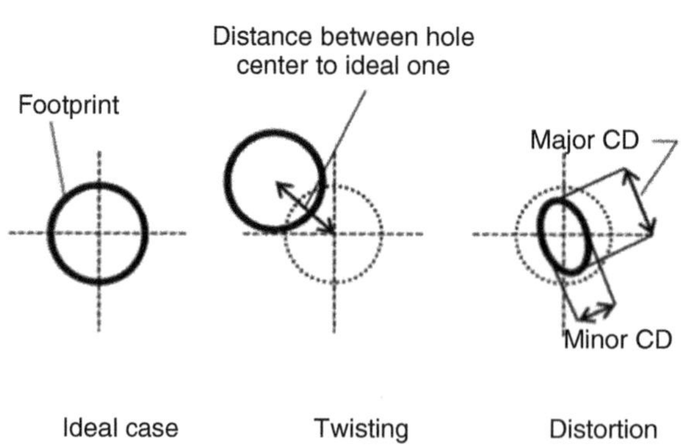

Figure 7.52 Definition of hole twisting and distortion. Source: Negishi et al. (2017). © 2017 AIP Publishing.

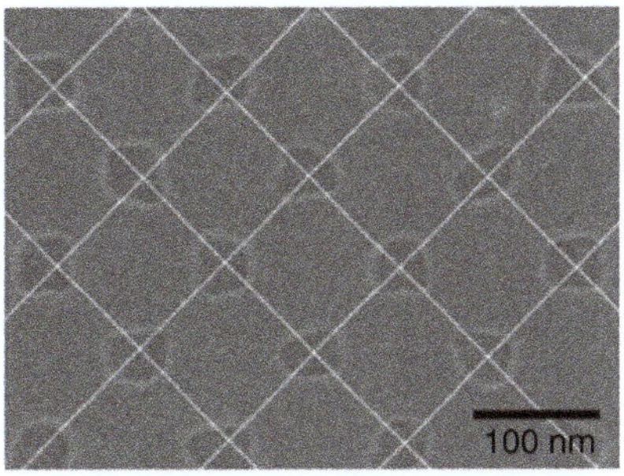

Figure 7.53 High aspect ratio bottom hole distortion and twisting. Source: Kim et al. (2015).

hence produce twisting. In this case, distortions and twisting may have the same root cause. However, twisting can be so severe that part or all of the circumference of the bottom share are outside the top shape. This case is shown in Figure 7.52. The bottom hole shape can also be distorted in this case, but the root cause may be different. Figure 7.53 shows experimental results for distortion and twisting. Twisting is visible as the shift of the hole center from the layout matrix lines (Kim et al. 2015).

One root cause hypothesis for twisting and distortions that is frequently discussed is asymmetric ion scattering. The idea is that the effect of ion scattering provides positive feedback for a hole that twists to one side. The ions would preferentially impact the less than vertical sidewall, get reflected, and impact the sidewall with more than 90° sidewall angle. This type of mechanism can be excluded. Firstly, sidewalls are nearly perfectly vertical, and the ion angular distribution is narrow (typically not more than 4° for high aspect ratio dielectric etching). Therefore, the flux of ions to the sidewall is small. Secondly, the number of scattered ions that undergo specular reflection without energy loss is small (see Figures 7.45 and 7.46). For these reasons, ion sputtering does actually provide negative feedback. The prevalence of direct line-of-sight sputtering (chemically enhanced, chemical and physical) is the reason why it is possible to etch high aspect ratio holes with reasonable repeatability.

There is evidence for two major root causes for twisting and distortions: asymmetric necking and sidewall charging. Sidewall charging is the only mechanism that can account for twisting of the profile beyond the line of sight of the ions.

Let us first discuss the evidence for necking as a root cause for twisting and bottom hole distortion. A first indication of asymmetric necking is visible in Figure 7.51. The 60 nm diameter carbon mask exhibits more redeposition on the right side (see circle). This is the result of stochastic effects and the interplay of deposition and removal.

Miyake et al. calculated the incident ion flux distributions at the hole bottom for axisymmetric and non-axisymmetric necking of a cylindrical hole (Miyake et al. 2009). In the case of non-axisymmetric necking, the imbalance of ion flux in the bottom of the hole brakes the etching symmetry in the bottom part of the hole, causing twisting as shown in Figure 7.54.

Miyake et al. found in experiments that the probability of twisting increases with more severe necking. The depth at which twisting started decreased with the minimum necking width for the same initial mask hole CD. The change in necking

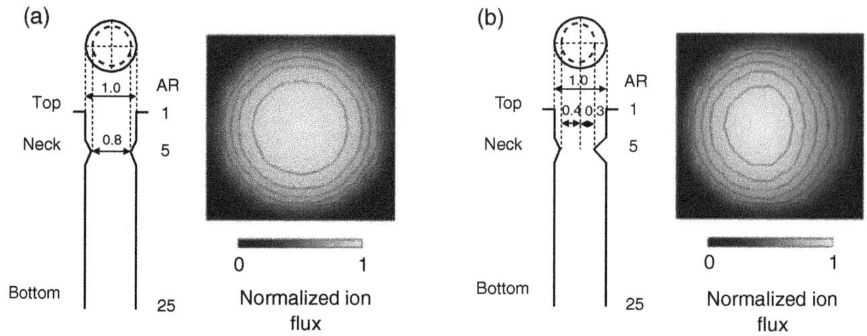

Figure 7.54 Calculated ion flux distributions at the hole bottom for axisymmetric and non-axisymmetric necking of a cylindrical hole. Source: Miyake et al. (2009). © 2009 Japan Society of Applied Physics.

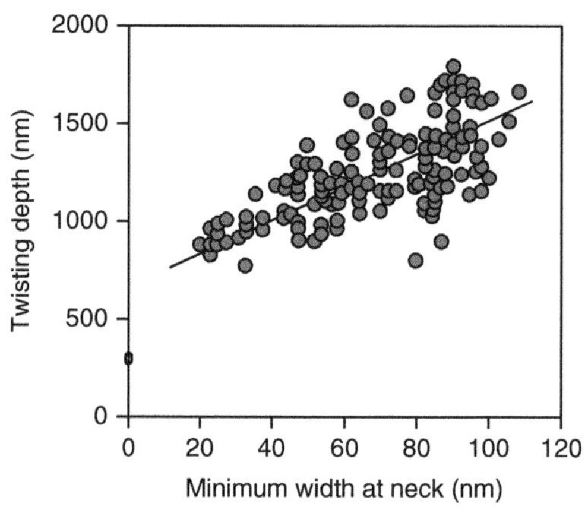

Figure 7.55 Twisting starting depth d as a function of minimum necking width w for a photoresist mask. Source: Miyake et al. (2009). © 2009 Japan Society of Applied Physics.

CD is caused by the amount of polymer precursors in the etching chemistry (see Figure 7.55). The results show that one has to be careful when increasing the flux of deposition precursors to reduce bowing as too much deposition can cause excessive necking and twisting. Multistep processes are a good approach to break the trade-off between bowing and twisting and distortions. Polymer removal steps have been demonstrated to reduce necking and to improve bottom hole distortions (Kim et al. 2015).

Sidewall roughness and its impact on pattern deformation have been studied directly with AFM on high aspect ratio trenches (see Figure 7.56). It was found that the lower spatial frequency component of the mask's sidewall roughness is amplified at the bottom region of the trench and that higher spatial frequency component of over 10 μm^{-1} disappears (Negishi et al. 2017). These results also confirm that mask deformation including necking should be considered as one of the root causes of bottom hole deformation.

Figure 7.57 confirms the relationship between necking widths for trench structures we observed for holes in Figure 7.55. The data in Figure 7.57 show the benefit of higher ion energies to extend the twisting free depth. Higher ion energies are therefore one of the main technology drivers for high aspect ratio dielectric etching.

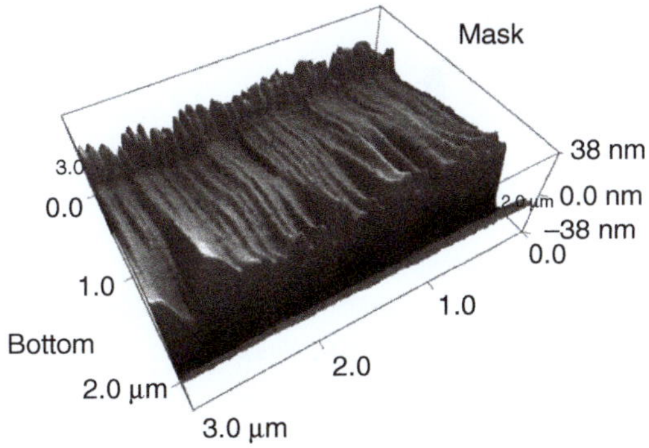

Figure 7.56 AFM measurements of high aspect ratio trench sidewall roughness from the mask to the trench bottom. Source: Negishi et al. (2017).

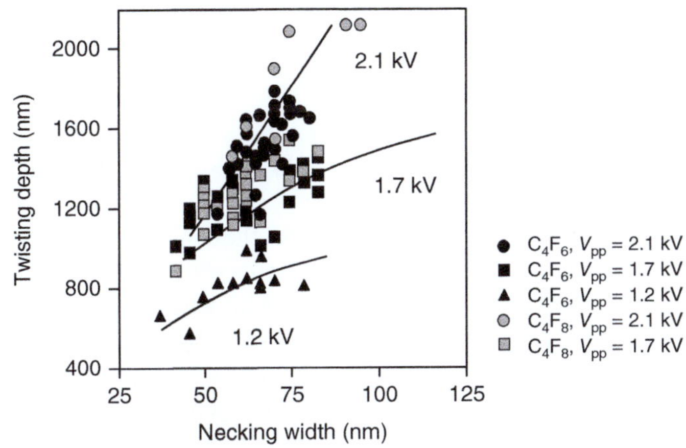

Figure 7.57 Twisting depth as a function of necking width and ion energy. Source: Negishi et al. (2017). © 2017 AIP Publishing.

Twisting and deformation out of line of sight of the ions can be explained by charging. The deflection of an ion in an electrical field is a function of the field strength and the time the ion is subjected to the electrostatic field. The deflection angle in a high aspect ratio trench can be calculated using the following equation (Berry et al. 2020):

$$\theta = a \tan\left(\frac{dV_{\text{SW}}}{wV_i}\right) \tag{7.11}$$

Here, d is the depth of the trench, w is the width of the trench, V_{SW} is the voltage on one sidewall, and V_i the acceleration voltage of the ion. According to Eq. (7.11), for an ion with 1000 eV energy, a trench with a depth of 1000 nm and a width of 50 nm, only 2 V charging of one of the sidewalls will deflect the ion by 2°. This angular deflection corresponds to 35 nm twisting of the 50 nm hole, which would be completely unacceptable for proper device performance.

Similar calculations can be conducted for charged holes. The differences between holes and trenches is that the charged and non-charged sidewall is shunted by the circular sidewall. The conductance of the sidewall polymer is therefore important

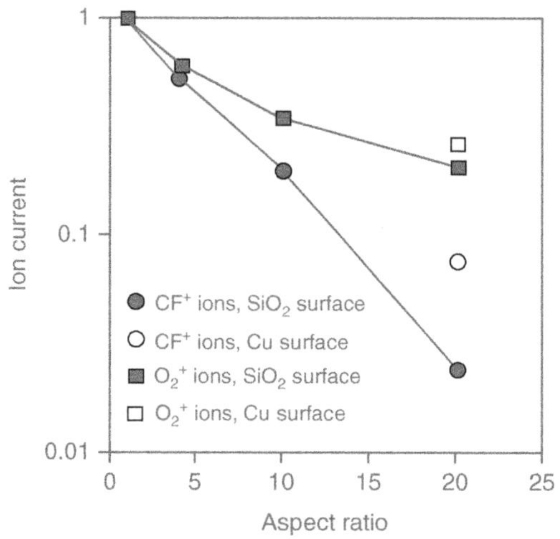

Figure 7.58 Normalized O_2^+ and CF^+ ion currents as a function of aspect ratio and sidewall coating. Source: Kurihara and Sekine (1996). © 1996 IOP Publishing.

for sidewall charging induced twisting of high aspect ratio holes in SiO_2 and other dielectric materials.

Literature sources provide inconclusive information about the conductance of CF_x polymers. Kurihara and Sekine conducted elegant experiments with capillary plates (Kurihara and Sekine 1996). They measured ion currents with lead glass capillary plates with aspect ratios of up to 20. These plates could be covered by copper to make the sidewall conductive. Figure 7.58 shows normalized O_2^+ and CF^+ ion currents as a function of aspect ratio and sidewall coating. The ions were generated by a C_4F_8 or oxygen CCP plasma above the capillary plate. The ion currents were measured below the capillary plate with a quadrupole mass spectrometer. The currents of both the non-polymerizing O_2^+ and the polymerizing CF^+ ions decrease with increasing aspect ratio and both increase when using the copper-coated capillary plate. The decrease in the ion current of the CF^+ ion is larger than that of the O_2^+ ion, which can be explained if the CF_x polymer is not conductive and enhances charging.

Shimmura et al. used an on-wafer monitoring device to measure the charging potential in SiO_2 contact holes (Shimmura et al. 2004). The potentials of the SiO_2 contact hole top and bottom surfaces were measured during argon plasma exposure with and without deposited fluorocarbon film in the holes. The fluorocarbon films were deposited by a 500 MHz CCP C_4F_8 plasma. The results revealed that the sidewall-deposited fluorocarbon film has high electric conductivity (Shimmura et al. 2004). After the experiments, the on-wafer monitor was removed from the plasma chamber and the current between the top and bottom electrodes across the holes in the monitor was measured for a voltage of 20 V. The results are depicted in Figure 7.59. They show that the current across the holes with fluorocarbon polymer coverage is 6 orders of magnitude larger than without.

Charging effects have been included in profile evolution models (Wang and Kushner 2010; Huang et al. 2019). Modeling results by Huang et al. are shown in Figure 7.60. The operating conditions were flux ratios of $Ar/C_4F_8/O_2 = 75/15/10$, a pressure of 25 mTorr, a total gas flow of 500 sccm, and RF powers of

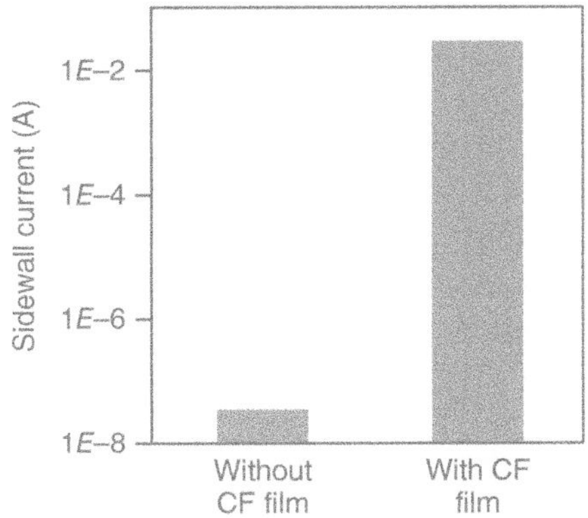

Figure 7.59 Current across contact holes for a 20 V voltage with and without deposited fluorocarbon sidewall film. Source: Shimmura et al. (2004). © 2004 AIP Publishing.

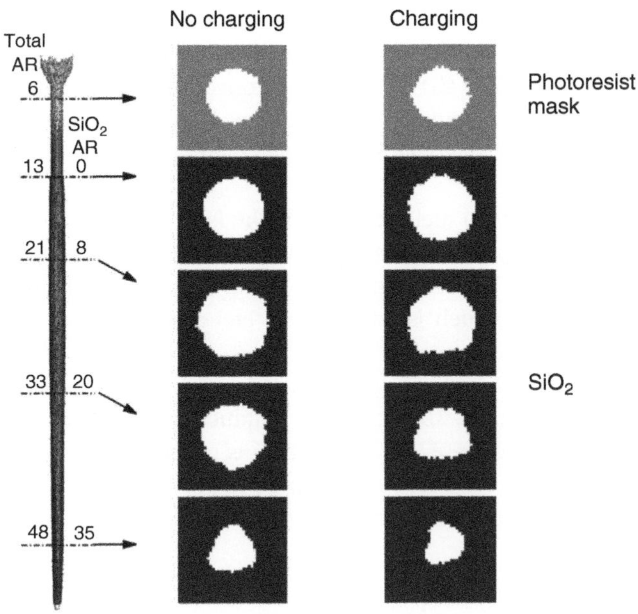

Figure 7.60 Horizontal slices through the final etching profiles with overetch, without and with charging. Source: Huang et al. (2019). © 2019 AIP Publishing.

80/10/5 MHz = 0.4/2.5/5 kW. Stochastic processes produce noncircular profiles deep in the feature. A small amount of roughness occurs on the walls of the PR due to the randomness and statistical nature of the grazing angle sputtering of the mask. This roughness results in a more diffusive scattering of grazing angle ions, which then leads to asymmetric bowing deeper in the feature. Without charging, the cross section of the final feature is largely circular with some statistical roughness for aspect ratios of up to 20.

For larger aspect ratios, the cross section of the feature deviates from a circular profile. An overetch tends to round previously noncircular features and removes the anomalies caused by the tapered etch front. The rounding of the feature by overetch is less effective with charging. Charging aggravates feature distortion by introducing more randomness and more persistence of randomness to the process (Huang et al. 2019).

7.3.3.2 High Aspect Ratio 3D NAND Etch

3D NAND devices are called vertically integrated because the memory cells are built inside a vertical channel, which is etched into a stack of alternating conductor and dielectric layers. The charges that represent the information are stored in a segment of the sidewall of this channel as shown in Figure 7.61.

Two strategies are deployed to scale the device, which means to increase memory density and to reduce the cost. The first strategy is conventional reduction of the dimensions in the *x*- and *y*-planes, which is called lateral scaling. The second approach is to scale in the vertical direction and to stack more device layers on top of each other. This drives the need for a number of critical high aspect ratio etches as shown in Figure 7.61. The most critical etch among them is the formation of the so-called channel or memory hole. The CD has to be controlled from top to bottom within a few nanometers or tens of a degree profile angle to avoid a broadening of the threshold voltage distribution of the devices along the memory hole (Oh et al. 2018).

The second critical high aspect ratio dielectric etch is the isolation trench etch, which separates blocks of memory cells from each other. The channel and the isolation trench are etched into the alternating layers. Finally, the devices have to be connected to the control logic, which is why contact etches are needed. This process forms holes in SiO_2.

There are two types of 3D NAND technologies, floating gate (FG) and replacement gate (RG) flash. In the case of a FG 3D NAND, the conducting layers are made of polysilicon and the isolating layers are SiO_2. The stack is called OPOP. RG 3D NAND devices use tungsten as the conductive layer and SiO_2 for isolation. These two materials cannot be etched together with meaningful aspect ratios. Therefore,

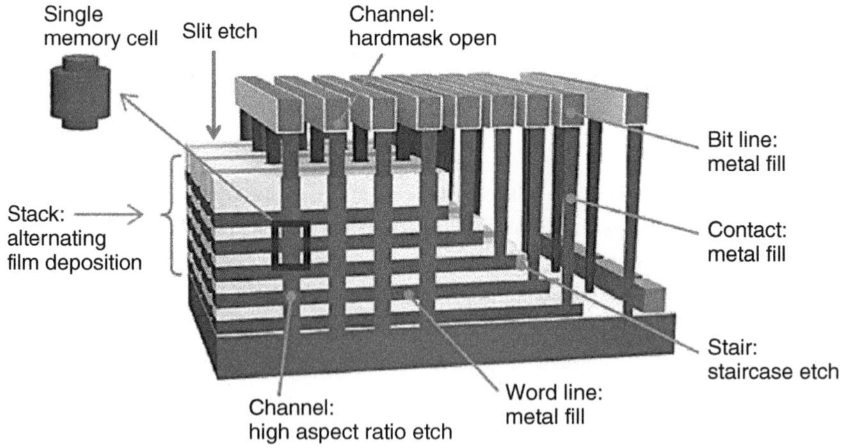

Figure 7.61 3D NAND memory architecture and critical processing steps. Source: Lill (2017).

the initial stack is SiO_2/Si_3N_4, which can be etched in situ. This stack is called ONON. After etching and some additional steps, the Si_3N_4 layer is removed via wet chemical etching and the open spaces are filled with tungsten. There is ongoing development to find a metal that has the right electrical properties and can be etched in situ with SiO_2. So far, such an OMOM stack remains elusive.

Thus, there are two types of critical channel etched in 3D NAND: OPOP and ONON channel etching. The obvious difference to the capacitor cell etch is the presence of polysilicon or Si_3N_4 layers, which drives changes in the etching chemistry. The other difference is that the CDs of the holes are different. The CDs are about 70 nm for OPOP and 100 nm for ONON while DRAM capacitor cells have a CD of 20–30 nm. The reason for larger CDs in 3D NAND devices is that there must be enough space to fit the device into the channel.

Despite significantly larger CDs, the final aspect ratios for 3D NAND are as large as for DRAM. The aspect ratios must be pushed to the limits to increase memory density. This means that the same etching challenges and solutions apply to 3D NAND channel holes as for DRAM capacitors: ARDE, mask selectivity, bowing, necking, twisting, and distortions. The OPOP and ONON stacks must be etched in situ because etching each layer and stopping on the next layer would be prohibitive for throughput and cost reasons.

OPOP channel hole etching requires a larger process change to accommodate etching of the polysilicon layers such as the addition of additional halogen-containing gases. The change from pure SiO_2 etching to ONON etching must accommodate etching of Si_3N_4. This is accomplished by adding hydrogen-containing gases to the gas mixture.

Despite process adjustments, the selectivity between the two types of layers is not always one after all other process results are dialed in. This raises the question whether etching selectivity between the two layers impacts twisting and distortions. The hypothesis is that if the etching front between a slowly and fast etching layer is broken though off center, the etching rate will accelerate in the faster etching layer. This should amplify the asymmetry and potentially introduce twisting and distortions.

Figure 7.62 shows Monte Carlo modeling results for etching of a stack of alternating materials with a selectivity of 3 : 1 (Berry et al. 2020). The ion energy is 1 keV, and the sputtering species is argon ions. Redepositing from the mask but not from the stack materials is considered. This is to emulate a chemical component of the etch. The mask is 150 nm thick and the alternating stack layers are 20 nm each. The mask CD is 20 nm. These dimensions are smaller than typical 3D NAND structures but allow the study of the trends.

The mask profile is tapered on one side by about 5° to introduce an initial asymmetry. After etching of two stack layers, the etching front is deeper on the opposite side from the tapered mask side. However, when the etching front reaches the third layer, the etching front tilt is reversed. As the etching front progresses deeper into the stack this tilt alternates stochastically. Twisting is not observed. The conclusion from this simulation is that differences in etching rate of the layers in the stack do not introduce twisting.

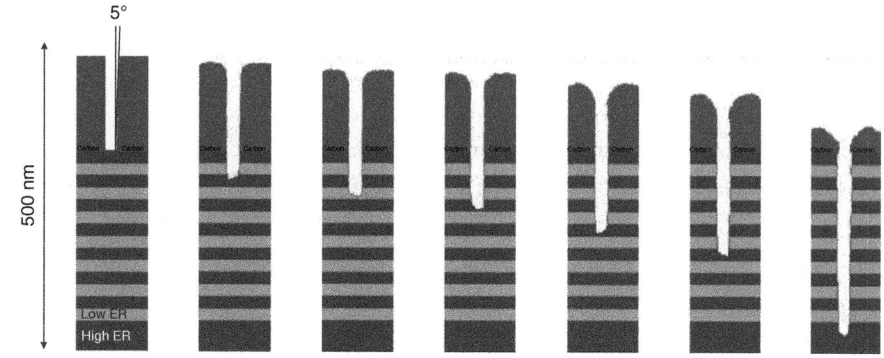

Figure 7.62 Effect of layer to layer selectivity on 3D NAND channel hole etching. Source: Berry et al. (2020).

Slit etching cuts through the same ONON or OPOP layers as the memory channel etching process. The aspect ratios are more relaxed. Twisting due to charging is a larger challenge because the opposite sidewalls are electrically not connected.

The high aspect ratio contact etch process patterns either the two dual material stack or SiO_2 depending on the type of contact. The contact to the staircase is patterned in SiO_2. Selectivity is the largest challenge because the etching depth varies step to step (see Figure 7.61).

7.3.4 Emerging Memories

One of the areas of intense research and development in the semiconductor industry is novel memories. The driving forces behind these efforts are performance and cost. Today, there is a gap between very fast and expensive DRAM and lower cost but slow 3D NAND. The information in DRAM can be addressed in bits while in 3D NAND, only entire pages or files can be moved. The information in DRAM must be refreshed every few tens of milliseconds while it is permanently stored in 3D NAND, which is considered a nonvolatile memory. The refresh cycles in DRAM consume energy and reduce the security of the data storage. Therefore, the industry is working on a bit addressable, nonvolatile memory with lower cost than DRAM. This memory will augment DRAM and 3D NAND flash and perform special tasks. This class of memory is called storage class memory (SCM).

At the same time, there are ideas to scale DRAM in the third dimension, which is impossible today because the DRAM uses transistors with single crystal silicon for speed and capacitors that are over one micrometer tall. Finally, new memory devices are needed to be embedded into logic chips. These memories are inserted into the many wiring layers and perform supportive functions. Their density must not be as compact as standalone memories. Magnetic random access memories (MRAMs) are being implemented as embedded memories replacing flash memory and static random access memory (SRAM). We will discuss patterning of MRAM devices in Chapter 8.

Table 7.1 shows a list of emerging memory devices and typical materials used in their implementation as well as their use cases.

Table 7.1 Emerging memory devices.

Emerging memory	Metals	Metal oxides
Phase change memory (PCM)	Chalcogenides ($Ge_xSb_yTe_z$, $As_xTe_yGe_z$)W	
Magnetic random access memory (MRAM)	Co, Fe, Ta, Ti, W, Pt	MgO
Resistive random access memory (ReRAM)	Chalcogenides (Ag, Cu)	TiO_2, NbO_x, HfO_x, Al_2O_3, TaO_x
Ferroelectric random access memory (FeRAM)	PZT, BST	HfO_x, $HfZr_xO_y$, $HfSi_xO_y$

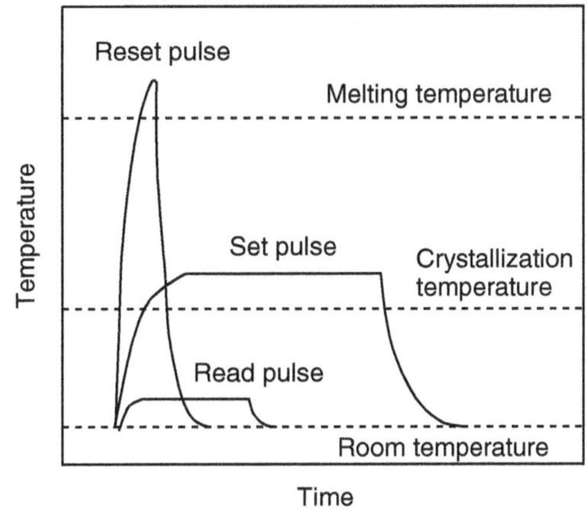

Figure 7.63 Schematic illustration of programming and reading of PCM cells. Source: Wong et al. (2010). © 2010 IEEE.

7.3.4.1 Phase Change Memory (PCM)

One of the first storage class memories that is produced in large volume is phase change memory (PCM). It is also called phase change random access memory (PCRAM). PCM leverages resistance differences between the amorphous and crystalline states in phase change materials to store information (Wong et al. 2010; Burr et al. 2010, 2016; Raoux et al. 2008). The most commonly used phase change material is an alloy of germanium, antimonium, and tellurium (GST). The material class is referred to as chalcogenides. Figure 7.63 illustrates the switching mechanism of a PCM cell. The memory is "set" by an electrical pulse that heats the material above the crystallization temperature. The crystalline material has a lower resistance than the amorphous state. To "reset" the material, a higher current is applied such that the material is melted. When the reset pulse is cut off abruptly, the material solidifies in the amorphous state, which has a high resistance.

Important material properties of the GST material are, among others, the resistance in the crystalline and amorphous state, the melting and crystallization temperatures, and the speed of these transitions. These properties depend on the composition of the GST. Etching can change the composition of the surface layers

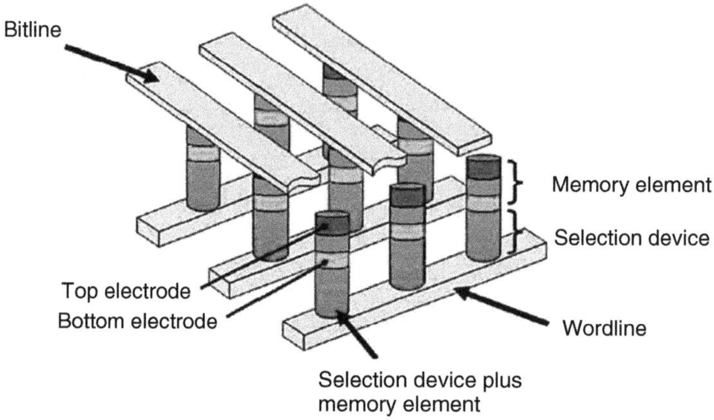

Figure 7.64 Cross-point implementation of PCM. Source: Wong et al. (2010). © 2010 IEEE.

via selective removal of volatile species and implantation of species from the plasma. Because germanium, antimony, and tellurium can oxidize in air, the handling of the wafers after etching is critical and different from other etching applications. Encapsulation with a hermetic film is needed (Shen et al. 2019).

A PCRAM product, called Optane®, with crossing wordlines and bitlines, was introduced in 2015. This cross-point architecture is schematically shown in Figure 7.64. It features a selector device in addition to the memory cell and can be stacked in the third dimension. The switching device is implemented to suppress the current through "half-selected" memory cells, which are connected to the same selected wordline or bitline. It ensures that only the memory cell at the crossing of the selected wordline and bitline is activated. There are many different selector devices under development for emerging memories (Burr et al. 2014).

The preferred switching device for PCM is called ovonic threshold switch (OTS). The composition of OTS switching materials is tailored to the specific needs, but all OTS devices are made of chalcogenides similar to the PCM cell material. One possible implementation is an alloy made of arsenic, tellurium, and germanium (Park et al. 2019). Because of the materials similarities, the challenges of etching OTS and GST without damage are comparable.

Table 7.2 shows the boiling points of halides and hydrides of elements used in phase change and ovonic threshold switching devices. All elements form volatile compounds with fluorine, chlorine, bromine, and hydrogen. The boiling points are the lowest for the hydrides and the highest for bromine. Bromine is used in RIE in the form of HBr where the effect of hydrogen dominates. The choice of the etching gas is therefore not determined by the volatility of the etching products but by the ability to preserve the stoichiometry of the chalcogenide.

The stoichiometry of GST and OTS must be preserved after etching because small changes can impact the switching temperature of GST and OTS as well as other electrical parameters. Even if only a few nanometers of the sidewall material is altered, the scalability to CDs of 20 nm and below can be at risk because diffusion

Table 7.2 Boiling points of halides and hydrides of elements used in phase change and ovonic threshold switching devices.

	Ge		Sb		Te		Se		As	
Etchant	Product	Boiling point (°C)	Product	Boiling point (°C)	Product	Boiling point (°C)	Product	Boiling point (°C)	Product	Boiling point (°C)
F	GeF_4	−36.5	SbF_5	141	TeF_4	196	SeF_4	100	AsF_5	−52.8
	GeF_2	130	SbF_3	345	TeF_6	−39			AsF_3	60.4
Cl	$GeCl_4$	86.6	$SbCl_3$	220	$TeCl_2$	328	$SeCl_4$	288	$AsCl_3$	130.2
	$GeCl_2$	450	$SbCl_5$	140	$[TeCl_4]_4$	387				
Br	$GeBr_4$	186	$SbBr_3$	286	$TeBr_2$	339	$SeBr_4$	N/A	$AsBr_3$	221
	$GeBr_2$	150			$[TeBr_4]_4$	414				
H	GeH_4	−88	SbH_3	−18	TeH_2	−1	SeH_2	−42	AsH_3	−62.5

Source: Kang et al. (2011a), Wikipedia.

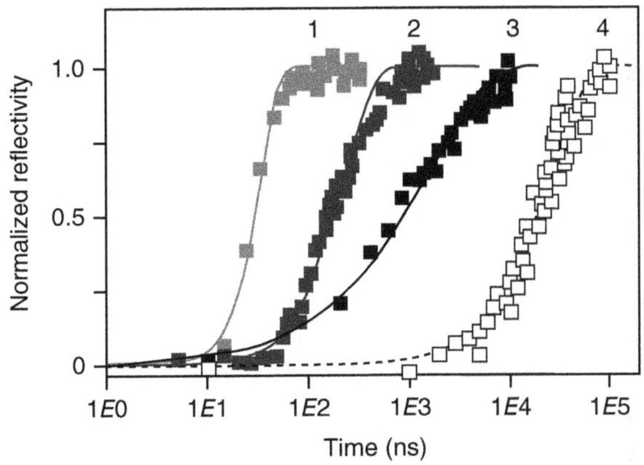

Figure 7.65 Fractional crystallization vs laser pulse heating time for nitrogen-doped GST samples as a function of surface treatment. Source: Washington et al. (2011). © 2011 AIP Publishing.

can change the bulk composition. The effect of a change in GST stoichiometry is shown in Figure 7.65. The graph shows the change in reflectivity of nitrogen-doped GST as a function of surface treatment by plasma etching. The reflectivity changes fastest for the unetched surface (1), followed by the sample etched with Ar, CHF_3, and Cl_2 (2), followed by etching the surface with the same plasma and cleaning it with a standard wet process (3). The slowest response was observed for the sample etched with the same process followed by an oxygen plasma, which represents a resist strip process (4). The treatments were applied to the horizontal surface and hence the laser probed the entire modified surface. The results illustrate empirical observations in processing of PCM devices: (i) GST and OTS materials are easily damaged by RIE processes. (ii) It is challenging to restore the surface by removing the damaged layer by means of wet cleaning because the latter introduces its own damage signature. (iii) Exposure of GST and OTS leads to surface oxidation and changes the device.

The use of halogens in RIE and neural beam etching of GST and OTS has been widely studied (Yoon et al. 2005; Kang et al. 2008, 2011a,b; Li et al. 2016; Park et al. 2019; Altieri et al. 2018; Canvel et al. 2019, 2020). A review of challenges and innovations of dry etching, wet cleaning, encapsulation, and the interaction between these processes for successful patterning of the PCM has been published by Shen et al. (2020). There are at least two approaches to etch PCM materials: (i) ensuring that etch by-products volatilities of the GST and OTS materials are as closely matched as possible. (ii) Using gas-phase passivation to protect the freshly etched sidewall from attack by reactive species (Shen et al. 2020).

Hydrides of germanium, antimony, tellurium, selenium, and arsenic have boiling points in a narrow range between −90 and 0 °C. Hydrogen can be delivered to the RIE reactor via methane and acetylene. This provides at the same time passivation of the sidewall with carbon (Altieri et al. 2018; Shen et al. 2019, 2020). The addition of argon provides a knob to control the amount of carbon passivation via dilution of the feed gas. Etching with hydrogen, methane, and acetylene does not leave halogen residues, which are known to enhance the corrosion of metals (Evans 1981).

Among the halogen gases, HBr was found to damage GST the least. This is most likely the result of bromine being less reactive and the etching effect of hydrogen. The highest halogen surface concentration was found for CF_4 (Kang et al. 2011a; Li et al. 2016). CF_4 gave more vertical profiles compared to chlorine because of the passivating effect of carbon (Kang et al. 2011a). Energy-dispersive X-ray spectroscopy (EDX) of etched GST lines reveals a high germanium concentration at the surface followed by a depleted area (Li et al. 2016). This is the result of germanium being more reactive with halogens than antimony and tellurium. Severe depletion of germanium and antimony after etching with a chlorine ICP plasma was also observed by Kang et al. as shown in Figure 7.66. This negative effect was suppressed when using a chlorine neutral beam. The reduced degradation of the neutral beam etched GST sidewall is believed to be due to the higher ratio of energetic vertical particle flux to random halogen radical flux compared to ICP plasma etching (Kang et al. 2011b). The neutral beam etching rate was 1 order of magnitude lower compared to RIE.

OTS faces similar damage challenges as GST. The preferred etching gases are methane and HBr. When etching OTS with fluorine- or chlorine-containing gases, halogenation of germanium was found while arsenic and tellurium remained in the metallic state (Park et al. 2019).

The etched GST and OTS surfaces are prone to oxidation in atmosphere. Figure 7.67 shows oxygen concentration on a GST surface as a function of time and oxygen pressure (Shen et al. 2019). The GST surface was cleaned with an argon ion

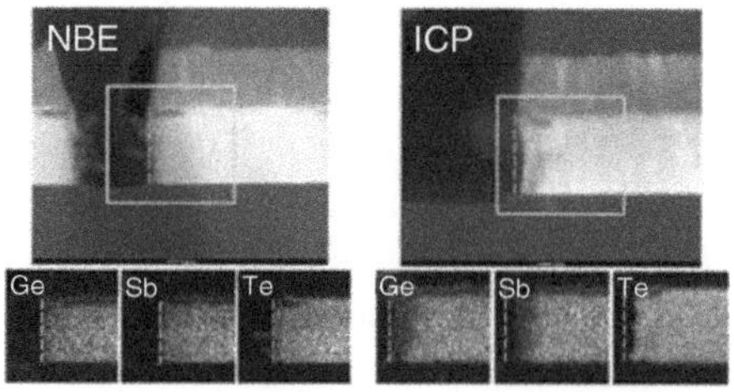

Figure 7.66 Cross-sectional images and the TEM/EDX images of the etched sidewall of GST etched with chlorine plasma and chlorine neutral beam. Source: Kang et al. (2011b).

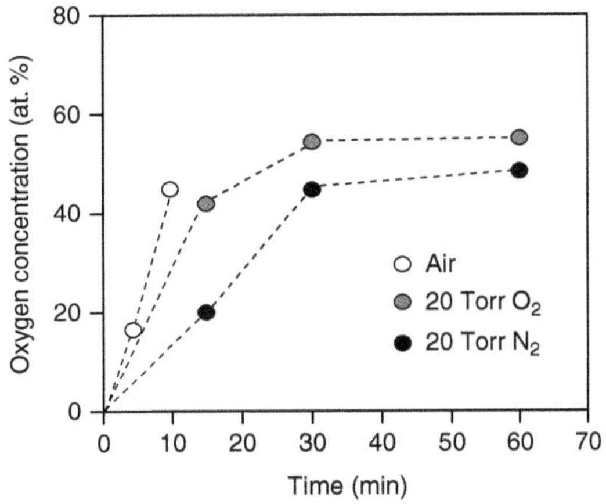

Figure 7.67 Oxygen concentration at the surface of a cleaned GST sample as a function of exposure time and oxygen pressure. Source: Shen et al. (2020).

beam. The data shows that GST oxidizes fastest in air, followed by 20 Torr oxygen and 20 Torr nitrogen. The background pressure of the experimental system was in the millitorr range, which is also the oxygen pressure in the nitrogen experiment. The conclusions from the experiment are as follows: (i) GST oxidizes very fast if exposed to even low oxygen concentrations. The fast saturation of the process suggests a CM oxidation mechanism (see Section 2.7). (ii) Nitrogen atmosphere provides some benefit in suppressing GST oxidation. Nitrogen purging of the wafers during transport after etching is a method used in production. (iii) Humidity in air accelerates oxidation.

The accelerating effect of humidity in air is illustrated in Figure 7.68 (Shen et al. 2019). When the humidity in air is increased and the exposure time is extended, the concentration of tellurium in the oxidized state increases. This is a sign of very deep oxidation as tellurium does not easily oxidize in dry oxygen (Gourvest et al. 2012; Yashina et al. 2008). Deep oxidation of GST and OTS by humid air can be effectively suppressed by handling of the wafers in a nitrogen atmosphere after etching (Shen et al. 2019).

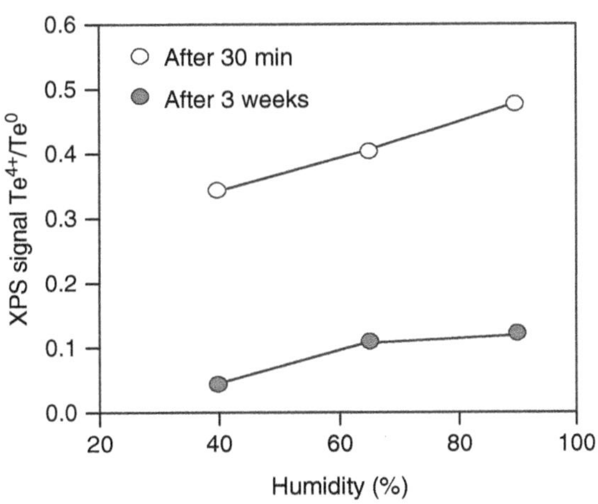

Figure 7.68 GST oxidation in air as function of humidity during the first 30 minutes of air exposure. Source: Shen et al. (2019). © 2018 IOP Publishing.

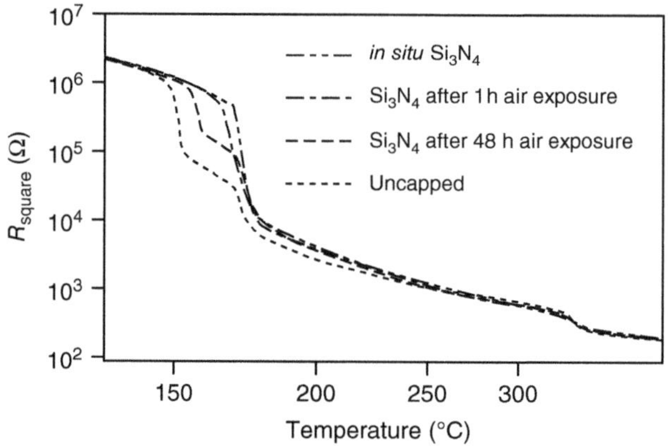

Figure 7.69 Resistance as a function of temperature in 100 nm thick $Ge_2Sb_2Te_5$ films left uncapped, capped in situ with Si_3N_4, or capped with SiN after exposure to air for either 1 or 48 hours. Source: Noe et al. (2018).

Oxidation shifts the onset of crystallization of amorphous GST to higher temperatures as shown in Figure 7.69. Furthermore, in the absence of oxidation, nucleation occurs inside the volume of the $Ge_2Sb_2Te_5$ film, while in oxidized films heterogeneous nucleation occurs at the oxidized surface. Therefore, oxidation of PCM devices must be avoided, which becomes a larger challenge with shrinking device sizes.

Capping or encapsulation with CVD Si_3N_4 is an effective method to freeze the composition. As PCM devices shrink, the method of choice becomes low-temperature Si_3N_4 ALD. Before encapsulation, the PCM device may be cleaned to remove etching residues (Shen et al. 2019, 2020).

7.3.4.2 ReRAM

Resistive random access memory (ReRAM) devices can be grouped into conductive bridge random access memory (CBRAM) and metal oxide resistive random access memory (OxRAM) (Yang et al. 2013). The switching mechanism of CBRAM devices is based on the mobility of electrochemically active metallic cations, for instance copper, in an electrolyte. Examples for electrolytes are sulfides (Ag-doped Ge_xS_x, As_2S_3, Cu_2S, $Zn_xCd_{1-x}S$), iodides (AgI, $RbAg_4I_5$), selenides (Ag-doped Ge_xSe_y), tellurides (Ge_xTe_y), and ternary chalcogenides ($GeSb_xTe_y$). The counter-electrode is usually an electrochemically inert metallic material, such as W, Pt, Au, Mo, Co, Cr, Ru, Ir, doped poly-Si, TiW, or TaN (Yang et al. 2013). The combination of hard-to-etch electrode materials and potentially damage-sensitive electrolytes makes etching of CBRAM stacks a formidable challenge. For large CD devices, such as embedded memories, argon sputtering in an RIE or IBE tool can be used. The typical thickness of a CBRAM memory stack is around 30 nm and sputtering will lead to tapered profiles and CD gain.

OxRAM devices are much more amiable to etching because the switching materials are metal oxides such as TiO_x, ZrO_x, HfO_x, VO_x, NbO_x, TaO_x and others. The fact that the metals are in the oxidized state suggests that they should generally react with halogens and form volatile by-products. The stack thickness is in the range of only 15–20 nm. Etching of OxRAM devices in a conventional planar device therefore does not create unsurmountable challenges. For many of these metal oxides, thermal ALE processes are already known (see Table 4.3). This opens the way to integrate OxRAM devices on the sidewalls of 3D structures using ALD and ALE using the concept shown in Figure 4.20.

Problems

P7.1 Explain the similarities and differences between chemical and chemically assisted sputtering.

P7.2 Explain the root causes for ARDE for hypothetical pure chemical and chemically assisted sputtering etching processes.

P7.3 Using Eq. (7.4), calculate a normalized etching rate as a function of the ratio between ion and neutral flux and discuss process sensitivity. Assume that the total flux is constant. What are the implications for ARDE?

P7.4 Does equation 7.4 contain contributions from pure neutral or radical etching?

P7.5 Explain the similarities and differences between directional ALE with a Cl_2/Ar^+ process and RIE with bias pulsing using a mixed Cl_2/Ar chemistry.

P7.6 Using Figures 7.3 and 7.4, discuss profile implications of the angular distribution of chemical sputtering.

P7.7 What is the implication of the formation of a mixed layer on intrinsic selectivity of RIE?

P7.8 Using Eq. (2.2), explain why temperature is an effective knob to compensate ERNU caused by neutral flux nonuniformities in RIE reactors.

P7.9 Explain the relationship between sidewall passivation etching resistance and CD microloading.

P7.10 Why is sidewall scattering a critical mechanism for RIE of high aspect ratio features?

References

Altieri, N.D., Chen, J.K.C., Minardi, L., and Chang, J.P. (2017). Review Article: Plasma-surface interactions at the atomic scale for patterning metals. *J. Vac. Sci. Technol., A* 35: 05C203 1–12.

Altieri, N.D., Chen, E., Fong, S. et al. (2018). Plasma processing of phase change materials for PCRAM. *AVS 65-th International Symposium and Exhibition*, Long Beach, CA, USA.

Arnold, J.C. and Sawin, H.H. (1991). Charging of pattern features during plasma etching. *J. Appl. Phys.* 70: 5314–5317.

Azarnouche, L., Pargon, E., Menguelti, K. et al. (2013). Benefits of plasma treatments on critical dimension control and line width roughness transfer during gate patterning. *J. Vac. Sci. Technol., B* 31: 012205 1–11.

Bartha, J.W., Greschner, J., Puech, M., and Maquin, P. (1995). Low temperature etching of Si in high density plasma using SF_6/O_2. *Microelectron. Eng.* 27: 453–456.

Bell, F. and Joubert, O. (1997). Polysilicon gate etching in high density plasmas. V. Comparison between quantitative chemical analysis of photoresist and oxide masked polysilicon gates etched in $HBr/Cl_2/O_2$ plasmas. *J. Vac. Sci. Technol., B* 15: 88–97.

Berry, I.L., Park, J.C., Kim, J.K. et al. (2017). Patterning of embedded STT-MRAM devices: challenges and solutions. *ECS and SMEQ Joint International Meeting 2018*, Cancun, Mexico.

Berry, I.L., Kanarik, K.J., Lill, T. et al. (2018). Applying sputtering theory to directional atomic layer etching. *J. Vac. Sci. Technol., A* 36: 01B105 1–7.

Berry, I.L., Kim, Y., and Lill, T. (2020). High aspect ratio etch challenges and co-optimized etch and deposition solutions. *Seminar at KAIST*, June 2020.

Burr, G.W., Breitwisch, M.J., Franceschini, M. et al. (2010). Phase change memory technology. *J. Vac. Sci. Technol., B* 28: 223–262.

Burr, G.W., Shenoy, R.S., Virwani, K. et al. (2014). Access devices for 3D crosspoint memory. *J. Vac. Sci. Technol., B* 32: 040802 1–22.

Burr, G.W., Brightsky, M.J., Sebastian, A. et al. (2016). Recent progress in phase-change memory technology. *IEEE J. Emerging Sel. Top. Circuits Syst.* 6: 146–162.

Canvel, Y., Lagrasta, S., Boixaderas, C. et al. (2019). Study of Ge-rich GeSbTe etching process with different halogen plasmas. *J. Vac. Sci. Technol., A* 37: 031302 1–9.

Canvel, Y., Lagrasta, S., Boixaderas, C. et al. (2020). Modification of Ge-rich GeSbTe surface during the patterning process of phase-change memories. *Microelectron. Eng.* 221: 111183 1–9.

Chang, J.P. and Sawin, H.H. (1997). Kinetic study of low energy ion-enhanced polysilicon etching using Cl, Cl_2, and Cl^+ beam scattering. *J. Vac. Sci. Technol., A* 15: 610–615.

Chang, J.P., Arnold, J.C., Zau, G.C.H., and Shin, H.S. (1997). Kinetic study of low energy argon ion-enhanced plasma etching of polysilicon with atomic / molecular chlorine. *J. Vac. Sci. Technol., A* 15: 1853–1863.

Chang, J.P., Mahorowala, A.P., and Sawin, H.H. (1998). Plasma-surface kinetics and feature profile evolution in chlorine etching of polysilicon. *J. Vac. Sci. Technol., A* 16: 217–224.

Cho, B.O., Hwang, S.W., Lee, G.R., and Moon, S.H. (2000). Angular dependence of SiO_2 etching in a fluorocarbon plasma. *J. Vac. Sci. Technol., A* 18: 2791–2798.

Choi, S.K. and Han, C.H. (1998). Low temperature copper etching using an inductively coupled plasma with ultraviolet light irradiation. *J. Electrochem. Soc.* 145: L37–L39.

Coburn, J.W. (1994a). Surface-science aspects of plasma-assisted etching. *Appl. Phys. A* 59: 451–458.

Coburn, J.W. (1994b). Ion-assisted etching of Si with Cl_2: the effect of flux ratio. *J. Vac. Sci. Technol., B* 12: 1384–1389.

Coburn, J.W. and Winters, H.F. (1979a). Ion- and electron-assisted gas-surface chemistry – an important effect in plasma etching. *J. Appl. Phys.* 50: 3189–3196.

Coburn, J.W. and Winters, H.F. (1979b). Plasma etching – a discussion of mechanisms. *J. Vac. Sci. Technol.* 16: 391–403.

Coburn, J.W. and Winters, H.F. (1989). Conductance considerations in the reactive ion etching of high aspect ratio features. *Appl. Phys. Lett.* 55: 2730–2732.

Cunge, G., Inglebert, R.L., Joubert, O., and Vallier, L. (2002). Ion flux composition in $HBr/Cl_2/O_2$ and $HBr/Cl_2/O_2/CF_4$ chemistries during silicon etching in industrial high-density plasmas. *J. Vac. Sci. Technol., B* 20: 2137–2148.

Desvoivres, L., Vallier, L., and Joubert, O. (2001). X-ray photoelectron spectroscopy investigation of sidewall passivation films formed during gate etch processes. *J. Vac. Sci. Technol., B* 19: 420–426.

Draeger, N. (2016). Tech Brief: FinFET Fundamentals. Lam Blog. https://blog.lamresearch.com/tech-brief-finfet-fundamentals (accessed 14 September 2020).

Eriguchi, K., Matsuda, A., Takao, Y., and Ono, K. (2014). Effects of straggling of incident ions on plasma-induced damage creation in "fin"-type field-effect transistors. *Jpn. J. Appl. Phys.* 52: 03DE02 1–6.

Evans, U.R. (1981). *An Introduction to Metallic Corrosion*, 3e. Edward Arnold Publishers Ltd.

Feil, H., Dieleman, J., and Garrison, B.J. (1993). Chemical sputtering of silicon related to roughness formation of a Cl-passivated Si surface. *J. Appl. Phys.* 74: 1303–1309.

Gerlach-Meyer, J.W. and Coburn, E.K. (1981). Ion-enhanced gas-surface chemistry: the influence of the mass of the incident ion. *Surf. Sci.* 103: 177–188.

Giapis, K.P., Scheller, G.R., Gottscho, R.A. et al. (1990). Microscopic and macroscopic uniformity control in plasma etching. *J. Appl. Phys.* 57: 983–985.

Goldfarb, D.L., Mahorowala, A.P., Gallatin, G.M. et al. (2004). Effect of thin-film imaging on line edge roughness transfer to underlayers during etch processes. *J. Vac. Sci. Technol., B* 22: 647–653.

Gottscho, R.A., Jurgensen, C.W., and Vitkavage, D.J. (1992). Microscopic uniformity in plasma etching. *J. Vac. Sci. Technol., B* 10: 2133–2147.

Gottscho, R.A., Cooperberg, D., and Vahedi, V. (1999). The black box illuminated. *Workshop on Frontiers in Low Temperature Plasma Diagnostics III (LTPD)*, Saillon, Switzerland.

Gou, F., Neyts, E., Eckert, M. et al. (2010). Molecular dynamics simulations of Cl^+ etching on a Si(100) surface. *J. Appl. Phys.* 107: 113305 1–6.

Gourvest, E., Pelissier, B., Vallee, C. et al. (2012). Impact of oxidation on $Ge_2Sb_2Te_5$ and GeTe phase-change properties. *J. Electrochem. Soc.* 159: H373–H377.

Graves, D.B. and Humbird, D. (2002). Surface chemistry associated with plasma etching processes. *Appl. Surf. Sci.* 192: 72–87.

Guo, W. and Sawin, H.H. (2009). Modeling of the angular dependence of plasma etching. *J. Vac. Sci. Technol., B* 27: 1326–1336.

Hayashi, H., Kurihara, K., and Sekine, M. (1996). Characterization of highly selective SiO_2/Si_3N_4 etching of high-aspect ratio holes. *Jpn. J. Appl. Phys.* 35: 2488–2493.

Helot, M., Chevolleau, T., Vallier, L. et al. (2006). Plasma etching of HfO_2 at elevated temperatures in chlorine-based chemistry. *J. Vac. Sci. Technol., A* 24: 30–40.

Huang, S., Huard, C., Shim, S. et al. (2019). Plasma etching of high aspect ratio features in SiO_2 using $Ar/C_4F_8/O_2$ mixtures: a computational investigation. *J. Vac. Sci. Technol., A* 37: 031304 1–26.

Huard, C.M., Zhang, Y., Sriraman, S. et al. (2017). Role of neutral transport in aspect ratio dependent plasma etching of three-dimensional features. *J. Vac. Sci. Technol., A* 35: 05C301 1–18.

Huard, C.M., Sriraman, S., Paterson, A., and Kushner, M.J. (2018). Transient behavior in quasi-atomic layer etching of silicon dioxide and silicon nitride in fluorocarbon plasmas. *J. Vac. Sci. Technol., A* 36: 06B101 1–25.

Hudson, E., Srivastava, A., Bhowmick, R. et al. (2014). Highly selective atomic layer etching of silicon dioxide using fluorocarbons. *AVS Symposium*, paper PS2+TF-ThM2, November 2014.

Iacopi, F., Choi, J.H., Terashima, K. et al. (2011). Cryogenic plasmas for controlled processing of nanoporous materials. *Phys. Chem. Chem. Phys.* 13: 3634–3637.

Ingram, S.G. (1990). The influence of substrate topography on ion bombardment in plasma etching. *J. Appl. Phys.* 68: 500–504.

Ito, H., Kuwahara, T., Higuchi, Y. et al. (2013). Chemical reaction dynamics of SiO_2 etching by CF_2 radicals: tight-binding quantum chemical molecular dynamics simulations. *Jpn. J. Appl. Phys.* 52: 026502 1–9.

Ito, H., Kuwahara, T., Kawaguchi, K. et al. (2014). Tight-binding quantum chemical molecular dynamics simulations of mechanisms of SiO_2 etching processes for CF_2 and CF_3 radicals. *J. Phys. Chem. C* 118: 21580–21588.

Jang, J.K., Taka, H.W., Yang, K.C. et al. (2019). Etch damage reduction of ultra low-k dielectric by using pulsed plasmas. *ECS Trans.* 89: 79–86.

Jin, W., Vitale, S.A., and Sawin, H.H. (2002). Plasma–surface kinetics and simulation of feature profile evolution in Cl_2+HBr etching of polysilicon. *J. Vac. Sci. Technol., A* 20: 2106–2114.

Joubert, O., Oehrlein, G.S., and Surendra, M. (1994). Fluorocarbon high density plasma. VI. Reactive ion etching lag model for contact hole silicon dioxide etching in an electron cyclotron resonance plasma. *J. Vac. Sci. Technol., A* 12: 665–670.

Joubert, O., Legouil, A., Ramos, R. et al. (2006). Plasma etching challenges of new materials involved in gate stack patterning for sub 45 nm technological nodes. *209th ECS Meeting*, Denver, CO, USA.

Kang, S.K., Oh, J.S., Park, B.J. et al. (2008). X-ray photoelectron spectroscopic study of $Ge_2Sb_2Te_5$ etched by fluorocarbon inductively coupled plasmas. *Appl. Phys. Lett.* 93: 043126 1–3.

Kang, S.K., Jeon, M.H., Park, J.Y. et al. (2011a). Etch damage of $Ge_2Sb_2Te_5$ for different halogen gases. *Jpn. J. Appl. Phys.* 50: 086501 1–4.

Kang, S.K., Jeon, M.H., Park, J.Y. et al. (2011b). Effect of halogen-based neutral beam on the etching of $Ge_2Sb_2Te_5$. *J. Electrochem. Soc.* 158: H768–H771.

Kiehlbauch, M.W. and Graves, D.B. (2003). Effect of neutral transport on the etch product lifecycle during plasma etching of silicon in chlorine gas. *J. Vac. Sci. Technol., A* 21: 116–126.

Kim, D., Hudson, E.A., Cooperberg, D. et al. (2007). Profile simulation of high aspect ratio contact etch. *Thin Solid Films* 515: 4874–4878.

Kim, J.K., Lee, S.H., Cho, S.I., and Yeom, G.Y. (2015). Study on contact distortion during high aspect ratio contact SiO_2 etching. *J. Vac. Sci. Technol., A* 33: 021303 1–6.

Kanarik, K.J., Tan, S., and Gottscho, R.A. (2018). Atomic Layer Etching: Rethinking the art of etching. *J. Phys. Chem. Lett.* 9: 4814–4821.

Kurihara, K. and Sekine, M. (1996). Plasma characteristics observed through high-aspect-ratio holes in C_4F_8 plasma. *Plasma Sources Sci. Technol.* 5: 121–125.

Layadi, N., Donnelly, V.M., and Lee, J.T.C. (1997). Cl_2 plasma etching of Si(100): nature of the chlorinated surface layer studied by angle-resolved X-ray photoelectron spectroscopy. *J. Appl. Phys.* 81: 6738–6748.

Lee, Y.H. and Chen, M.M. (1986). Silicon doping effects in reactive plasma etching. *J. Vac. Sci. Technol., B* 4: 468–475.

Lee, C.G.N., Kanarik, K.J., and Gottscho, R.A. (2014). The grand challenges of plasma etching: a manufacturing perspective. *J. Phys. D: Appl. Phys.* 47: 273001 1–9.

Li, J., Xia, Y., Liu, B. et al. (2016). Direct evidence of reactive ion etching induced damages in $Ge_2Sb_2Te_5$ based on different halogen plasmas. *Appl. Surf. Sci.* 378: 163–166.

Lill, T. (2017). Critical challenges and solutions in 3D NAND volume manufacturing. *Flash Memory Summit*, Santa Clara, CA, USA.

Lill, T., Yuen, S., Ameri, F. et al. (2001). Advanced gate etching: stopping on very thin gate oxides. *Proceeding of the 1st International Conference on Semiconductor Technology*, vol. 2, Shanghai, May 2001, pp. 548–557.

Lill, T., Kamarthy, G., Eppler, A. et al. (2014). Advanced patterning: plasma etch challenges and solutions. *Presentation at the SPIE Advanced Lithography Conference*, San Jose, CA, USA.

Lill, T., Kanarik, K.J., Tan, S. et al. (2018). Benefits of atomic layer etching (ALE) for material selectivity. *ALD/ALE conference 2018*, Incheon, Korea.

Lionti, K., Volksen, W., Magbitang, T. et al. (2015). Toward successful integration of porous low-k materials: strategies addressing plasma damage. *ECS J. Solid State Sci. Technol.* 4: N3071–N3083.

Lutker-Lee, K.M., Lu, Y.T., Lou, Q. et al. (2019). Low-k dielectric etch challenges at the 7 nm logic node and beyond: continuous-wave versus quasiatomic layer plasma etching performance review. *J. Vac. Sci. Technol., A* 37: 011001 1–9.

Marchak, N. and Chang, J.P. (2011). Perspectives in nanoscale plasma etching: what are the ultimate limits? *J. Phys. D* 44: 174011 1–11.

Martin, R.M., Blom, H.-O., and Chang, J.P. (2009). Plasma etching of Hf-based high-*k* thin films. Part II. Ion-enhanced surface reaction mechanisms. *J. Vac. Sci. Technol., A* 27: 217–223.

Mayer, T.M., Barker, R.A., and Whitman, L.J. (1981). Investigation of plasma etching mechanisms using beams of reactive gas ions. *J. Vac. Sci. Technol.* 18: 349–352.

Metzler, D., Bruce, R.L., Engelmann, S. et al. (2014). Fluorocarbon assisted atomic layer etching of SiO_2 using cyclic Ar/C_4F_8 plasma. *J. Vac. Sci. Technol., A* 32: 020603 1–4.

Metzler, D., Li, C., Engelmann, S. et al. (2016). Fluorocarbon assisted atomic layer etching of SiO_2 and Si using cyclic Ar/C_4F_8 and Ar/CHF_3 plasma. *J. Vac. Sci. Technol., A* 34: 01B101 1–10.

Miyake, M., Negishi, N., Izawa, M. et al. (2009). Effects of mask and necking deformation on bowing and twisting in high-aspect-ratio contact hole etching. *Jpn. J. Appl. Phys.* 48: 08HE01 1–5.

Mizotani, K., Isobe, M., and Hamaguchi, S. (2015). Molecular dynamics simulation of damage formation at Si vertical walls by grazing incidence of energetic ions in gate etching processes. *J. Vac. Sci. Technol., A* 33: 021313 1–6.

Negishi, N., Miyake, M., Yokogawa, K. et al. (2017). Bottom profile degradation mechanism in high aspect ratio feature etching based on pattern transfer observation. *J. Vac. Sci. Technol., B* 35: 051205 1–9.

Noe, P., Vallee, C., Hippert, F. et al. (2018). Phase-change materials for non-volatile memory devices: from technological challenges to materials science issues. *Semicond. Sci. Technol.* 33: 013002 1–32.

Oehrlein, G.S. and Lee, Y.H. (1987). Reactive ion etching related Si surface residues and subsurface damage: their relationship to fundamental etching mechanisms. *J. Vac. Sci. Technol., A* 5: 1585–1594.

Oehrlein, G.S., Phaneuf, R.J., and Graves, D.B. (2011). Plasma-polymer interactions: a review of progress in understanding polymer resist mask durability during plasma etching for nanoscale fabrication. *J. Vac. Sci. Technol., B* 29: 010801 1–34.

Oh, Y.T., Kim, K.B., Shin, S.H. et al. (2018). Impact of etch angles on cell characteristics in 3D NAND flash memory. *Microelectron. J.* 79: 1–6.

Pargon, E., Martin, M., Thiault, J. et al. (2008). Linewidth roughness transfer measured by critical dimension atomic force microscopy during plasma patterning of polysilicon gate transistors. *J. Vac. Sci. Technol., B* 26: 1011–1020.

Pargon, E., Menguelti, K., Martin, M. et al. (2009). Mechanisms involved in HBr and Ar cure plasma treatments applied to 193 nm photoresists. *J. Appl. Phys.* 105: 094902 1–11.

Park, J.W., Kim, D.S., Lee, W.O. et al. (2019). Etch damages of ovonic threshold switch (OTS) material by halogen gas based-inductively coupled plasmas. *ECS J. Solid State Sci. Technol.* 8: P341–P345.

Posseme, N., Pollet, O., and Barnola, S. (2014). Alternative process for thin layer etching: application to nitride spacer etching stopping on silicon germanium. *Appl. Phys. Lett.* 105: 051605 1–4.

Raley, A., Mack, C., Thibaut, S. et al. (2018). Benchmarking of EUV lithography line/space patterning versus immersion lithography multipatterning schemes at equivalent pitch. *Proceedings of SPIE 2018*, Volume 10809, 1080915-1.

Raoux, S., Burr, G.W., Breitwisch, M.J. et al. (2008). Phase-change random access memory: a scalable technology. *IBM Res. Dev.* 52: 465–479.

Regis, J.M., Joshi, A.M., Lill, T., and Yu, M. (1997). Reactive ion etch of silicon nitride spacer with high selectivity to oxide. *1997 IEEE/SEMI Advanced Semiconductor*

Manufacturing Conference and Workshop ASMC 97 Proceedings, September 1997, Cambridge, MA, USA.

Reynolds, G.W., Taylor, J.W., and Brooks, C.J. (1999). Direct measurement of X-ray mask sidewall roughness and its contribution to the overall sidewall roughness of chemically amplified resist features. *J. Vac. Sci. Technol., B* 17: 3420–3425.

Rudolph, U., Weikmann, E., Kinne, A. et al. (2004). Extending the capabilities of DRAM high aspect ratio trench etching. *2004 IEEE/SEMI Advanced Semiconductor Manufacturing Conference and Workshop* (IEEE Cat. No. 04CH37530), Boston, MA, USA, pp. 89–92.

Schaepkens, M., Standaert, T.E.F.M., Rueger, N.R. et al. (1999). Study of the SiO_2-to-Si_3N_4 etch selectivity mechanism in inductively coupled fluorocarbon plasmas and a comparison with the SiO_2-to-Si mechanism. *J. Vac. Sci. Technol., A* 17: 26–37.

Schwartz, G.C. and Schaible, P.M. (1983). Reactive ion etching of copper films. *J. Electrochem. Soc.* 130: 1777–1779.

Shaqfeh, E.S.G. and Jurgensen, C.W. (1989). Simulation of reactive ion etching pattern transfer. *J. Appl. Phys.* 66: 4664–4675.

Shen, M., Lill, T., Hoang, J. et al. (2019). Meeting the challenges in patterning phase change materials for next generation memory devices. *AVS Symposium 2019*, Columbus, Ohio, USA.

Shen, M., Lill, T., Altieri, N. et al. (2020). A review on recent progress in patterning phase change materials. *J. Vac. Sci. Technol., A* 38: 060802 1–17

Sherpa, S.D. and Ranjan, A. (2017). Quasi-atomic layer etching of silicon nitride. *J. Vac. Sci. Technol., A* 35: 01A102 1–6.

Shimmura, T., Suzuki, Y., Soda, S. et al. (2004). Mitigation of accumulated electric charge by deposited fluorocarbon film during SiO_2 etching. *J. Vac. Sci. Technol., A* 22: 433–436.

Shin, H., Zhu, W., Donnelly, V.M., and Economou, D.J. (2012). Surprising importance of photo-assisted etching of silicon in chlorine-containing plasmas. *J. Vac. Sci. Technol., A* 30: 021306 1–10.

Stan, G., Ciobanu, C.V., Levin, I. et al. (2015). Nanoscale buckling of ultrathin low-k dielectric lines during hard-mask patterning. *Nano Lett.* 15: 3845–3850.

Tachi, S. and Okudaira, S. (1986). Chemical sputtering by F^+, Cl^+, and Br^+ ions: reactive spot model for reactive ion etching. *J. Vac. Sci. Technol., B* 4: 459–467.

Tachi, S., Tsujimoto, K., and Okudaira, S. (1988). Low-temperature reactive etching and microwave plasma etching of silicon. *Appl. Phys. Lett.* 52: 616–618.

Tan, S., Yang, W., Kanarik, K.J. et al. (2015). Highly selective directional atomic layer etching of silicon. *ECS J. Solid State Technol.* 4: N5010–N5012.

Toyoda, N. and Ogawa, A. (2017). Atomic layer etching of Cu film using gas cluster ion beam. *J. Phys. D: Appl. Phys.* 50: 184003 1–5.

Toyoda, H., Morishima, H., Fukute, R. et al. (2004). Beam study of the Si and SiO_2 etching processes by energetic fluorocarbon ions. *J. Appl. Phys.* 95: 5172–5179.

Vallier, L., Foucher, J., Detter, X. et al. (2003). Chemical topography analyses of silicon gates etched in $HBr/Cl_2/O_2$ and $HBr/Cl_2/O_2/CF_4$ high density plasmas. *J. Vac. Sci. Technol., B* 21: 904–911.

Vitale, S.A. and Smith, B.A. (2003). Reduction of silicon recess caused by plasma oxidation during high-density plasma polysilicon gate etching. *J. Vac. Sci. Technol., B* 21: 2205–2211.

Vitale, S.A., Chae, H., and Sawin, H.H. (2001). Silicon etching yields in F_2, Cl_2, Br_2, and HBr high density plasmas. *J. Vac. Sci. Technol., A* 19: 2197–2206.

Wang, M. and Kushner, M.J. (2010). High energy electron fluxes in dc-augmented capacitively coupled plasmas. II. Effects on twisting in high aspect ratio etching of dielectrics. *J. Appl. Phys.* 107: 023309 1–11.

Washington, J.S., Joseph, E.A., Raoux, S. et al. (2011). Characterizing the effects of etch-induced material modification on the crystallization properties of nitrogen doped $Ge_2Sb_2Te_5$. *J. Appl. Phys.* 109: 034502 1–7.

Winters, H.F. and Haarer, D. (1987). Influence of doping on the etching of Si(111). *Phys. Rev. B* 36: 6613–6623.

Winters, H.F., Coburn, J.W., and Kay, E. (1977). Plasma etching – a "pseudo-black-box" approach. *J. Appl. Phys.* 48: 4973–4983.

Winters, H.F., Coburn, J.W., and Chuang, T.J. (1983). Surface processes in plasma-assisted etching environments. *J. Vac. Sci. Technol., B* 1: 469–480.

Wise, R. and Shamma, N. (2018). Low roughness EUV lithography. US Patent 9,922,839.

Wong, P., Raoux, S., Kim, S.B. et al. (2010). Phase change memory. *Proc. IEEE* 98: 2201–2227.

Wu, B., Kumar, A., and Pamarthy, S. (2010). High aspect ratio silicon etch: a review. *J. Appl. Phys.* 108: 051101 1–20.

Yang, J.J., Strukov, D.B., and Stewart, D.R. (2013). Memristive devices for computing. *Nat. Nanotechnol.* 8: 13–24.

Yashina, L.V., Püttner, R., Neudachina, V.S. et al. (2008). X-ray photoelectron studies of clean and oxidized α-GeTe(111) surfaces. *J. Appl. Phys.* 103: 094909 1–12.

Yokohama, S., Ito, Y., Ishihara, K. et al. (1995). High-temperature etching of PZY/Pt/TiN structure by high-density ECR plasma. *Jpn. J. Appl. Phys.* 34: 767–770.

Yoon, S.-M., Lee, N.-Y., Ryu, S.-O. et al. (2005). Etching characteristics of $Ge_2Sb_2Te_5$ using high-density helicon plasma for the nonvolatile phase-change memory applications. *Jpn. J. Appl. Phys.* 44: L869–L872.

8

Ion Beam Etching

So far, we have been exploring chemically assisted and chemical etching processes. Material can also be removed by pure physical sputtering. This technology is called ion beam etching (IBE). It is an important technology for patterning of nonvolatile and chemically sensitive materials.

8.1 Mechanism and Performance Metrics of Ion Beam Etching

IBE is a physical sputtering process and can be described by the mechanisms and equations 2.5, 2.6, 2.7, 2.8, and 2.9 in Section 2.5. The etching rate is a function of the ion flux, the sputtering yield, and the density of the material. The sputtering yield is a function of the ion energy, the mass of the incoming ion and the surface atoms, the impact angle, and the binding energy of the solid (see Eq. (2.8)).

To achieve the same sputtering rate in all points on the wafer, the ion flux uniformity should be good. This can be achieved by scanning an ion beam with a relatively small cross section across the wafer or by using a source with a diameter that is larger than the wafer, which has a large number of beamlets (a typical number is several thousand beamlets for a 300 mm IBE system). The latter approach has the advantage of high productivity because the entire wafer surface is etched simultaneously. The angular distribution within the beamlets must be slightly divergent to create overlapping impact areas on the wafer. This is called source divergence and its value is typically a few degrees. The design of IBE grids with good ion flux and divergence uniformity is considered an art. Besides the grid design, divergence uniformity is also influenced by the uniformity of the plasma density behind the grids. To improve ion flux uniformity, the wafer is also rotated.

To achieve vertical profiles, the wafer must be tilted and rotated. Without rotation, the profile will be asymmetric, which is typically not desired. The combined effect of tilting and rotation is that feature sidewalls that point toward the wafer edge are exposed to the plasma when they are closest to the ion source while sidewalls that point to the center of the wafer are exposed at the point of the largest distance to the source. Because the beamlets are slightly divergent, a so-called "inboard/outboard" effect can be observed. The outer feature sidewalls etch slightly

Atomic Layer Processing: Semiconductor Dry Etching Technology, First Edition. Thorsten Lill.

faster than the inner ones. This can be corrected by increasing the distance between the source and the wafer and by making sure the average angular distribution of all beamlets is as narrow as possible. This drives the need for very low vacuums to avoid gas-phase collisions between ions and neutrals and electron beam neutralization to avoid space charge effects.

Etching selectivities are generally low in IBE. They are governed by Eq. (2.8). It is possible to increase selectivity for some metals by adding oxygen neutrals to the process. This approach works if one of the metals forms preferentially an oxide with higher binding energy and lower sputtering rate than the non-oxidized metal.

IBE is prone to aspect ratio dependent etching (ARDE) because the transport of the sputtered species out of the features is a function of their aspect ratio. The sputtered species are typically atoms and are nonvolatile because there is no chemistry involved in the process. They redeposit upon impact on feature sidewalls. We discussed this effect in the context of directional atomic layer etching (ALE) in Section 6.1.1.

8.2 Applications Examples

One class of applications that remains inaccessible for reactive ion etching (RIE) is patterning of magnetic random access memory (MRAM) devices. While it is possible to oxidize the metals in typical MRAM stack with halogens (Chen et al. 2017; Altieri et al. 2019), the RIE reaction products have such high boiling points that they redeposit on the feature sidewalls causing tapered instead of vertical profiles. They also deposit on the reactor walls where they cause particles and plasma stability issues. Table 8.1 shows the boiling points of selected MRAM materials such as cobalt, iron, and nickel to illustrate this point.

What makes the application of RIE to MRAM patterning even more challenging is that the tunneling dielectric, MgO, is very sensitive to chemical damage. The

Table 8.1 Boiling points of selected MRAM materials.

Element	Compound	Boiling point (°C)
Cobalt	CoF_2	1400
	$CoCl_2$	1049
	$CoBr_2$	n/a
Iron	FeF_3	1327
	$HfCl_4$	315
	$HfBr_4$	n/a
Nickel	NiF_2	1450
	$NiCl_2$	1001
	$NiBr_2$	n/a

Source: Boiling points of selected MRAM materials, Wikipedia.

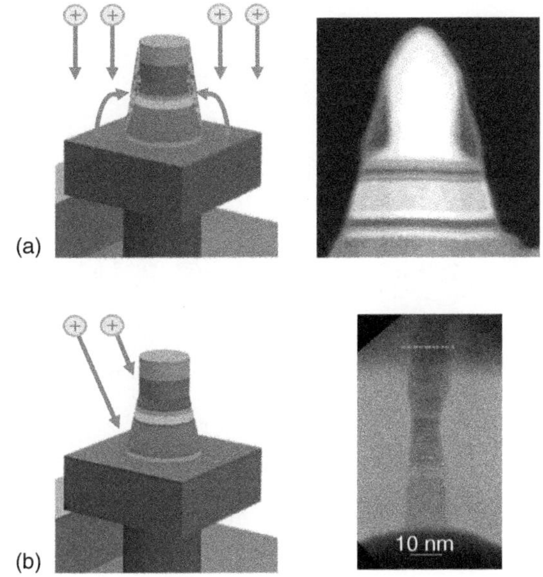

Figure 8.1 Comparison of MRAM profiles etched with RIE (a) and IBE (b). Source: IBM Corporate Website (2016).

use of halogens is not recommended for device performance reasons. Unless this requirement can be relaxed, this creates a large barrier for the use of RIE and ALE technologies. In the absence of synergy from chemical reactions, a sputtering process must be used.

The angular dependence of the sputtering yield predicts a tapered profile for normal ion incidence angle (see Figure 2.9). When the beam is tilted with respect to the wafer, the sidewall can be sputtered, and vertical profiles can be achieved. Figure 8.1 shows a comparison of MRAM profiles patterned with RIE and IBE.

IBE has a much smaller process parameter space than RIE. The parameters that can be selected are ion mass, energy, impact angle, and rotation speed. Ion energies above 1000 eV provide the best profile performance for MRAM patterning. However, this leads to intermixing of stack materials as shown in Figure 8.2.

Intermixing leads to device shorts across the MgO layer. For 100 eV Ar ions, the depth of intermixing in the middle of the MgO layer is 1 nm vs 4 nm for 1000 eV in the main etch step (see Figure 8.2). The best performance can be obtained with multistep processes, which optimize profile and short performance (Song et al. 2016).

IBE requires line of sight access to the bottom of the feature. This limits the useful wafer tilt and ion impact angles for high aspect ratio features. This will become a challenge as MRAM migrates from embedded to standalone devices in the future. While new approaches must be found for patterning of high-density MRAM devices, IBE has established itself in the semiconductor industry as the preferred approach for embedded MRAM devices. The ability to change ion impact angles and to control the ion angular distribution and energy over a wide range is attractive for other potential applications as well.

To become more widely used, IBE chambers must be able to handle reactive background gases and reactive ions to introduce chemical effects (Hrbek 1977). This is

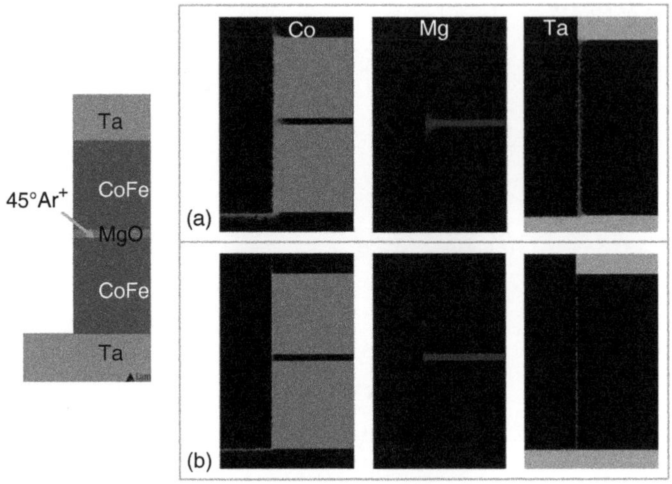

Figure 8.2 Simulated elemental mapping of intermixing after IBE for 1000 eV (a) and 100 eV (b) ion energy. Source: Lill et al. (2019).

called chemically assisted ion beam etching (CAIBE) or reactive ion beam etching (RIBE). It is interesting to note that research on CAIBE and RIBE was quite active in the 1980s just before RIE was widely implemented in the semiconductor industry (Chinn et al. 1983; Revell and Goldspink 1984). If RIBE and CAIBE find application in advanced device patterning, it would have been a long way in the making. Fundamental challenges such as grid erosion in reactive chemistries and wafer contamination by materials sputtered from the grid must be overcome to make this happen.

Problems

P8.1 Why is it necessary to tune the ion impact angle in an IBE tool? Why is it possible in IBE but not in RIE?

P8.2 Why does surface oxidation decrease the sputtering rate of some metals? Using Eq. (2.2), explain the condition for this approach to work.

P8.3 Why do IBE tools operate at very low pressures?

References

Altieri, N.D., Chen, J.K.C., and Chang, J.P. (2019). Controlling surface chemical states for selective patterning of CoFeB. *J. Vac. Sci. Technol., A* 37: 011303 1–6.

Chen, J.K.C., Altieri, N.D., Kim, T. et al. (2017). Directional etch of magnetic and noble metals. I. Role of surface oxidation states. *J. Vac. Sci. Technol., A* 35: 05C304 1–6.

Chinn, J.D., Adesida, I., and Wolf, E.D. (1983). Chemically assisted ion beam etching for submicron structures. *J. Vac. Sci. Technol., B* 1: 1028.

Hrbek, J. (1977). Sputtering of metals in the presence of reactive gases. *Thin Solid Films* 42: 185–191.

IBM Corporate Website (2016). Researchers celebrate 20th anniversary of IBM's invention of Spin Torque MRAM by demonstrating scalability for the next decade. A new mechanism is proposed for exciting the magnetic state of a ferromagnet. https://www.ibm.com/blogs/research/2016/07/ibm-celebrates-20-years-spin-torque-mram-scaling- 11-nanometers/ (accessed 4 September 2019).

Lill, T., Vahedi, V., and Gottscho, R. (2019). Etching of semiconductor devices. In: *Materials Science and Technology*. Wiley-VCH. 1–25.

Revell, P.J. and Goldspink, G.F. (1984). A review of reactive ion beam etching for production. *Vacuum* 34: 455–462.

Song, Y.J., Lee, J.H., Shin, H.C. et al. (2016). Highly functional and reliable 8Mb STT-MRAM embedded in 28 nm logic. *Proceedings of 2016 IEDM Conference*, San Francisco, CA, USA.

9

Etching Species Generation

In this chapter, we will establish a link between the etching mechanisms we studied so far and the various approaches to generate the reactive species needed to realize these etching processes.

The etching species used in thermal etching and thermal atomic layer etching (ALE) are neutral gases. They are delivered to the etching reactor from gas or liquid sources and delivery systems. We will not discuss them here.

Radical etching, reactive ion etching (RIE), and directional ALE require radicals and ions that are efficiently created by plasmas. The ions in IBE are also extracted from plasma. Therefore, this chapter provides the basics of plasmas and how they can be tuned to reach the process conditions needed for the different.

Outstanding textbooks about plasma processing with focus on plasma physics already exist. Excellent examples are "Principles of plasma discharges and materials processing" by Lieberman and Lichtenberg (2005) and "Physics of radio-frequency plasmas" by Chabert and Braithwaite (2011). It would have been fine to end this book here and to defer the reader who is interested in how etching species are generated to these textbooks. However, I believe that it is useful to summarize the plasma fundamentals into a number of essential insights and to show what they mean from an etching performance point of view. Aspects of plasma physics that are less important for etching of semiconductor devices will be omitted. Hopefully, this will allow engineers who practice dry etching to link their etching challenge with the etching mechanism and the choice of plasma generation and tuning.

9.1 Introduction of Low-Temperature Plasmas

A plasma is a quasi-neutral gaseous particle system in the form of free ions, electrons, and neutral particles. Why is plasma used in dry etching? This question seems a tautology since dry etching is frequently also called plasma etching. After plasma was discovered as the fourth state of matter, it was found that is can sputter solid-state surfaces. Later, chemically active gases were added with the result of faster etching rates. In this narrative, plasma was a tool looking for an application, which the semiconductor industry provided.

Atomic Layer Processing: Semiconductor Dry Etching Technology, First Edition. Thorsten Lill.

When one approaches dry etching from the surface mechanism point of view, it appears desirable to have a tool with independent ion, radical, and neutral beams to generate exactly the right combination of species fluxes needed to achieve the required process results. The main reason why this is not practical is cost. Dedicated sources would be needed for the different etching species. More importantly, by generating the species remotely and transporting them to the wafer surface, the fluxes would attenuate with the consequence of diminished etching rates. In contrast, the highest possible flux of all species can be obtained when bringing the plasma in direct contact with the wafer. Exceptions are remote sources that are being used in the semiconductor industry for radical and ion beam etching. We will cover these sources in Section 9.6. The separation of the plasma from the wafer is achieved via grids or tubes.

RIE is implemented with a reactive plasma in direct contact with the wafer. This maximizes the species fluxes but also introduces enormous complexity. It makes RIE a highly convoluted system with complicated surface mechanisms (see Eq. (7.4)) and strong feedback via the plasma. One of these feedback mechanisms is the reintroduction of etching products into the plasma generation region where they may get ionized and dissociated to be accelerated toward the wafer surface again (Kiehlbauch and Graves 2003). In this chapter, we will introduce other fundamental complexities that arise when a plasma is in contact with a wafer. It is very important for etching engineers to know them in order to develop processes efficiently.

The type of plasma that is used in dry etching is called low-temperature or nonequilibrium plasma (Graves and Humbird 2002; Graves and Brault 2009; Oehrlein and Hamaguchi 2018). Nonthermal or nonequilibrium plasmas emerge as a small number of positive ions and negative electrons in a sea of neutral gas. The density of neutral gas atoms, n_n, is 2–6 orders of magnitude higher than the ion and electron density, n_i and n_e, respectively (Graves and Humbird 2002). Other important parameters for typical nonequilibrium plasmas are listed in Table 9.1.

One defining characteristic of this type of plasma is that the charged species have much higher energies than neutral species, and hence the term nonequilibrium. These plasmas are generated in confined spaces with walls. Charged species are lost to the wall faster than they are able to transfer via collisions with neutrals the energy they receive from an external energy source. Hence, the neutrals are colder than the ions – they have a temperature of about 300–2000 K (Graves and Humbird 2002). In contrast, the temperatures in equilibrium plasmas range from 4000 K for easily ionizable elements such as cesium to 20 000 K for elements such as helium (Lieberman and Lichtenberg 2005). Table 9.1 lists the surface impact characteristics of nonequilibrium plasmas. While the peak power density per ion impact is very high (10^8–10^{12} W/cm^2) and causes collision cascades and sputtering, the average power delivered to the wafer is low (10^4 W/cm^2 for an ion energy of 100 eV and an ion current density of 10 mA/cm^2). Therefore, the wafer surface can be etched without excessive thermal impact (Graves and Humbird 2002). The heat generated by the plasma on the wafer must still be removed with a wafer cooling mechanism, typically an electrostatic chuck (ESC).

Table 9.1 Characteristics of nonequilibrium plasmas.

Characteristic property	Value
Gas pressure	1–1000 mTorr
Gas temperature	300–2000 K, or 0.03–0.15 eV per atom
Degree of ionization (N_i/N_n)	10^{-6} to 10^{-2}
Average electron energy	1–10 eV
Average ion energy in plasma	0.05–1 eV
Average ion energy impacting surface	10 to several 1000 eV
Typical plasma dimension	0.1–1 m
Radius of surface area impacted by ion (A_{im})	25 Å for 200 eV ion
Ion flux to surface	In the order of 100 A/m^2 or 10^{21} ion/m^2 s
Approximate time between ion impacts within A_{im}	10^{-4} s
Complete energy dissipation from single ion impact	Within 10^{-12} s
Peak power density in impacted area	10^8 to 10^{12} W/cm^2
Average power delivered to surface	104 W/cm^2 for 100 eV and 10 mA/cm^2

Source: Graves and Humbird (2002). © 2020 Elsevier.

The pressure range for nonequilibrium plasmas is 1–1000 mTorr. The reason is that the mean free path of particles decreases for higher pressures. As the pressures increases, the probability of charged particle recombination raises. This is different for equilibrium plasmas where all particles, ion, electrons, and neutrals have such a high kinetic energy that collisions do not lead to recombination and loss of charged particles.

The second reason to use low pressures in directional dry etching is the need to accelerate ions in one preferred direction. This means after they are generated in the gas phase, they need to travel to the wafer without colliding with other species. If the pressure is too high, the distance they can travel becomes too small.

The average electron energy in such a nonequilibrium plasma is 1–10 eV. Most bond energies are of a few electron volts, and ionization energies can be higher than 10 eV. The first ionization energy for helium, for instance, is 24.6 eV. The electron energy E_e is generally lower than the threshold energies for molecule dissociation, E_{diss}, and for atom and molecule ionization, E_{iz}. Despite this lack of energy, dissociation and ionization occur in nonequilibrium plasmas. The reason is that electrons have a distribution with some very high energy electrons. This distribution function is schematically shown in Figure 9.1. Here, $N_e(E_e)dE_e$ is the number of electrons per unit volume with energies between E_e and $E_e + dE_e$. Electrons in the high-energy tail can have energies higher than E_{diss} and E_{iz} and contribute to dissociation and ionization of the neutral species. The distribution in Figure 9.1 is drawn as a Maxwellian distribution of the bulk electron energy E_e, but this may not always reflect reality.

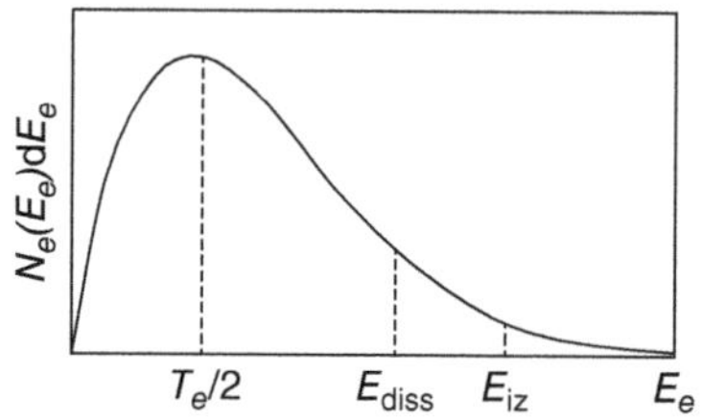

Figure 9.1 Electron energy distribution function (EEDF) for a nonequilibrium plasma.

Deviations from a Maxwellian distribution are common due to other effects in the plasma.

The energy of an electron is a function of its temperature, T_e. The relationship for a single electron is given in the following equation (compare with Eq. (7.7), which is the analog for atoms and molecules):

$$E_e = k_B T_e \tag{9.1}$$

The mean energy of an assembly of electrons with Maxwellian distribution is

$$< E_e >= \frac{3}{2} T_e \tag{9.2}$$

The electron temperature is an important plasma parameter. It characterizes the density of charged particles in the plasma or in short, the plasma density. It also determines the lowest possible ion energy that can be achieved in a plasma. This is important for the intrinsic selectivities of directional ALE and RIE. We will discuss the relationship between electron temperature and ion energy below.

Figure 9.1 shows that the ionization energy is higher than the dissociation energy for molecules. This has critical implications for what kind of chemistry is possible in a plasma. Generally, because molecules dissociate at lower energies than they ionize, molecules fragment in the plasma at the same time as they get ionized. The underlying processes are called electron impact dissociation and ionization. For instance, an oxygen molecule can dissociate into two oxygen radicals under electron impact via the following pathway:

$$O_2 + e \rightarrow O^* + O^* + e \tag{9.3}$$

At higher electron energies, electron impact can also lead to ionization:

$$O_2 + e \rightarrow O_2^+ + e \tag{9.4}$$

Dissociation and ionization can also occur simultaneously. This process is called dissociative ionization:

$$O_2 + e \rightarrow O^+ + O^* + e \tag{9.5}$$

Because ionization occurs at higher energies than dissociation and because ionization is needed to sustain a plasma, surface chemistry using complex molecules with many functional groups is hardly possible in RIE. Therefore, RIE process engineers think frequently in terms of the ratio of certain elements in the given feed gases rather than about their structure and chemical properties. For instance, the choice between C_4F_6 and C_4F_8 is largely driven by the carbon to fluorine ratio needed for a given process. The fact that C_4F_8 has a cyclic molecule while C_4F_6 is linear is less important due to dissociation in the plasma. However, what plasma offers are very reactive radicals, which are formed during electron-stimulated dissociation. The full range of chemistry possibilities can be leveraged in thermal etching and thermal ALE.

Large and complex ionic molecules can be formed via gentler electron attachment ionization. In this type of ionization, the electron is captured by the molecule to form a negative ion. For this ionization pathway, the electron energies must not exceed an energy of a few electron volts. If the energy is too high, electron attachment can also lead to dissociation. Negative ions are formed only under very special conditions, which we will discuss in Section 9.5.

Because nonequilibrium plasmas that are used in dry etching operate at low pressures, they must be confined in a vacuum chamber with a vacuum pump and walls. Owing to loss of ions and electrons at these chamber walls, the generation of new charged particles requires input of energy. This energy is typically provided by electric and magnetic fields ranging from direct current (DC) to microwave. Most industrial etching tools use radio frequency (RF) fields.

Under the influence of the electric and magnetic fields, the neutral gas can be ionized and becomes conductive due to the free electrons in the gas. By passing currents through the gas, the charged species are heated, thereby sustaining the plasma. The peak neutral gas temperature can vary from near room temperature to over 2000 K, but even when the peak gas temperature is high, the low heat capacity of a gas at low pressure means that the gas rapidly cools near walls.

A plasma that is confined by walls is called a bounded plasma system (BPS). When a plasma is bounded or confined by a solid body, the quasi-neutrality is distorted in the vicinity of the walls. This region is called space charge sheath or sheath. The formation of this region is the result of the enormous difference in mobility between electrons and ions due to their mass difference. Electrons are very fast, and they escape the plasma and move toward the walls. Ions are too slow to follow the electrons and hence a region devoid of electrons is formed, the sheath. This mechanism is schematically illustrated in Figure 9.2.

The sheath retards the electron motion out of the plasma and accelerates positive ions toward surfaces. The fact that plasmas form sheaths of thickness on the order of

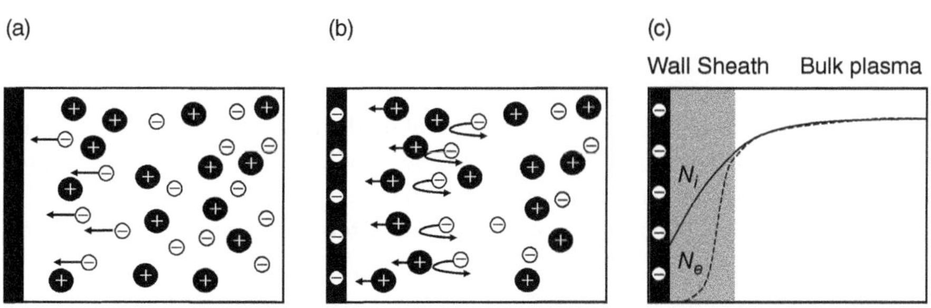

Figure 9.2 Sheath formation in a bounded plasma. (a) Electrons escape the plasma and move towards the walls. (b) The sheath retards electrons and accelerates positive ions. (c) Electron and ion density in the sheath.

hundreds of microns means that ions can be accelerated across a nearly collision-less sheath to impact a surface with little angular scattering at pressures of tens to hundreds of mTorr (Graves and Humbird 2002). This means that the ion energy toward the surface is obtained in the sheath and that ion energies are tuned by manipulating the sheath. The pre-sheath is yet another plasma region that extends beyond the sheath. It is quasi-neutral but both electron and ion densities are lower than in the bulk (Lieberman and Lichtenberg 2005).

The Debye length λ_D is the parameter that defines the width of the shield and the voltage drop across it. It is the length over which the plasma shields itself from an applied electric field. The Debye length is a function of the electron density and temperature (Lieberman and Lichtenberg 2005):

$$\lambda_D = 743\sqrt{\frac{T_e}{N_e}} \tag{9.6}$$

The Debye length equals the sheath thickness when no additional potentials are applied to the wall or electrode. For an electron temperature of 4 V and an electron density of 10^{10} cm^{-3}, the Debye length is 0.14 mm (Lieberman and Lichtenberg 2005).

According to Eq. (9.4), the sheath without applied potential is wider for plasmas with higher temperatures. This makes sense because the higher the electron temperature, the larger a potential they can overcome to escape the bulk plasma. The sheath is thinner for plasmas with higher electron densities. This means that for the hypothetical case that the plasma can be sustained via supply of external energy without generating additional potentials on the walls, increase in density tends to diminish the sheath voltage while increase in electron temperature increases it. The behavior of the electron temperature as a function of applied RF power depends on how this power is delivered to the plasma and will be discussed later.

The electron temperature also determines the sheath voltage V_{sh} (Lieberman and Lichtenberg 2005):

$$V_{\mathrm{sh}} = \frac{T_e}{2}\ln\left(\frac{M_i}{2\Pi M_e}\right) \tag{9.7}$$

Here, M_i and M_e are the mass of the ion and electron, respectively. For argon, Eq. (9.7) simplifies to (Lieberman and Lichtenberg 2005)

$$V_{\mathrm{sh,Ar}} \approx 4.7\,T_e \tag{9.8}$$

For an electron temperature of 4 eV, which is common in RIE reactors, the sheath voltage for argon is 16.4 V. This value is also called the floating potential or self-bias. It is an important parameter as it defines how low an ion energy an etching tool can achieve without powering the electrode in which the wafer rests. This is called a source-power-only regime. We will explain the terms source and bias power in Sections 9.2 and 9.3.

Self-biasing gives the argon ion an energy of 16.4 eV, which is much lower than the lower boundary of the ideal ALE window for directional ALE of silicon with Cl_2/Ar^+ (about 50 eV). It is marginal for the same ALE process when it is applied to germanium, which is about 30 eV as shown in Figure 6.7.

According to Eq. (9.8), the self-bias can be reduced by lowering the electron temperatures, which depends on the way the external energy is coupled to the plasma. Another method is to remove the plasma generation region as much as possible from the wafer surface. In other words, larger gap reactors have lower electron temperatures above the wafer because the electrons cool via inelastic collisions as they travel from the plasma generation region through the bulk plasma to the wafer. An extreme embodiment of this idea is the electron beam plasma sources (Walton et al. 2015). Walton et al. demonstrated electron temperatures between 0.3 and 1.0 eV and ion energies in the range of 1–5 eV.

Another implication of Eq. (9.8) is that the sheath voltage cannot be zero. Ion masses are larger than the mass of an electron and the electron temperature must by larger than zero to sustain a plasma. This means that any plasma will lead to ion bombardment when being in contact with the wafer. Tools for application that cannot tolerate ion bombardment because of the need for high intrinsic selectivity therefore remove the plasma even further away from the wafer. These plasmas are called remote. They feed only the neutral gas and radicals via tubes to the reactor. Another engineering embodiment is the internal grids between the plasma and the wafer. Here, the sheath is formed between the plasma and the grid and the neutrals and radicals travel through the openings in the grid toward the wafer surface. These grids must be designed properly to avoid ions and electrons from entering the lower chamber.

Most plasma processes, however, require high-energy ions for etching. For these applications, the self-bias and hence electron temperature are not very important.

9.2 Capacitively Coupled Plasmas

Let us now study how RF power is coupled into the plasma and how this changes the electron temperature, sheath thickness, and voltage. The case of driving a plasma with RF power is important for semiconductor manufacturing because wafers are generally electrically insulating due to many dielectric layers. Thus, exciting the plasma with RF provides the option to apply the RF signal to the electrode on which the wafer is placed. The electrical circuit is closed via a grounded electrode, which is electrically isolated from the powered electrode in the absence of plasma. The resulting sheath voltages in the plasma and at the electrodes for the times of the most negative and positive parts of the RF cycle are shown in Figure 9.3. The voltage of the bulk plasma with respect to ground is called plasma voltage or potential V_p. The sheath voltage on the powered electrode is the difference between the RF voltage and the sheath potential. It is largest for the negative RF cycle ($V_{sh,1}$) and close to zero for the positive cycle. We show in Figure 9.3 a very small potential for the positive and the neutral part of the cycle labeled $V_{sh,2}$. This is the self-bias as discussed in chapter 9.1 for a BSP. It can be neglected in practical applications because it is much smaller than $V_{sh,1}$.

The plasma sheath acts like a capacitance. This allows to calculate the sheath voltage, which is induced by the RF signal (in addition to the self-bias we

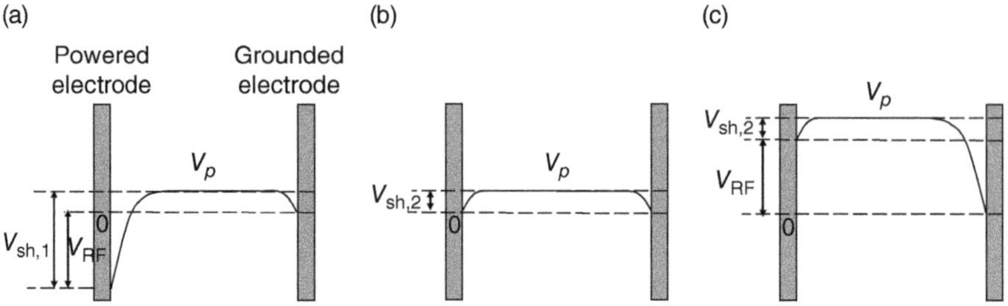

Figure 9.3 Plasma chamber with external RF voltage. (a) Most negative part of RF cycle, (b) neutral part of RF cycle, and (c) most positive part of RF cycle.

discussed previously):

$$V_{\text{sh}} = X_c I \tag{9.9}$$

where X_c is the capacitive reactance of the sheath and I is the current through the plasma. This means that higher currents that are generated by higher RF power generate larger sheath voltages and ion energies. Some of the RF energy is also used to increase the plasma density.

The capacitive reactance is a function of frequency ω and capacitance C:

$$X_c = \frac{1}{\omega C} \tag{9.10}$$

Equation (9.10) has two very important practical implications. Firstly, lower frequencies produce higher sheath voltages and hence higher ion energies. This means that the plasma frequency is a useful parameter to drive ion energies. Secondly, the sheath voltage depends on the capacitance and hence on the electrode area or more precisely on the relative sizes of the two electrodes.

Figure 9.4 illustrates another important frequency effect on sheath voltage. For excitation frequencies of 100 kHz and lower, the sheath collapses and the plasma potential approaches zero when the voltage on the RF voltage is zero or negative. Only a very small self-bias $V_{\text{sh, 2}}$ remains, which can be neglected. Such a sheath is called a resistive sheath. For frequencies above 100 kHz, the RF voltage changes too fast for the sheath and the plasma potential to follow. In this case, a frequency-dependent sheath voltage $V_{\text{sh, 3}}$ is formed. Since the latter represents a DC voltage it is typically called V_{DC}. Because the sheath voltage accelerates the ions, the temporal behavior of the sheath voltages translates into ion energy distributions. These distributions exhibit two peaks. The width of the ion energy distribution or energy dispersion, ΔE_i, is wider for the resistive sheath.

In most RIE tools, frequencies over 100 kHz are used and the ions are accelerated by V_{DC} of a capacitive sheath. Resistive sheaths are interesting because they resemble the ion bombardment behavior of a plasma where the bias is pulsed. They are, however, not used in high-volume semiconductor production to date.

So far, we established that for a capacitively coupled plasma, the plasma and sheath voltages are higher the lower the frequency is (see Eq. (7.18)). When the frequency is higher than the inverse of the transit time of the ions, a DC voltage is

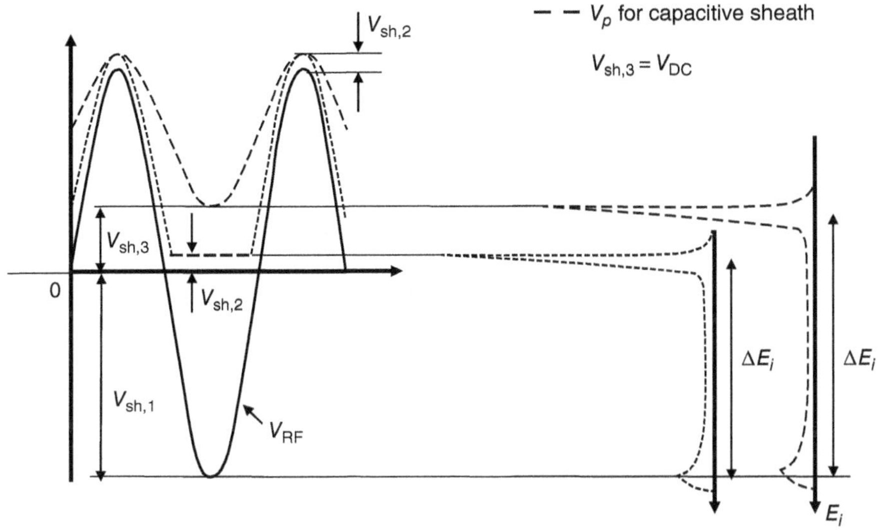

Figure 9.4 Time-dependent sheath voltage for resistive and capacitive sheaths and ion energy distributions for electrodes with equal size. Source: Modified from Koehler et al. (1985a).

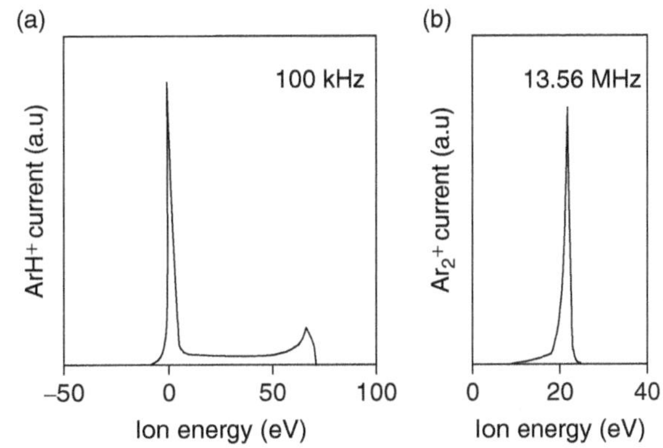

Figure 9.5 Energy distributions of ions extracted through the ground plane for a capacitively coupled argon plasma at 50 mTorr for 100 kHz (a) and 13.56 MHz (b). Source: Modified from Koehler et al. (1985a).

generated. Under these conditions, sheath voltage will be between the DC voltage at the low end and the plasma potential at the high end (see Figure 9.4). The higher the RF frequency is, the narrower should be the ion energy distribution. If the RF frequency is high enough, the ion energy distribution should become a single peak.

This is indeed the case as shown in Figure 9.5. For 100 kHz, the ArH^+ energy distribution has a low-energy peak at zero volts and a high energy peak at about 70 V. The energy distribution of $Ar_2{}^+$ ions in the case of 13.56 MHz is monoenergetic (Koehler et al. 1985a). The reason is that the saddle in the plasma voltage for negative RF half-cycles, which can be seen in Figure 9.4, is almost completely filled in at this frequency for $Ar_2{}^+$.

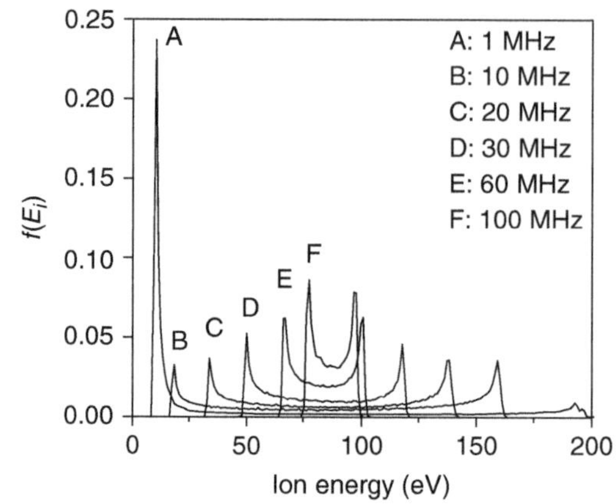

Figure 9.6 PIC simulation of ion energy distributions for a capacitively coupled helium with symmetric electrodes. Source: Kawamura et al. (1999). © 1999 IOP Publishing.

This bimodal behavior for low frequencies has been modeled using particle-in-cell (PIC) models (Kawamura et al. 1999). Simulations allow the investigation of a wide range of frequencies. The results, which are shown in Figure 9.6, illustrate the effect of larger peak ion energies but also wider energy distributions (IED) for lower excitation frequencies. The results for 1 MHz show that the magnitude of the low-energy peak is much larger than that of the high-energy peak. This is caused by the sheath voltage staying at the minimum value for a longer part of the cycle than at a maximum value.

Okamoto and Tamagawa derived an analytical expression for the energy dispersion ΔE_i (Okamoto and Tamagawa 1970):

$$\Delta E_i \propto \frac{1}{\omega}\left(\frac{M_i}{e}\right)^{-1/2} V_p \tag{9.11}$$

This relationship shows that the energy is inversely proportional to the circular frequency of the RF field ω, and inversely proportional to the square root of the atomic mass and proportional to the amplitude of the time varying component of the plasma potential.

According to Eq. (9.11), lower mass ions have a larger ΔE_i or, in other words, a wider IED. In Figure 9.7, experimental results for Eu^+ ($M = 152$ amu), H_2O^+ ($M = 18$ amu), and H_3^+ ($M = 3$ amu) are compared. The data was collected in a capacitively coupled reactor.

The IED for a gas mixture is the superposition for the IED of the ionic species. If the gases are molecular, dissociation products have their own IED. This leads to IEDs with a series of peaks. Figure 9.8 depicts the measured IED for a capacitively coupled CF_4 plasma, which exhibits peaks for the fragmentation products F^+, CF^+, CF_2^+, and CF_3^+.

Figure 9.9 shows experimental results for the relationship between the ion mass M_i and the ion energy dispersion ΔE_i (Coburn and Kay 1972). The experimental result confirms that the ion energy dispersion is inversely proportional to the square root of the ion mass as predicted by Eq. (9.11). The corresponding IEDs for the highlighted data points in Figure 9.9 are shown in Figure 9.7.

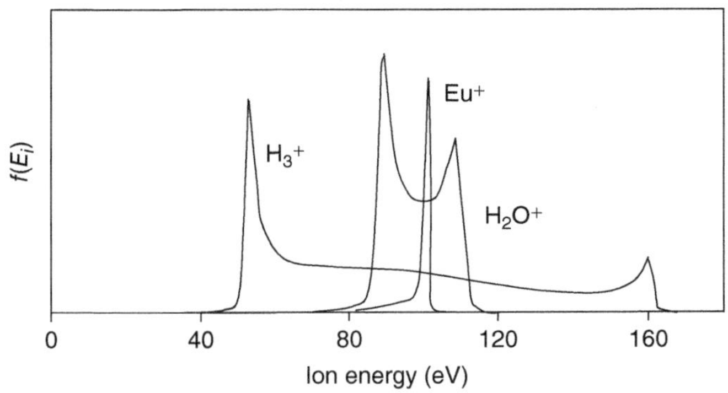

Figure 9.7 Energy distributions of H_3^+, H_2O^+, and Eu^+ for 13.56 MHz at 75 mTorr pressure. Source: Coburn and Kay (1972). © 1972 AIP Publishing.

Figure 9.8 Measured ion energy distribution for a capacitively coupled CF_4 plasma at 3 mTorr and 3 kW (13.56 MHz RF power). Source: Kuypers and Hopman (1990). © 1990 AIP Publishing.

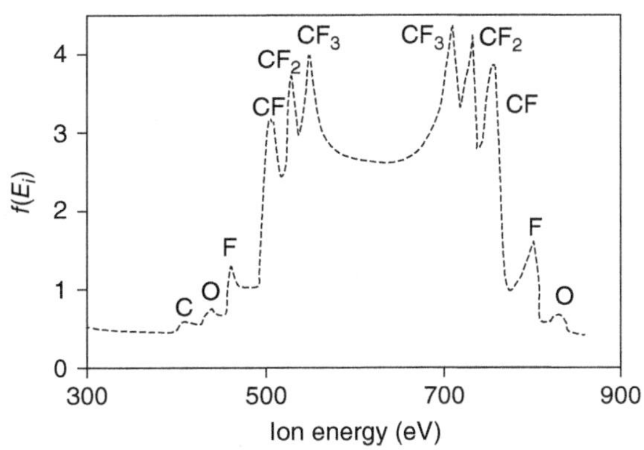

Figure 9.9 Full width ion energy dispersions as a function of the atomic mass of the ions. The data correspond to the ion energy distributions in Figure 9.7. Source: Coburn and Kay (1972). © 1972 AIP Publishing.

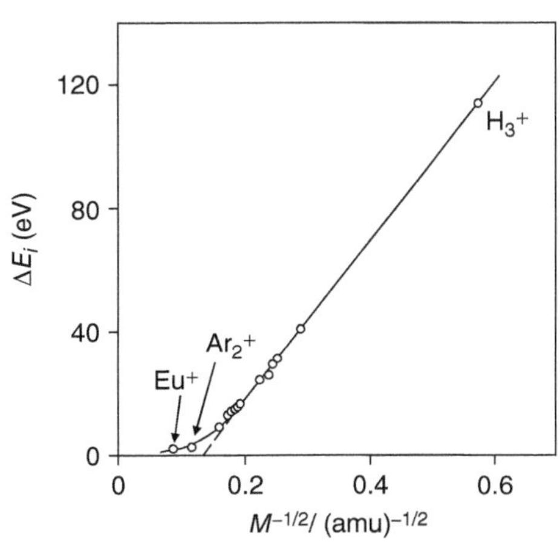

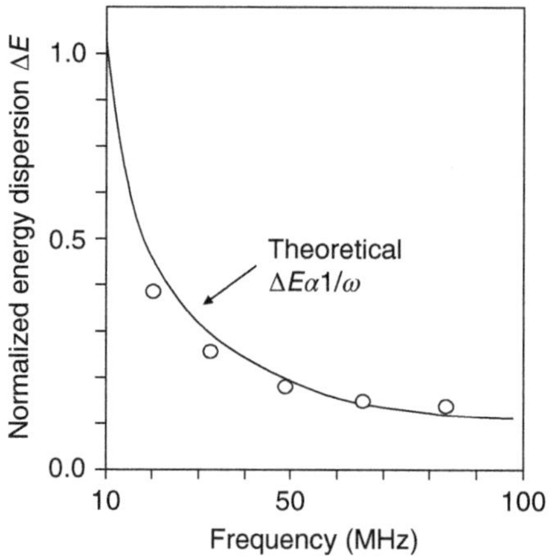

Figure 9.10 Dependence of the ion energy dispersion ΔE_i on RF frequency. Source: Okamoto and Tamagawa (1970). © 1970 Physical Society of Japan.

Okamoto and Tamagawa confirmed the relationship between ΔE_i and plasma frequency in a capacitively coupled reactor with two ring electrodes wrapped around a dielectric tube (Okamoto and Tamagawa 1970). The results in Figure 9.10 are normalized to unity for 10 MHz, which is the lowest frequency they probed. The molecular mass of 28 a.m.u. represents N_2^+ ions. The pressure is around 1 mTorr. The process parameters were tuned to obtain a mean ion energy of 102 eV.

The frequency and mass dependence of the IED can be explained using the framework of an ion transient time τ_i through the sheath (Kawamura et al. 1999):

$$\tau_i = 3d_{sh}\sqrt{M_i/2eV_{sh}} \tag{9.12}$$

Here, d_{sh} is the sheath thickness, M_i is the ion mass, e is the elementary charge, and V_{sh} is the sheath voltage.

The critical parameters that control ion modulation in an RF sheath are the product of the ion transient time τ_i and the circular frequency ω of the applied RF signal. When $\omega\tau_i \ll 1$, ions traverse the sheath in a short time compared to the field oscillations. These ions experience the instantaneous sheath voltage at the time the ion enters the sheath. This condition is fulfilled for low RF frequencies (with a resistive sheath, see Figure 9.4). The condition is also met for thin sheaths due to low sheath voltage or small Debye length or small ion mass. For the case of $\omega\tau_i \ll 1$, the energy dispersion can be expressed as

$$\Delta E_i \approx 2V_{RF} \tag{9.13}$$

Here, V_{RF} is the sinusoidal component of the excitation voltage. In this case, the energy dispersion is not frequency dependent. In industrial RIE etching tools, frequencies of 400 kHz, 1 MHz, 2 MHz, 13.56 MHz, 27 MHz, 60 MHz, and 162 MHz are used. The case of $\omega\tau_i \ll 1$ is valid for frequencies below 400 kHz. In other words, it is not typically used in semiconductor manufacturing yet.

The other extreme case is $\omega\tau_i \gg 1$ where ions experience many sheath oscillations while traveling through the sheath. In this case, ions will respond to the time-averaged sheath potential, and the IED function will exhibit a single peak:

$$\Delta E_i \approx 2V_{RF}/\omega\tau_i \tag{9.14}$$

Equation (9.14) implies that the energy dispersion is frequency dependent in the high frequency regime. For typical RIE process pressures and RF powers, this is the case for RF frequencies above 27 MHz. Frequencies of 27 MHz and higher are used in all capacitively coupled plasma (CCP) etching tools to drive the plasma densities. They are considered "source frequencies." Source frequencies can power the top electrode (in which case the name "source frequency" is more intuitive) or the cathode on which the wafer is placed.

RIE reactors with only one electrode, which are not used in semiconductor manufacturing anymore, used preferentially an intermediate frequency of 13.56 MHz because it allowed to create plasma density and to accelerate the ions meaningfully. Dual frequency CCP reactors use a second RF signal to accelerate the ions. The preferred frequencies are between 400 kHz and 13.56 MHz, which are not covered by Eqs. (9.13, 9.14). Panagopoulos and Economou developed a model that covers the transitional regime between the two extremes where $\omega\tau_i \sim 1$ (Panagopoulos and Economou 1999). The model is based on prior work by Miller and Riley (1997). It can be used to calculate the IEDs from the actual voltage signal at the cathode on an RIE reactor.

So far, we assumed that ions travel through the sheath without colliding with neutrals. This is not always the case. In nonequilibrium plasmas, the neutral density is several orders of magnitude higher than the ion and electron density. These neutrals have the same density in the sheath as in the bulk plasma because they do not respond to electric fields. Therefore, ion collisions with neutrals during acceleration in the sheath are becoming increasingly likely with increasing pressure.

The physics of the elastic collision is the same as in the case of solid-state sputtering. The change in energy is described by Eq. (2.5). This means that sheath collisions will completely change the energy distribution and dramatically widen the angular distribution of the ions. This regime must be avoided for directional etching to be effective. This insight was one of the driving forces for the development of high-density plasma tools in the early 1990s. High-density plasma reactors operate at low pressure (long mean free path) and the plasma has a small Debye length (thin sheath), which results in collision-free sheaths.

The mean free path is a measure of the distance an ion can travel in a gas without collisions. It is defined as the distance over which an un-collided ion beam decreases by $1/e$ from the initial value. The mean free path is a function of the neutral gas density and the cross section of the collision σ:

$$\lambda_i = \frac{1}{N_n \sigma} \tag{9.15}$$

According to Eq. (9.15) the mean free path of an accelerated ion in the sheath is inversely proportional to the pressure. This is important to consider in process development. When a process engineer attempts to increase the process pressure to change the plasma chemistry, there is a risk of creating sheath collisions. Sheath collisions will manifest themselves as increased profile bowing due to the wider ion

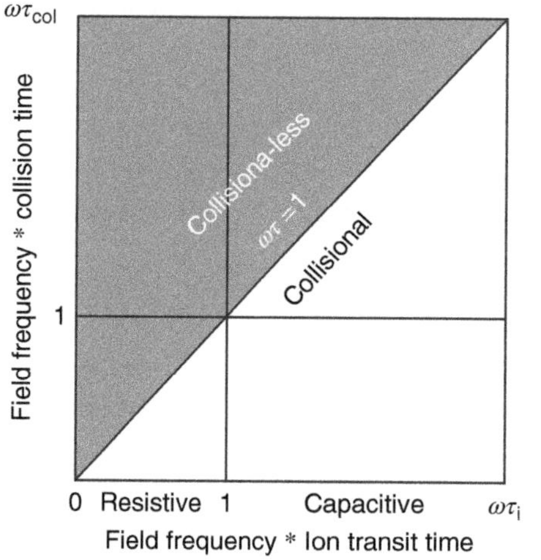

Figure 9.11 Generalized sheath diagram for time-dependent sheaths. Source: Panagopoulos and Economou (1999). © 1999 AIP Publishing.

angular distribution (IAD). This effect can also be used, for instance, for cleaning of residues from sidewalls where a wide ion energy distribution is needed.

Unwanted sheath collisions can also be created at constant pressure by increasing the low-frequency RF power. This widens the sheath thickness, which increases the distance an ion must travel. The time between collisions, which is referred to as collision time τ_{col}, depends on the mean ion velocity:

$$\tau_{col} = \lambda_i / V_i \tag{9.16}$$

The existence of sheath collisions can be evaluated by comparing the ion transient time τ_i and the collision time τ_{col}. For $\tau_i < \tau_{col}$, the sheath is considered collision free.

Panagopoulos and Economou combined the concepts of ion transient time vs collision time and the ion transient time vs RF frequency into a generalized sheath diagram, which is depicted in Figure 9.11 (Panagopoulos and Economou 1999). The generalized sheath diagram exhibits four major regimes depending on the product of the applied RF frequency and the ion collision frequency $\omega\tau_{col}$ (y axis) as well as the ion transit time through the sheath $\omega\tau_i$ (x axis): collision-less resistive, collision-less capacitive, collisional resistive, and collisional capacitive. The diagonal ($\tau_i = \tau_{col}$) separates the collision-less from the collisional sheath regimes (Panagopoulos and Economou 1999).

Because the number of sheath collisions depends on the ion transient time τ_i, it is proportional to the sheath thickness and inversely proportional to the square root of the sheath voltage (see Eq. (9.12)). This means that when the RF power is increased to produce higher ion energies, the sheath thickness increases, which in turn increases the probability of ion–neutral collisions in the sheath. The sheath thickness of RF-powered plasmas is governed by the Child–Langmuir law:

$$J_i = \frac{4}{9}\varepsilon_0 \left(\frac{2e}{M_i}\right)^{1/2} \frac{V_{sh}^{2/3}}{d^2} \tag{9.17}$$

This law describes the space charge limited current in a plasma between planar walls or electrodes separated by a distance d. The existence of the sheath is caused by the difference in electron and ion mobilities due to their difference in mass. Space charge, which is the result of ions repelling each other while traveling through the sheath, lowers the ion mobility. Considering this effect, the thickness of a Child–Langmuir sheath can be written as (Lieberman and Lichtenberg 2005)

$$d = \frac{\sqrt{2}}{3}\lambda_D\left(\frac{2V_{\text{sh}}}{T_e}\right)^{3/4} \tag{9.18}$$

The thickness of Child–Langmuir sheaths can be of the order of hundreds of Debye length or several tens of millimeters, while the thickness of a non-powered plasma as discussed in Section 9.1 is around one Debye length.

In summary, capacitive coupling of RF power using planar electrodes is the preferred method of ion acceleration in RIE and directional ALE reactors. Lower frequencies provide higher ion energies but also wider ion energy distributions. The IED has two maxima for low frequencies, for single mass ion plasmas and in the absence of sheath collisions. Typical plasmas used in semiconductor device etching have several feed gases that can dissociate. This generates a range of ions and creates additional peaks in the IED. This means that the IED in real applications constitutes a nearly continuous distribution with several peaks. The goal of the process engineer is to shape this distribution to achieve the best possible results. The gas pressure must be low enough to avoid sheath collisions. This is increasingly important for etching with very high ion energies (for instance high aspect ratio dielectric etch) and wide sheaths.

Next, we will discuss how the IED depends on the geometry of the etching reactor. A capacitively coupled plasma has two electrodes that are characterized by their own capacitances. The equivalent circuit of a two-electrode RF plasma system is shown in Figure 9.12 (Koehler et al. 1985b). The capacitive sheath approximation assumes that the resistive components of this circuit are negligible ($R_1 = 0$; $R_2 = 0$; $R_p = \infty$).

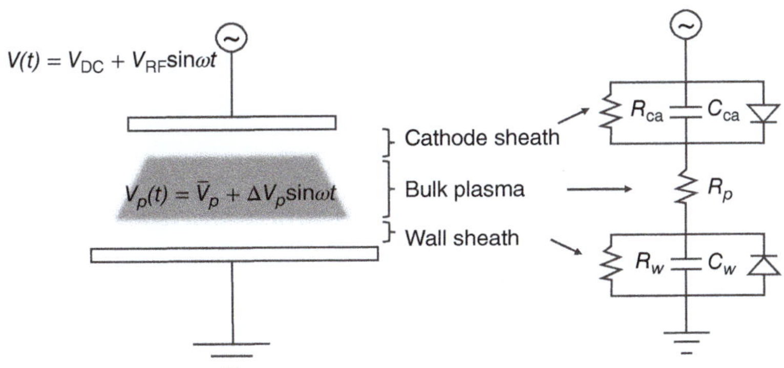

Figure 9.12 Equivalent circuit in capacitively coupled two-electrode RF plasma system. Source: Modified from Koehler et al. (1985b).

This equivalent circuit model represents a voltage divider with two capacitive resistances and the voltage on the excitation electrode is

$$V(t) = V_{DC} + V_{RF} \sin \omega t \quad (9.19)$$

The plasma potential of a capacitively coupled plasma that is powered by a sinusoidal RF voltage can be written as

$$V_p(t) = \overline{V_p} + \Delta V_p \sin \omega t \quad (9.20)$$

When the RF power is capacitively coupled to the excitation electrode, the capacitive sheath approximation predicts a time-averaged plasma potential $\overline{V_p}$ of the form (Koehler et al. 1985a)

$$\overline{V_p} = \frac{1}{2}(V_{RF} + V_{DC}) \quad (9.21)$$

This is an important expression that can be used to estimate the averaged plasma potential when using the RF voltage that is supplied by the generator and the DC voltage that can be measured at the powered electrode.

The DC bias voltage V_{DC} is related to the RF amplitude V_{RF} by the expression (Koehler et al. 1985b)

$$V_{DC} = V_{RF}\left(\frac{C_{ca} - C_w}{C_{ca} + C_w}\right) \quad (9.22)$$

Here, C_{ca} and C_w are the capacitances of the powered cathode and the grounded wall, respectively. These capacitances are proportional to their area. The sheath potential that accelerates the ions is

$$|V_{sh}| = |\overline{V_p}| + |V_{DC}| \quad (9.23)$$

Equations (9.22, 9.23) imply that the cathode with the wafer and the wall will be bombarded with the same ion energy if the areas are the same ($V_{DC=0}$ in Eq. (9.22)). If the areas are different, the ion energy will be larger on the smaller electrode. Figure 9.13 depicts time-averaged potentials on the powered and grounded electrodes for the case of equal and non-equal electrode sizes.

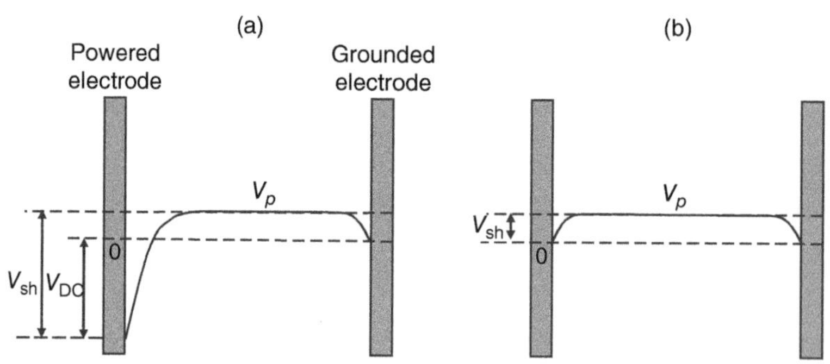

Figure 9.13 Time-averaged potentials on the powered and grounded electrodes for the case of equal and non-equal electrode sizes. (a) Unequal area electrodes and (b) Equal area electrodes.

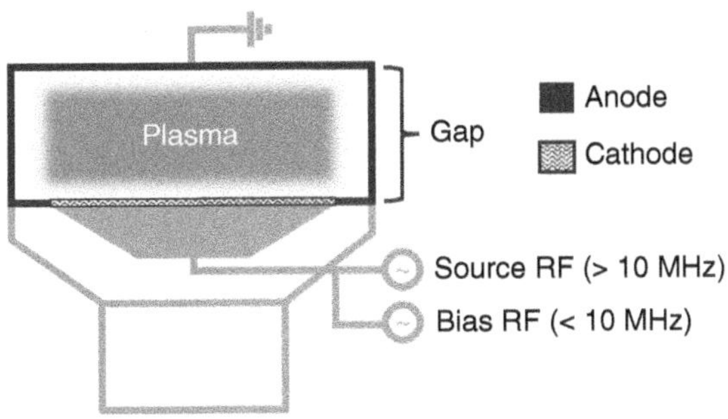

Figure 9.14 Cathode and anode areas for a generic CCP RIE reactor.

The dependence of the sheath voltage on the relative sizes of the electrodes is important for the design of capacitively coupled plasma etching reactors (Horwitz 1983). Figure 9.14 illustrates the cathode and anode areas for a generic CCP reactor. The electrode that holds the wafer is called the cathode. The grounded area that includes the grounded upper plate, the reactors walls, and potential pumping screens that are touched by the plasma is called the anode. This terminology can be confusing because the plasma is powered by RF. In a hypothetical reactor where the areas of the electrodes are the same, the electrodes would be electrically indistinguishable. However, in real reactors the electrode that holds the wafer is always smaller and hence the smaller electrode has a larger negative potential with respect to the bulk plasma, and hence the name cathode.

The gap between the cathode and the upper electrode (anode) is an important design parameter. In reactors with variable gap, it is also a process parameter. When the gap is increased, the anode to cathode ratio and the ion energy on the wafer increase while the ion energy on the upper electrode decreases. This is desirable, for instance, for high aspect ratio dielectric etch applications (see Section 7.3.3). A lower ion energy on the upper electrode or anode reduces erosion, increases the useful lifetime of the parts, and reduces the cost of operation.

Despite the beneficial effect of increasing the anode to cathode ratio for reaching higher ion energies, CCP etching tools for high aspect ratio etching have relatively small gaps around several tens of millimeters. So why is it not advisable to increase the gap as much as possible to maximize the ion energy one can obtain for a give RF power? The reason is that narrow gap reactors have benefits in terms of polymer management because less of an area must be cleaned after etching a wafer. Also, a small plasma volume allows for faster mixed mode pulsing (MMP). In addition, shorter residence times reduce the degree of fragmentation because the time molecules spend in the plasma excitation region is reduced. Finally, the plasma density for a given RF power is higher for a smaller plasma volume.

So far, we discussed the mechanism of how capacitive coupling interacts with the plasma sheath. From a practical point of view, this explains the origin of ion energy. It is important to point out that all RIE and directional ALE reactors use capacitive coupling to accelerate ions. What differentiates the different RIE reactors is how the ion and radical densities and fluxes are generated.

In CCP reactors, both ion energy and flux are driven by capacitive coupling of RF power. The simplest implementation of a CCP RIE reactor utilizes an RF-powered

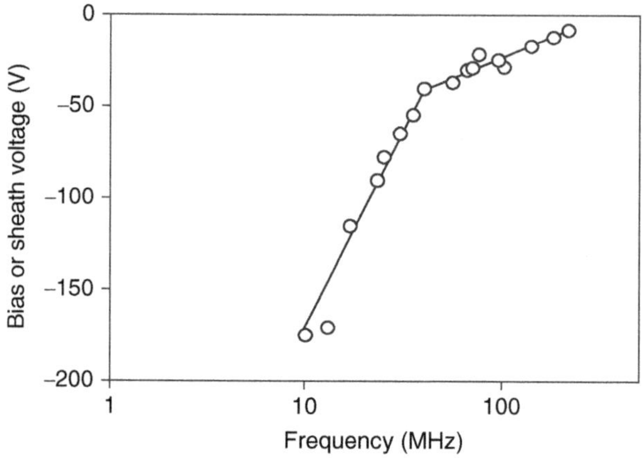

Figure 9.15 Bias voltage as a function of RF frequency for a capacitively coupled argon plasma for 100 W RF power at 7 mTorr. Source: Goto et al. (1992). © 1992 AIP Publishing.

cathode and grounded walls. The power that is applied to the cathode will sustain the plasma *and* accelerate the ions. The challenge with such a design is that both the ion flux and the ion energy will change when the RF power is changed. To reach regimes with low ion flux and high ion energy or high ion flux and low ion energy, it is desirable to have one parameter or "process knob" that tunes the ion flux and one to tune the ion energy. It turns out that the use of two different RF frequencies allows to decouple ion energy and ion flux (Goto et al. 1992). Capacitive coupling with frequencies higher than 13.56 MHz drives primarily the plasma density and ion flux while RF frequencies of 13.56 MHz and below are well suited to tune the ion energy.

We already discussed that ions can follow the RF potential on the powered electrode at low frequencies, which results in wider IEDs with higher maximum ion energies. In completely capacitive sheaths, ions cannot follow the electrode potential and the ion energy is determined by an average sheath or DC voltage. Figure 9.15 shows measured average ion energies for frequencies between 10 and 100 MHz, which is in the regime of capacitive sheaths (Goto et al. 1992). The experiments were conducted with argon in a parallel plate reactor with equally sized electrodes at a constant RF power of 100 W and a pressure of 7 mTorr. The data shows a reduction of the bias voltage across the entire frequency region. The data can be fitted with a logarithmic relationship. Two distinct regions between 10 and 40 MHz and 40 and 100 MHz are visible.

Figure 9.16 depicts the corresponding electron densities. Here, the electron densities increase from 10 to 40 MHz and drop beyond 40 MHz. The data indicates that the driving frequency influences excitation of the plasma and impacts the electron and plasma density. This effect was brought to the attention of the plasma etching community by Flamm (1986). His calculations showed that the RF frequency alters not only the energy of ions impinging on surfaces, but also the electron energy distribution function (EEDF). In general, frequency changes are associated with a defined transition in discharge characteristics when the RF excitation frequency is close to the characteristic frequencies of critical electrical or chemical processes in the plasma. The relationship between the RF frequency and the transition time of

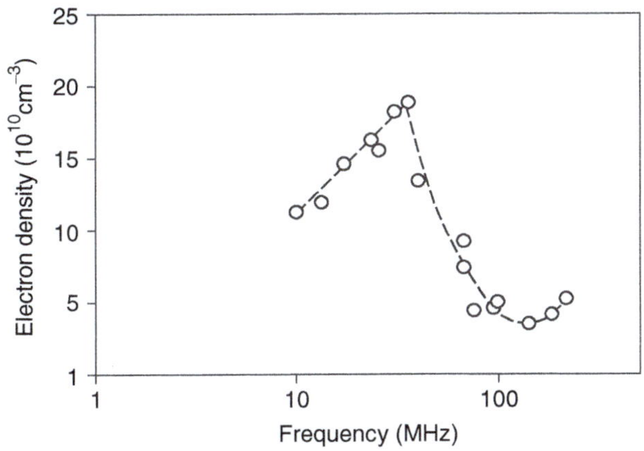

Figure 9.16 Electron density dependence on RF frequency for the experiments shown in Figure 9.15. The measurements were conducted 20 mm from the axial center and 15 mm from each electrode. Source: Goto et al. (1992). © 1992 AIP Publishing.

ions through the sheath and the plasma frequency, which we discussed in the context of the origin of IEDs, is a good example. Flamm considered in his analysis over ten time-dependent processes.

Furthermore, Flamm predicted that the RF frequency impacts the spatial distribution of species and electrical fields across the discharge, which means the plasma uniformity across the wafer can change. This is a very important point which limits the use of very high frequencies in etching if semiconductor devices (Flamm 1986).

Surendra and Graves conducted PIC Monte Carlo simulations of RF discharges driven at various frequencies and their results predict that the plasma density, ion current, and power scale as the square of the RF frequency, for a constant voltage (Surendra and Graves 1991). They also found that raising the frequency decreases the sheath thickness, thereby increasing ion directionality in the sheath at constant pressure. Their work predicted superior profile performance for CCP reactors when operating at high frequencies (100 MHz and higher) and low pressures (50 mTorr and lower). These insights sparked considerable interest in CCP reactors with very high frequencies in the late 1990s. The research work was focused on the impact of the RF frequency on the species fluxes and ion energies and nonuniformities caused by very high frequencies.

Hebner et al. detected increasing electron densities for higher RF frequencies in a parallel plate reactor for 300 mm wafers (Hebner et al. 2006). Their results are shown in Figure 9.17. Within the frequency range from 10 to 120 MHz, the data shows a linear increase. However, there are significant departures from the linear behavior. Above 120 MHz, the increase appears to follow a frequency-squared dependence. At the two highest frequencies investigated, 163 and 189 MHz, the authors found a shift in the spatial distribution from uniform to highly peaked in the center. Thus, the density inflection at approximately 100 MHz appears to be due to changes in both the spatial distribution and the peak density.

Zhu et al. conducted theoretical and experimental studies of the electron density as a function of RF frequency (Zhu et al. 2007). They confirmed the finding by Surendra and Graves that the power that is absorbed by the plasma increases with the square

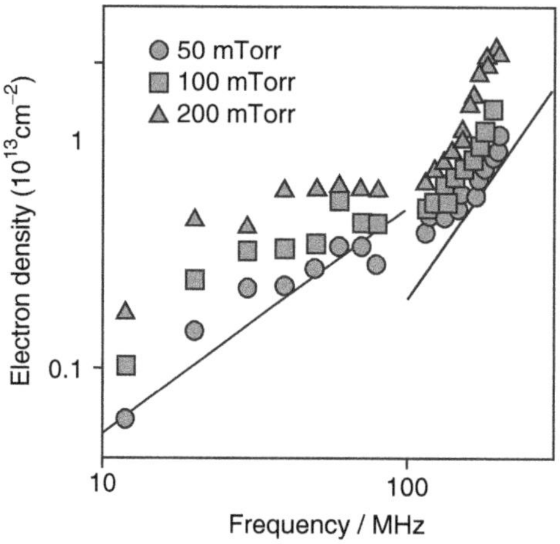

Figure 9.17 Electron density dependence on RF frequency for pressures of 50, 100, and 200 mTorr at constant 300 W power. The solid lines represent the slope for a linear and quadratic increase in density with increased frequency. Source: Hebner et al. (2006). © 2006 IOP Publishing.

of the driving frequency (Surendra and Graves 1991). Ahn et al. found in experiments that as the driving frequency increases, the electron temperature increases while the electron density is almost constant or decreases. They explain this to be the result of an enhancement of collisional heating in the bulk plasma (Ahn et al. 2006).

To summarize, experimental (Goto et al. 1992; Hebner et al. 2006; Ahn et al. 2006; Zhu et al. 2007) and theoretical (Surendra and Graves 1991; Zhu et al. 2007) investigations show overall a trend that the electron and plasma densities increase with RF frequency. The discrepancies between these studies with respect to the concrete functional relationship could be explained by spatial effects for frequencies over 100 MHz and the interaction with the reactor geometry. The emergence of these spatial nonuniformities is in part explained by the fact that the dimensions of the RF-driven electrodes may become a significant fraction of the RF wavelength for frequencies above 100 MHz. This can lead to the so-called standing waves. Therefore, the majority of CCP reactors used in the manufacturing of 300 mm wafers utilize source frequencies below 100 MHz.

Most commercial CCP reactors utilize at least two frequencies, 13 MHz or lower to drive the ion energy and a high frequency (for example, 27 and 60 MHz) to drive the plasma density and ion flux. The use of such high frequencies effectively decouples to bias and source effects (Kitajima et al. 2000). The source frequency can be applied to the upper electrode of the reactor but also to the cathode, which holds the wafer. Because high frequencies induce low ion energies, the cathode to anode ratio difference between the two implementations is negligible. The choice of where to apply the source power is mostly driven by engineering considerations. Adding the source power to the upper electrode complicates the gas delivery through the showerhead while adding to the bottom electrode complicates the design of the cathode with the bias RF power, the ESC with helium cooling, and the wafer lift mechanism.

Many CCP reactors use two bias frequencies, for instance, 2 and 13.56 MHz, to create the so-called dual frequency sheaths (Shannon et al. 2005). With this, the width of ion energy distribution functions can be tuned by mixing the higher and lower

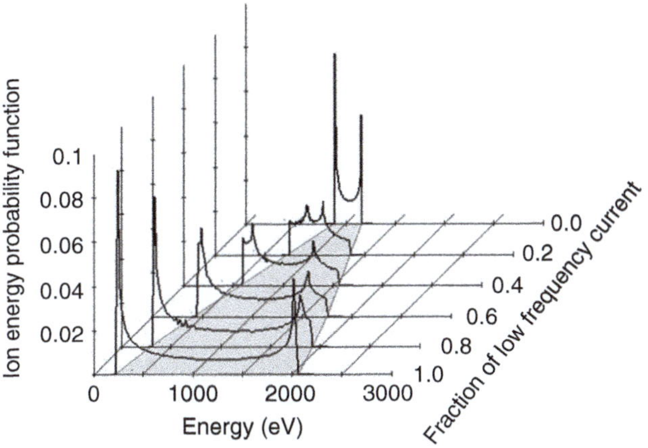

Figure 9.18 Calculated IEDs as a function of frequency mixing for an argon plasma. Source: Shannon et al. (2005). © 2005 AIP Publishing.

frequency RF components. Computational results by Shannon et al. are depicted in Figure 9.18. The graphs illustrate how the upper and lower energy peaks can be moved by changing the ratio of the lower and higher RF powers.

The mechanism that allows to change electron and plasma densities by changing RF power is called electron heating. The latter is responsible for the existence of the EEDF that drives dissociation and ionization in the plasma (see Figure 9.1). In capacitively coupled plasmas, electrons receive energy from the electric field via ohmic and stochastic heating. Ohmic heating is the result of elastic electron collisions with the neutral background gas. It is interesting to note that the electrons obtain only negligible energies from the electric field in the plasma body. The maximum energy electrons can obtain from the electric field in the bulk plasma is orders of magnitude smaller than the thermal energy of the gas, which is a few electron volts. Hence, the electrons thermalize when colliding with the background gas. Ohmic heating requires a high enough pressure to be effective. Because the effect is thermal, the EEDF for ohmic heating resembles a Maxwellian distribution.

Stochastic electron heating is the effect of energy transfer to electrons via the sheath. When electrons from the plasma reach the sheath, they get decelerated unless they are fast enough to escape to the electrode. As discussed before, the sheath moves with the RF frequency and the amplitude depends on the applied voltage and on the frequency. Electrons that are reflected from the moving sheath obtain additional energy. This effect is called stochastic heating. The efficiency of stochastic heating is dominant for high RF frequencies but not dependent on pressure. The effects of ohmic and stochastic heating are illustrated in Figure 9.19. Stochastic heating manifests itself as a high-energy tail in the EEDF resulting in a bi-Maxwellian distribution.

The temperature of the background gas is determined by the temperatures of the wall and the gas delivery system of the reactor. The charged particles do not heat the background gas meaningfully in nonequilibrium plasmas (hence the term "low-temperature plasma"). Ion bombardment heats the reactor walls. This effect

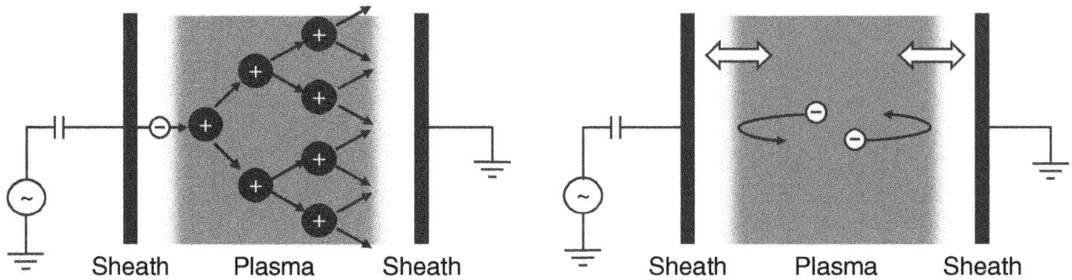

Figure 9.19 Schematic illustration of (a) ohmic and (b) stochastic heating.

is proportional to the ion flux and ion energy. For sources with small volume, such as downstream radical sources with an applicator tube, the walls can heat significantly and the gas exiting the source can reach several hundred degrees, which can contribute to the heating of the wafer.

The gas temperature determines the energy component parallel to the sheath and hence contributes to the ion angular distribution (IAD). Because there are no known practical engineering solutions to reduce the temperature of the background gas, the approach to narrow the IAD is to increase the vertical ion energy via tuning of the sheath voltage. This has limitations for directional ALE, which relies on ion energies below the sputtering threshold of the bulk materials (see Sections 6.1.2 and 6.1.3).

In summary, the dependencies of the ion energy, IED, IAD, and ion flux on RF power and frequency, ion mass, pressure, and electrode are complex and nonlinear. Advanced CCP reactors utilize at least two RF frequencies to decouple ion energy and flux. For processes that use very high RF power, low frequency, and a mixture of multiple molecular gases, the IED is a broad continuous distribution where only a small fraction of the ions reaches the peak energy. This is the situation in high aspect ratio SiO_2, alternating SiO_2 and Si_3N_4 (ONON), and alternating SiO_2 and poly silicon (OPOP) etching processes that are technology enabling. It is possible that ions with a variety of ion energies are needed for successful high aspect ratio etching; however, it is unknown what the perfect IED would look like because it is nearly impossible to tune the IED without changing other plasma parameters. There are technologies that allow narrowing the IED and we will discuss them in Section 9.4.

9.3 Inductively Coupled Plasmas

Another approach to decouple ion energy and flux is to use an RF magnetic field to power the plasma. This is the case in the so-called inductively coupled plasma (ICP) or transformer coupled plasma (TCP) reactors. Here, a magnetic field is generated by a coil that is situated on top or on the sides of a ceramic window. It generates an RF electric field inside the reactor, which transfers energy to the charged species in the plasma efficiently. The term "Transformer coupled" was coined because the resulting electron currents in the plasma resemble the secondary wiring of a transfer while the coil is the primary one. In the equivalent circuit of an ICP/TCP source the inductive components are represented by a transformer.

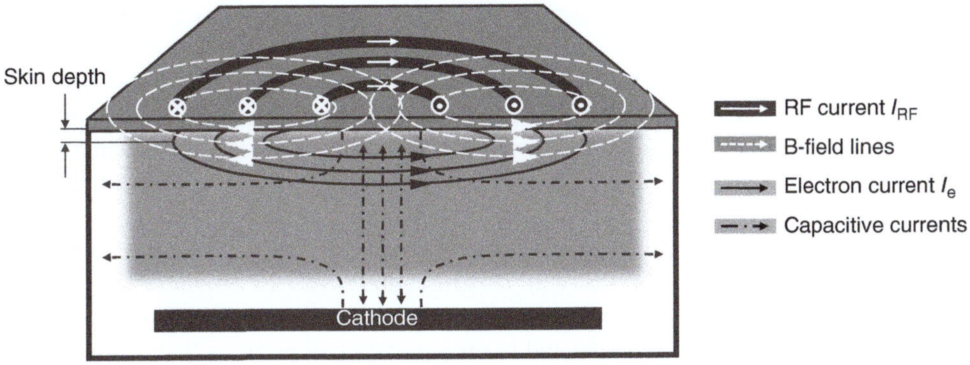

Figure 9.20 Schematic illustration of electric and magnetic fields in of ICP/TCP plasma reactors.

Other embodiments, for instance, helicon sources (Perry and Boswell 1989), are possible but are less frequently used in the manufacturing of semiconductor devices.

The driving force for the development of ICP/TCP sources was the need for processing at pressures below 10 mTorr to avoid sheath collisions and loss of directionality. This became important in the early 1990s when feature sizes reached 1 μm and below. Low-pressure etching is practical if the reactants can be produced in concentrations high enough to maintain high etching rates. High plasma densities require more efficient use of the electron energy in the ionization process (Carter et al. 1993).

The electric and magnetic fields in a typical ICP/TCP reactor with planar coil configuration are schematically shown in Figure 9.20. The coil induces electric and magnetic fields that penetrate the dielectric window on the top of the process chamber. In the presence of a plasma, an azimuthal electric field and an associated current are induced within the plasma according to Faraday's law. The plasma current has the opposite direction of the coil current and is confined to a layer near the surface of the dielectric window (Lieberman and Lichtenberg 2005).

Most ICP/TCP tools in the semiconductor industry use RF frequencies of 2 or 13.56 MHz. The RF frequency exceeds the electron collision frequency and the electron heating is determined the effective electric field underneath the dielectric window.

The power is absorbed by the plasma in a ring-shaped region. This leads to low ion flux densities in the center of the wafer. Because ion flux densities drop toward the edge of the wafer, the result is the so-called "M-shaped" etch rate profile across the wafer. Azimuthal and side-to-side asymmetries of the coil will correlate to similar asymmetries in the inductively coupled electric field and ion production rates (Kushner et al. 1996). These effects can be suppressed if the spacing between the top window and the wafer is large enough for diffusion of charged species to smoothen out the ion flux uniformity. In reactors for processing of 300 mm wafers, the ion flux uniformities reach satisfactory levels for gaps larger than 10 cm.

This is a disadvantage for fast gas switching in comparison to smaller gap CCP reactors.

Besides producing large ion fluxes even at low pressures below 10 mTorr, ICP/TCP sources have the additional benefit of a relatively straightforward implementation of radial ion flux uniformity tuning. Two concentric coils can be powered either individually or from one RF generator using a power spitting device (Collins et al. 2000; Long et al. 2015). The uniformity can be further improved when the currents in the concentric coils are phase shifted (Banna et al. 2015). This concept can also be applied to three concentric coils (Banna et al. 2019).

The inductive coil also couples capacitively to the plasma. This can lead to sputtering of the ceramic window and to the generation of particles that impact device yield. This effect can be reduced in planar coils with a configuration called a balanced transformer, which places a virtual ground in the middle of the coil and reduces the maximum coil-to-plasma voltage by a factor of 2. An electrostatic shield placed between the coil and the plasma can further reduce the capacitive coupling (Lieberman and Lichtenberg 2005).

A certain degree of capacitive coupling, however, is needed to strike and to maintain the plasma. For low powers, the ICP/TCP source can under certain conditions couple entirely capacitively to the plasma. When the power to the coil is increased, the coupling mode can switch to inductive. This transition is called $\mathcal{E}$ to $\mathcal{H}$ mode transition. This can lead to process instabilities when the source power is running in the transition region for a given process.

The electron temperatures in ICP/TCP reactors are typically higher than in CCP reactors with source frequencies below 100 MHz. This does not necessarily mean that ICP/TCP reactors generate higher self-biases than CCP reactors. According to Eq. (9.4), the self-bias is indeed proportional to the electron temperature. However, CCP coupling increases the sheath voltage due to the RF voltage, which is applied to the driving electrode according to Eqs. (9.20–9.22). This effect is reduced for higher RF frequencies.

This has important implications for directional ALE, which requires ion energies of 50 eV and lower in the removal step. Both ICP/TCP reactors and CCP reactors with sufficiently high RF frequency (40 MHz and higher) can deliver such low ion energies under conditions when only the plasma source is powered. Of course, as soon as the bias power is applied, the ion energies are determined by it.

Figure 9.21 depicts typical ICP/TCP reactors that are used in the manufacturing of semiconductor devices. These reactors differ in the shape of the dielectric window (flat or domed), the shape of the coil (flat or vertical), and the number of coils. Reactors with a ceramic dome have typically only one coil because the tuning effect of two coils is diminished by the large plasma volume that is created by the dome. Vertical coils allow moving the points of high voltage in the coil away from the ceramic window. However, this can introduce plasma stability challenges that can be remedied by routing higher voltage segments of the coil closer to the window. Flat coils on the other hand can be moved away from the window to reduce sputtering. The thickness of the dielectric window also impacts sputtering.

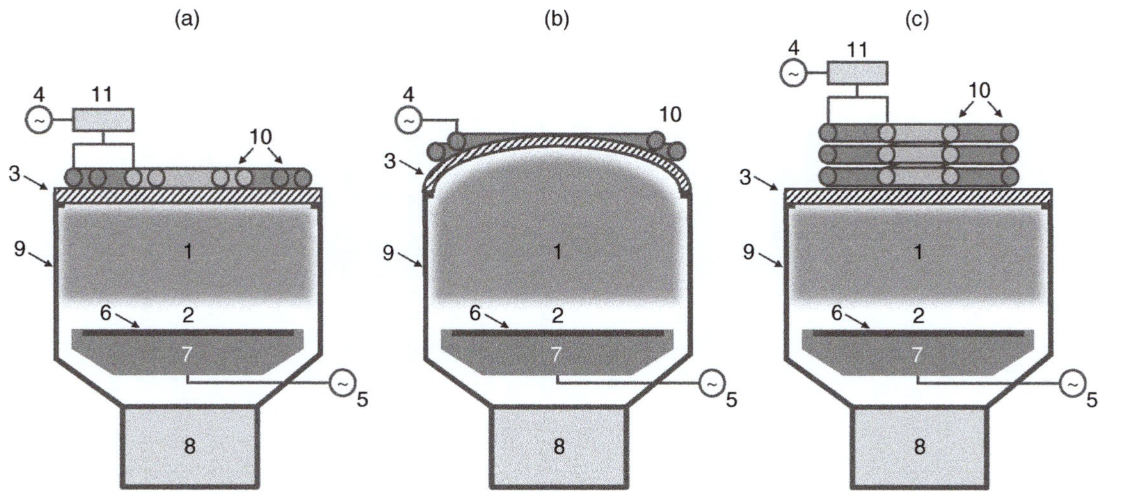

Figure 9.21 Schematic illustration of typical ICP/TCP reactors. (a) Reactor with flat window and planar dual coil. (b) Reactor with domed window and single coil. (c) Reactor with flat window and vertical dual coil. The following main features are highlighted: (1) plasma body; (2) plasma sheath; (3) dielectric window or dome; (4) source RF system; (5) bias RF system; (6) wafer; (7) cathode with electrostatic chuck; (8) turbo pump; (9) reactor walls; (10) ICP/TCP coil; and (11) current divider.

9.4 Ion Energy Distribution Modulation

The ion energy of both ICP/TCP and CCP reactors is driven by the bias power, which uses frequencies of 13.5 MHz and lower. Lower frequencies provide higher peak ion energies but also wider IEDs as shown in Figures 9.5–9.10. Ideally, process engineers would like to work with narrow IEDs because for any given etching mechanism, there should be one best solution. For high aspect ratio etching, for instance, we know that ions that reach the bottom of the feature directly without collisions dominate the etching process (Huang et al. 2019). Their etching yield increases with ion energy. Ions that scatter off the sidewall lose a lot of the energy. Their remaining energy also increases with the incoming energy. Because of these considerations, an IED with a narrow maximum at the highest energies would deliver superior performance. To passivate the sidewall in the mid-section of the trench around 10 : 1, it might be beneficial to have ions with somewhat lower energies as well. In other words, there should be one preferred IED to achieve the best possible HAR dielectric profile with one or more peaks to drive certain surface reactions (etching on the bottom, passivating the mask, passivating the sidewall). The challenge is that the IEDs for typical RIE reactors represent a continuous distribution, the width of which is determined by the lowest RF frequency that is capacitively coupled into the reactor.

In this chapter, we will discuss solutions to narrow IEDs caused by capacitive coupling of RF power. Conceptually, such narrow IEDs can be stitched together to create the desired multi-peak IEDs in the time domain using plasma pulsing. We will investigate plasma pulsing in Section 9.5.

The fundamental reason why IEDs are broad is that ion energy is controlled by varying the amplitude of an RF sinusoidal bias voltage applied to the wafer electrode. The sheath follows the applied voltage and ions are subjected to varying electric fields. Depending on the point in time when they enter the sheath, the ion will see different acceleration voltages. Only for the highest RF frequencies of 40 MHz and higher will the sheath transition time τ_i be large enough that the ion will be subjected to an average sheath voltage. This is the reason why high-frequency plasmas have narrow IEDs as discussed in Section 9.2.

However, high-frequency RF power also increases the plasma density. A hypothetical reactor with only high-frequency RF would firstly not allow to decouple ion energies and fluxes. Secondly, it would not deliver high enough ion energies. In addition, the IEDs are mass dependent (see Figure 7.60 and Eq. (9.11)). While using higher frequencies may narrow the IEDs for high-mass ions, low-mass ions may still have a wide IED.

This creates the challenge to generate mass-independent, narrow IEDs for RF frequencies of 13.56 MHz and below. Wang and Wendt proposed to replace the sinusoidal waveform with a periodic bias voltage waveform (Wang and Wendt 2000). This approach is known as "tailored waveform bias" (TWB) or "shaped waveform bias" (Bruneau et al. 2016). Wang and Wendt modeled a short voltage spike in combination with a longer slow voltage ramp. This waveform produces a constant sheath potential during the ramp period, which constitutes most of each cycle, resulting in

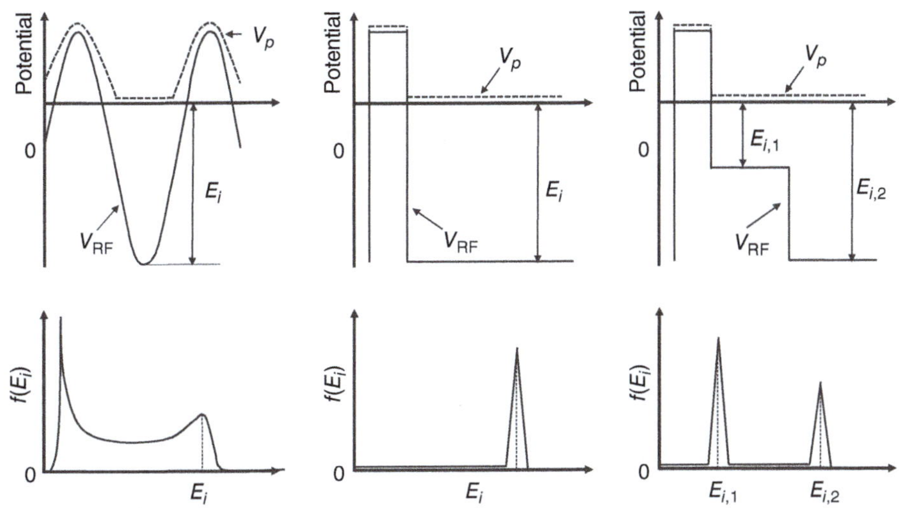

Figure 9.22 Schematic illustration of the effect of tailored bias waveforms on the ion energy distribution. Plasma potential and RF potential on the cathode for a sinusoidal RF waveform (a), for a square-shaped RF waveform (b), and a square-shaped waveform with two levels (c).

a narrow IED. Short high-voltage pulses are needed to prevent charge accumulation on the substrate surface (Wang and Wendt 2000). To produce a constant sheath potential, the voltage at the cathode must be ramped to compensate for charging.

The signal to the cathode is produced by a waveform generator that combines sinusoidal signals with different frequencies to a desired signal shape in an inverse Fourier transformation. This signal must be amplified to reach the desired voltages. The cost of this amplification is a challenge for the wide adoption of this technology in the semiconductor industry especially for high aspect ratio (HAR) dielectric etching, which requires very high ion energies.

Figure 9.22 depicts the mechanism of generating near monoenergetic IEDs by means of waveform shaping. The first panel represents the plasma potential and RF potential on the cathode for a sinusoidal RF waveform (see also Figure 9.4). The resulting IED is bimodal and broad because the ions see varying potentials depending on the time of their entry into the sheath region.

The middle panel shows a square-shaped signal at the cathode. Here, the ions see a constant voltage for most of the time. The signal is interrupted by a short positive pulse to attract electrons and to compensate the positive charge from the ion flux. During the positive pulse, the ions do not react with the wafer and therefore do not contribute to the IED. This results in a near monoenergetic IED. The RF signal shown in Figure 9.22 represents the voltage on the electrode or wafer surface. The actual signal from the waveform generator will be different to account for charging effects and for the other RF components such as transmission lines and the RF match.

The right panel of Figure 9.22 illustrates that bimodal IEDs with two distinct peaks can be produced using the TWB approach. Here, the RF signal has two levels of

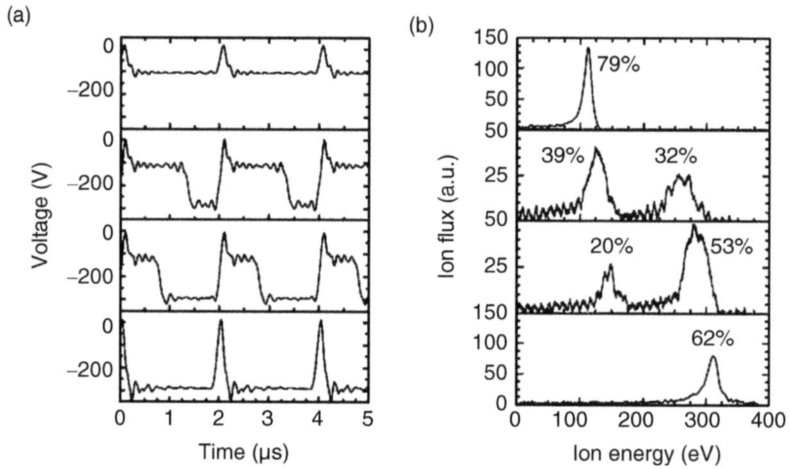

Figure 9.23 Two-peak tailored bias waveforms (a) and corresponding IEDs (b). The displayed percentage by each peak indicates the area of each peak as a fraction of the total area. Source: Qin et al. (2010). © 2010 IOP Publishing.

negative potential. The relative duration of the two levels determines the relative flux of the high- and low-energy ions.

Figure 9.23 shows experimental results that were obtained by Qin et al. (2010). The ion energy measurements were conducted in an argon high-density plasma at a pressure of 10 mTorr. The substrate bias power supply consisted of an arbitrary waveform generator and a broadband RF power amplifier. The tailored waveforms in Figure 9.23a have voltage plateaus at 100 and 300 V with variable durations. The corresponding IEDs in Figure 9.23b show two peaks near the ion energies expected of 100 and 300 eV, with a progression of ion flux shifting from the 100 eV peak to the 300 eV peak reflecting the duration of the 100 and 300 V voltage plateaus.

Agarwal and Kushner investigated the effect of tailored, non-sinusoidal bias waveforms on profile evolution using computational methods (Agarwal and Kushner 2005). When the positive voltage spike was sufficiently short, the average sheath potential was close to the quasi-DC sheath potential during the negative voltage portion of the cycle resulting in a near monoenergetic IED (Agarwal and Kushner 2005). Their calculations demonstrated that the width of the ion energy distribution, ΔE_i, is nearly independent of the ion mass, which is not the case for sinusoidal RF (see Eq. (9.11)). Mass independence of the energy distribution is a very important advantage of TWB.

The authors found that the silicon oxide to silicon selectivity in fluorocarbon plasmas can be controlled by adjusting the width and energy of the IED via tuning of the short positive voltage pulse. Tailoring of the bias waveform allows tuning the average ion energy and the width of the IED to fit the process window. The results can be interpreted based on the etching mechanism for silicon oxide as discussed in Section 6.1.3 in the context of directional ALE of SiO_2 with C_4F_8/Ar and for silicon oxide RIE with fluorocarbon gases. The process involves the deposition of a reactive C_xF_y layer on the surface. Ion bombardment leads to the intermixing of this layer with SiO_2, which yields volatile products such as SiF_4 and CO_2. Selectivity to silicon exists due to the absence of carbon consumption via CO_2. To etch silicon, the ion energies must be high enough to cause bulk sputtering. This means that for a given thickness of the C_xF_y layer, an ion energy can be dialed in precisely with TWB where silicon oxide etches while silicon does not etch.

9.5 Plasma Pulsing

We discussed various source and bias pulsing combinations in the context of how radical and ion species fluxes can be manipulated in RIE in Section 7.1.1 (see Figure 7.1). Bias pulsing is by far the most widely uses implementation of plasma pulsing in the semiconductor industry because it allows to alternate between deposition/activation (bias off) and etching/sputtering (bias on). The mechanism of bias pulsing of an etching reactor with decoupled source and bias powers can be understood as a sequence of very short processing steps with and without bias power. Thus, the effect of bias pulsing can be understood based on fundamentals of capacitively coupled plasmas, self-bias, and sheath voltage (see Sections 9.1 and 9.2).

For pulsing frequencies in the kilohertz range, the duration of these steps is in the order of several milliseconds. Because gases cannot be exchanged that fast, the changes in plasma chemistry can come only from variations of the electron temperature and dissociation pattern when the source power is pulsed. The fundamentals of source pulsing are based on RF power coupled capacitively or inductively into the plasma, on electron heating, and its effect on dissociation and ionization (see Sections 9.2 and 9.3).

Plasma pulsing is not limited to the on and off states of source and bias power. Multi-level pulsing is rapidly advancing. The various levels on multilevel pulsing constitute different combinations of source and bias power. Process engineers design these levels knowing the etching mechanism, what species are needed to realize the mechanism, and what source and bias RF power settings are needed to generate these species. In that sense multilevel pulsing is not that different from conventional process development except that the steps are much shorter and that the feed gases are not changing.

Bias pulsing can be combined with tailored waveform bias, which changes the shape of a pulse. This is schematically shown in Figure 9.24. The combination of TWB and two-level bias pulsing generates alternating near-monoenergetic ions on a timescale of milliseconds. TWB with two voltage levels gives monoenergetic ions on a nanosecond scale. For comparison, in a typical ICP/TCP reactor, saturation of the surface by radicals and ions occurs on a timescale of milliseconds and seconds, respectively (Kanarik et al. 2018). Therefore, bias pulsing is a method to implement cyclic processes akin to multistep ALE and MMP. Two-level TWB, however, results in quasi-simultaneous ion fluxes with two energies.

Plasma pulsing goes beyond simply turning on and off the source and bias power when considering the transitions between steps (Midha and Economou 2000). New conditions such as ion–ion plasmas (IIPs) can be produced in these

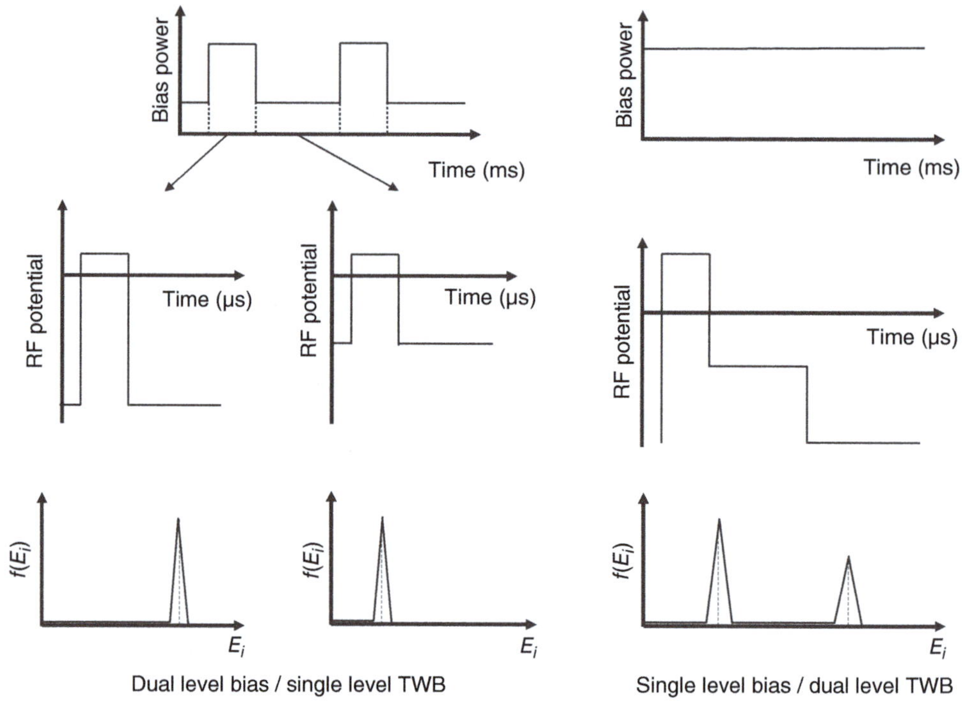

Figure 9.24 Schematic illustration of IED for combined bias pulsing and TWB. (a) Combination of single level TWB and two level bias pulsing. (b) Dual level TWB and single level bias bias pulsing.

transition periods (Midha and Economou 2001; Economou 2007). The pulsing frequencies must be high for these transitions to contribute to the overall process in a meaningful way. Next, we will give an overview of the important additional effects that can be generated with plasma pulsing. Review articles by Economou and Banna are recommended for further reading (Banna et al. 2012; Economou 2014).

Most plasmas that are used in etching of semiconductor devices energize gas mixtures, which contain electronegative gases, for instance, halogens and oxygen. This can be understood from the discussion of the mechanism of chemically assisted etching (see Chapter 2 and Figure 2.1). Electronegative radicals and neutrals are needed to adsorb at the surface and to form strong bonds. This leads to the weakening of the bonds between the surface atoms and the bulk material. These bonds can then be broken with less energy. In the plasma, electronegative gas plasmas form negative ions via electron attachment. These negative ions cannot enter the sheath since the negative ion energy is far less than the sheath potential. Therefore, they are trapped in the plasma. Source pulsing provides a means to having these negative ions reach the wafer.

Figure 9.25 shows the evolution of species wall fluxes predicted by a model of pulsed chlorine plasma between two parallel plates (Midha and Economou 2000). Each cycle is separated into four time segments: (1) early active glow, (2) late active glow, (3) early afterglow, and (4) late afterglow. The simulations show a drop in

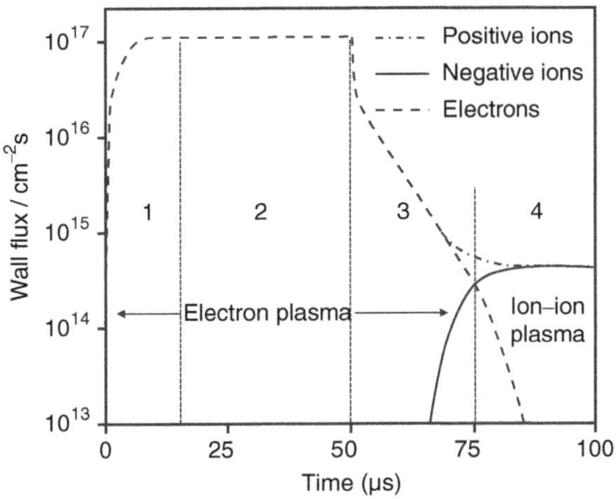

Figure 9.25 Evolution of species wall fluxes predicted by a model of pulsed chlorine plasma between two parallel plates. Source: Midha and Economou (2000). © 2000 IOP Publishing.

electrons reaching the reactor walls (and wafer) in the early afterglow. At the same time, negative ions emerge. At the beginning of the late afterglow, the negative ion flux equals the electron flux. An IIP forms, which decays by ion–ion recombination and diffusion to the walls (Midha and Economou 2001). The electron flux is negligible, making the positive and negative ion fluxes equal. The peak magnitude of the negative ion flux in the calculations for a chlorine plasma is about 2 orders of magnitude less than the electron flux in the active glow (Midha and Economou 2000).

Negative ions that are generated by source pulsing can provide useful effects for etching. Source pulsing is, for instance, used to reduce ion energies below the self-bias of continuous plasmas. We discussed the origin of self-bias in Section 9.1 for a plasma in contact with a grounded wall. Self-bias is the result of the difference in mobility between electrons and ions due to their large difference in mass, which leads to the formation of a sheath. Because there are only positive and negative ions in an IIP, the ion energies can be extremely low, as low as a few electron volts and lower compared to 10 eV and higher for a non-biased electron–ion plasma. This means source pulsing produces a bimodal ion energy distribution with the high energy being the self-bias of the active glow and the low energy corresponding to an IIP in the afterglow. The ratio between the low- and high-energy peaks is determined by the duty cycle as shown in Figure 9.26. Thus, source pulsing offers a way to achieve very low ion energies.

This can be used, for instance, for directional ALE such as silicon ALE with Cl_2 plasma and argon ion removal. To prevent etching during the chlorine plasma modification step, the source power can be pulsed. Pulsing during the removal step would not be needed in this example because the lower limit of the ideal ALE window is about 50 eV and can be easily produced with an ICP/TCP source without pulsing. For materials with very low ion energy thresholds of the ideal ALE window such as germanium (see Figure 6.12), pulsing the source power during the argon ion removal step may improve synergy. Source pulsing will create a bimodal ion energy distribution though. Low duty cycles will create near monoenergetic, low-energy

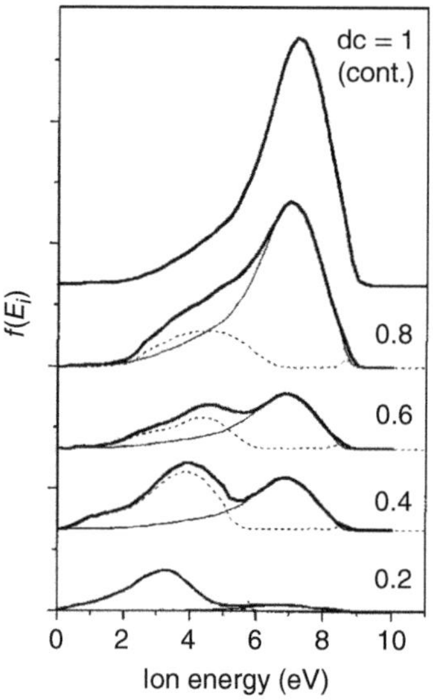

Figure 9.26 Ion energy distribution functions in an argon microwave plasma with a peak power of 960 W pulsed at 1 kHz for varying duty cycles. Source: Zabeida and Martinu (1999). © 1999 AIP Publishing.

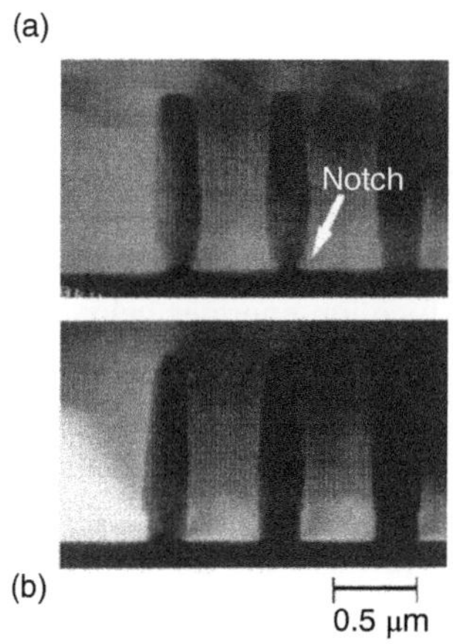

Figure 9.27 SEM cross-sectional polysilicon gate profiles etched in continuous and source pulsing pulsed modes. (a) CW mode and (b) pulsed-power mode. Source: Ahn et al. (1996).

IEDs but also reduce the ion flux, which may impact the throughput of the process negatively.

Another useful effect of the afterglow is wafer surface charge neutralization. During the late afterglow period, the collapsed sheath during this phase of the pulse allows negative ions to reach the bottom of the feature (Banna et al. 2012). This effect is illustrated in Figure 9.27 (Ahn et al. 1996). Finally, source pulsing has been shown to improve ERNU across the wafer (Subramonium and Kushner 2004; Banna et al. 2009; Tokashiki et al. 2009). The uniformity improvement is attributed to ion and neutral relaxation during the afterglow.

IIPs can form downstream of continuous wave discharges in strongly electronegative gases. This is called a spatial afterglow (Economou 2007). Grids between an inductive source and the wafer can be used to generate IIPs (Singh et al. 2019). An electron–ion plasma is generated in the upper sub-chamber. Electrons that diffuse to the lower sub-chamber are cooled as they pass through the grid. Under certain conditions, an IIP can form in the lower chamber via attachment of low-temperature electrons. Because the electron temperature is lower, dissociation and redeposition of reaction products is reduced in IIPs. This can reduce CD microloading via gas-phase deposition (see Section 7.1.4).

9.6 Grid Sources

Ion beam etching (IBE) and radical etching are examples of technologies adjacent to RIE, which cover specialized, all chemical (see Chapter 5) or all physical etching (see Chapter 8). IBE and radical etch reactors can be viewed as derivatives of RIE reactors where some of the species are being suppressed. Figure 9.28 shows schematic depictions of RIE, IBE, and radical etching chambers (Lill et al. 2019). For better contrast of the key differences, only ICP/TCP source technology is shown.

What becomes immediately clear from Figure 9.28 is that in IBE and radical etch tools, grids are inserted between the source and the wafer. Plasma grid technology is essential to enable these technologies. Grids allow separation of ions and radicals. In the case of ion sources, they also introduce an independent ion energy control knob.

Let us first review IBE reactors in more detail. The definition of "beam" as a verb is to "transmit in a specified direction." Plasma directs ions in only one direction, i.e. normal to the sheath and wafer surface. To be able to control the ion impact angle, the sheath must be separated from the wafer via a grid. With a grid, the wafer can be tilted to change the ion impact angle. Because this creates asymmetric profiles, the wafer must also be rotated. The plasma is confined above the grid. Below the grid is high vacuum good enough to prevent gas collisions with residual gas molecules, which would broaden the IAD. This plasma is typically at least 1 order of magnitude lower than the pressures used in RIE. Pressures of 0.1 mTorr are typical. The ions are accelerated via a DC voltage applied to the grid. The wafer is at ground with respect to the acceleration grid.

Advanced IBE sources use a three-grid design to allow for acceleration of the beamlets, which are formed in each of the several hundred openings in the grid. A schematic of such a source is depicted in Figure 9.29. The first grid sets the acceleration voltage and is called the plasma electrode. The second grid and the grounded third grid form an ion optic to shape the beamlets. The second grid is called the suppressor electrode because it suppresses the backflow of electrons into the ion source. The third grid is called the ground electrode. It is held at the same potential as the wafer and the chamber surrounding the wafer.

To avoid spreading of the IAD due to space charging and to avoid charge up of the wafer, the ions are neutralized by electron injection on their path from the source to the wafer. Because the ions are accelerated by the grids, there is no RF power applied to the wafer. The cathode is simpler in the electrical design but more complex mechanically because of the need for tilting and rotation in the presence of a temperature-controlled electrostatic chuck.

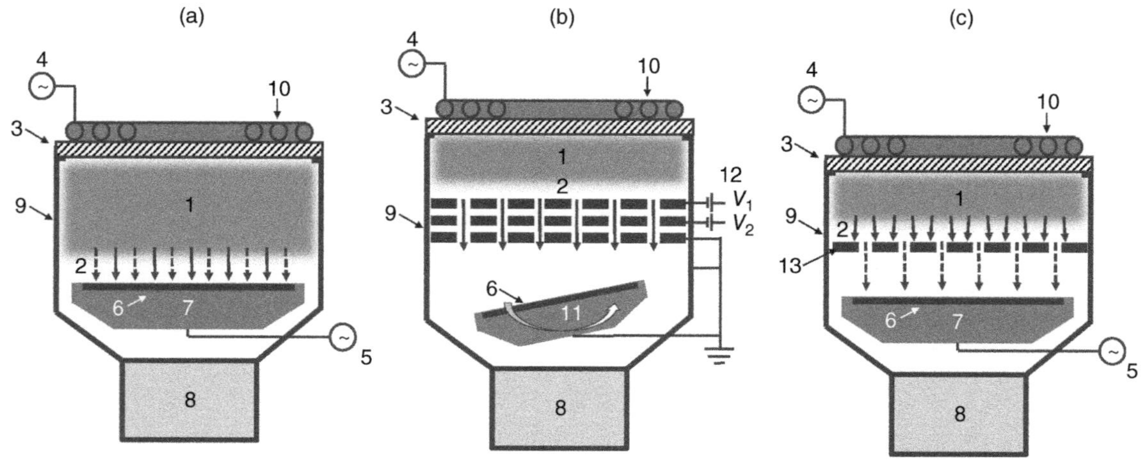

Figure 9.28 Comparison of an RIE or directional ALE reactor (a), an ion beam reactor (b), and a radical etching reactor (c). The plasma source is an ICP/TCP source in all cases. Ions are represented by solid arrows, radicals by dashed arrows. The following main features are highlighted: (1) plasma body; (2) plasma sheath; (3) dielectric window; (4) source RF system; (5) bias RF system; (6) wafer; (7) cathode with electrostatic chuck; (8) turbo pump; (9) reactor walls; (10) ICP/TCP coil; (11) rotating and tilting, grounded sample holder with electrostatic chuck; (12) ion extraction system with multiple biased grids; and (13) grounded ion blocking grid. Source: Lill et al. (2019).

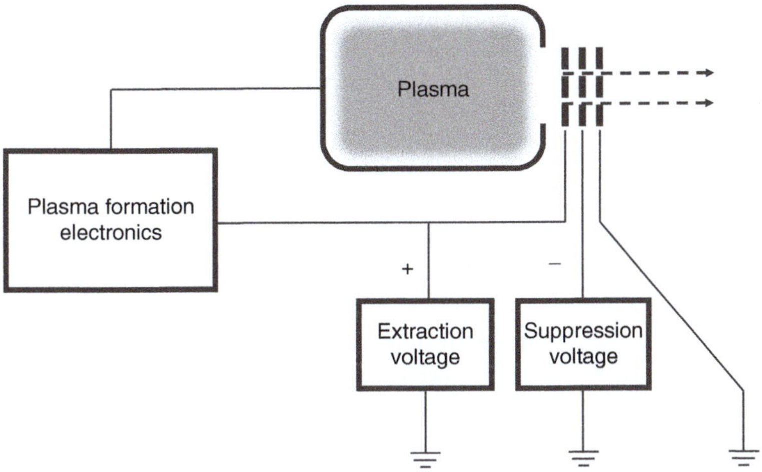

Figure 9.29 Schematic illustration of an ion beam source with three grids.

An in-depth review of ion beam sources can be found in "The physics and technology of ion sources" edited by Brown (2004).

The grid in radical etching reactors is there to block the ions from impinging onto the wafer. In the simplest case, both grid and wafer are grounded. Bias power can be applied to the cathode to ignite a plasma in the lower chamber. This technique is used to break through surface oxides that may prevent radical etching. In addition to radicals, neutrals can be injected into the lower chamber bypassing dissociation in the source to enrich the chemistry with neutral reactions. Depending on the lifetime of the radicals, the plasma source can be removed from the etching reactor completely. The radicals are then delivered into the etching chamber via tubes. In this case, grids are not needed because the ions are lost in wall collisions inside the radical delivery system.

Other use cases of plasma grids have been reported. Grids can be modified to neutralize the extracted ions, which allows the formation of fast neutral beams. The ions are neutralized by passing through either apertures with high aspect ratios (Samukawa 2006) or slightly tilted apertures (Park et al. 2005). The plasma can be generated by ICP/TCP or CCP sources. The ions neutralize during grazing incidence ion scattering (see Section 2.8.2). The benefit of fast neutrals in comparison to ions is the absence of surface charging, which can lead to aspect ratio dependent etching (ARDE), profile distortions, and device damage (Ohori et al. 2019). An additional benefit of blocking the plasma is the suppression of high-energy photons that can damage the surface. Neutral beam sources have not yet found application in the semiconductor industry due to their complexity, cost, and potential contamination of the wafer with materials from the grid.

Grids have been combined with plasma pulsing of a capacitively coupled plasma to extract nearly monoenergetic ion beams. During the pulsing, the electron temperature decays in the afterglow, resulting in nearly uniform plasma potential and minimal energy spread of ions. The ion energy was controlled by a DC bias on an electrode in contact with the plasma (Xu et al. 2005; Nam et al. 2007).

Problems

P9.1 Using Table 9.1, discuss the relationship between the temperatures of the neutral gas, the electrons, and the wafer surface at the ion impact site.

P9.2 Using Eq. (9.7), explain the significance of the electron temperature for the ion energy in non-biased plasmas (for instance, source-only TCP/ICP plasmas). How does this change when the wafer is biased by RF power?

P9.3 Why are low frequencies used to accelerate ions and high frequencies to create plasma densities in CCP reactors?

P9.4 Why does the anode to cathode ratio in CCP reactors impact the ion energy?

P9.5 What are the effects that contribute to the wide, continuous IEDs in plasma etch reactors?

P9.6 Why are TCP/ICP plasmas efficient in generating high-density plasmas?

P9.7 Why are the IEDs for TWB mass independent?

P9.8 What are the differences in the surface processes for combining single-level TWB with dual-level bias pulsing vs a combination of dual-level TWB and single-level bias pulsing?

References

Agarwal, A. and Kushner, M.J. (2005). Effect of non-sinusoidal bias waveforms on ion energy distributions and fluorocarbon plasma etch selectivity. *J. Vac. Sci. Technol., A* 23: 1440–1449.

Ahn, T.H., Nakamura, K., and Sugai, H. (1996). Negative ion measurements and etching in a pulsed-power inductively coupled plasma in chlorine. *Plasma Sources Sci. Technol.* 5: 139–144.

Ahn, S.K., You, S.J., and Chang, H.Y. (2006). Driving frequency effect on the electron energy distribution function in capacitive discharge under constant discharge power condition. *Appl. Phys. Lett.* 89: 161506 1–3.

Banna, S., Agarwal, A., Tokashiki, K. et al. (2009). Inductively coupled pulsed plasmas in the presence of synchronous pulsed substrate bias for robust, reliable, and fine conductor etching. *IEEE Trans. Plasma Sci.* 37: 1730–1746.

Banna, S., Agarwal, A., Cunge, G. et al. (2012). Pulsed high-density plasmas for advanced dry etching processes. *J. Vac. Sci. Technol., A* 30: 040801 129.

Banna, S., Chen, Z., and Todorow, V. (2015). Inductively coupled plasma source with phase control. US Patent 8,933,628.

Banna, S., Bishara, W., Giar, R. et al. (2019). High efficiency triple-coil inductively coupled plasma source with phasecontrol. US Patent 10271416

Brown, I.G. (ed.) (2004). *The Physics and Technology of Ion Sources*, 2e. Wiley-VCH.

Bruneau, B., Lafleur, T., Booth, J.P., and Johnson, E. (2016). Controlling the shape of the ion energy distribution at constant ion flux and constant mean ion energy with tailored voltage waveforms. *Plasma Sources Sci. Technol.* 25: 025006 1–8.

Carter, J.B., Holland, J.P., Peltzer, E. et al. (1993). Transformer coupled plasma etch technology for the fabrication of subhalf micron structures. *J. Vac. Sci. Technol., A* 11: 1301–1306.

Chabert, P. and Braithwaite, N. (2011). *Physics of Radio-Frequency Plasmas*, 1e. Cambridge University Press.

Coburn, J.W. and Kay, E. (1972). Positive-ion bombardment of substrates in rf diode glow discharge sputtering. *J. Appl. Phys.* 43: 4965–4971.

Collins, K., Rice, M., Trow, J. et al. (2000). Parallel plate electrode plasma reactor having an inductive antenna and adjustable radial distribution of plasma ion density. US Patent 6,054,013

Economou, D.J. (2007). Fundamentals and applications of ion–ion plasmas. *Appl. Surf. Sci.* 253: 6672–6680.

Economou, D.J. (2014). Pulsed plasma etching for semiconductor manufacturing. *J. Phys. D: Appl. Phys.* 47: 303001 1–27.

Flamm, D.L. (1986). Frequency effects in plasma etching. *J. Vac. Sci. Technol., A* 4: 729–738.

Goto, H.H., Loewe, H.D., and Ohmi, T. (1992). Dual excitation reactive ion etcher for low energy plasma processing. *J. Vac. Sci. Technol., A* 10: 3048–3054.

Graves, D.B. and Brault, P. (2009). Molecular dynamics for low temperature plasma–surface interaction studies. *J. Phys. D: Appl. Phys.* 42: 194011 1–27.

Graves, D.B. and Humbird, D. (2002). Surface chemistry associated with plasma etching processes. *Appl. Surf. Sci.* 192: 72–87.

Hebner, G.A., Barnat, E.V., Miller, P.A. et al. (2006). Frequency dependent plasma characteristics in a capacitively coupled 300 mm wafer plasma processing chamber. *Plasma Sources Sci. Technol.* 15: 879–888.

Horwitz, C.M. (1983). RF sputtering–voltage division between two electrodes. *J. Vac. Sci. Technol., A* 1: 60–68.

Huang, S., Huard, C., Shim, S. et al. (2019). Plasma etching of high aspect ratio features in SiO_2 using $Ar/C_4F_8/O_2$ mixtures: a computational investigation. *J. Vac. Sci. Technol., A* 37: 031304 1–26.

Kanarik, K.J., Tan, S., and Gottscho, R.A. (2018). Atomic layer etching: rethinking the art of etching. *J. Phys. Chem. Lett.* 9: 4814–4821.

Kawamura, E., Vahedi, V., Lieberman, M.A., and Birdsall, C.K. (1999). Ion energy distributions in rf sheaths; review, analysis and simulation. *Plasma Sources Sci. Technol.* 8: R45–R64.

Kiehlbauch, M.W. and Graves, D.B. (2003). Effect of neutral transport on the etch product lifecycle during plasma etching of silicon in chlorine gas. *J. Vac. Sci. Technol., A* 21: 116–126.

Kitajima, T., Takeo, Y., Petrovic, Z.L., and Makabe, T. (2000). Functional separation of biasing and sustaining voltages in two-frequency capacitively coupled plasma. *Appl. Phys. Lett.* 77: 489–491.

Koehler, K., Coburn, J.W., Horne, D.E. et al. (1985a). Plasma potentials of 13.65-MHz rf argon glow discharges in a planar system. *J. Appl. Phys.* 57: 59–66.

Koehler, K., Horne, D.E., and Coburn, J.W. (1985b). Frequency dependence of ion bombardment of grounded surfaces in rf argon glow discharges in a planar system. *J. Appl. Phys.* 58: 3350–3355.

Kushner, M.J., Collinson, W.Z., Grapperhaus, M.J. et al. (1996). A three-dimensional model for inductively coupled plasma etching reactors: azimuthal symmetry, coil properties, and comparison to experiments. *J. Appl. Phys.* 80: 1337–1344.

Kuypers, A.D. and Hopman, H.J. (1990). Measurement of ion energy distributions at the powered rf electrode in a variable magnetic field. *J. Appl. Phys.* 67: 1229–1240.

Lieberman, M.A. and Lichtenberg, A.J. (2005). *Principles of Plasma Discharges and Materials Processing*, 2e. Wiley.

Lill, T., Vahedi, V., and Gottscho, R. (2019). Etching of semiconductor devices. In: *Materials Science and Technology*. Wiley-VCH. 1–25.

Long, M., Marsh, R., and Paterson, A. (2015). TCCT match circuit for plasma etch chambers. US Patent 9,059,678.

Midha, V. and Economou, D.J. (2000). Spatio-temporal evolution of a pulsed chlorine discharge. *Plasma Sources Sci. Technol.* 9: 256–269.

Midha, V. and Economou, D.J. (2001). Dynamics of ion–ion plasmas under radio frequency bias. *J. Appl. Phys.* 90: 1102–1114.

Miller, P.A. and Riley, M.E. (1997). Dynamics of collisionless rf plasma sheaths. *J. Appl. Phys.* 82: 3689–3709.

Nam, S.K., Economou, D.J., and Donnelly, V.M. (2007). Particle-in-cell simulation of ion beam extraction from a pulsed plasma through a grid. *Plasma Sources Sci. Technol.* 16: 90–96.

Oehrlein, G.S. and Hamaguchi, S. (2018). Foundations of low-temperature plasma enhanced materials synthesis and etching. *Plasma Sources Sci. Technol.* 27: 023001 1–21.

Ohori, D., Fujii, T., Noda, S. et al. (2019). Atomic layer germanium etching for 3D Fin-FET using chlorine neutral beam. *J. Vac. Sci. Technol., A* 37: 021003 1–5.

Okamoto, Y. and Tamagawa, H. (1970). Energy dispersion of positive ions effused from an RF plasma. *J. Phys. Soc. Jpn.* 29: 187–191.

Panagopoulos, T. and Economou, D.J. (1999). Plasma sheath model and ion energy distribution for all radio frequencies. *J. Appl. Phys.* 85: 3435–3443.

Park, S.D., Lee, D.H., and Yeom, G.Y. (2005). Atomic layer etching of Si(100) and Si(111) using Cl_2 and Ar neutral beam. *Electrochem. Solid-State Lett.* 8: C106–C109.

Perry, A.J. and Boswell, R.W. (1989). Fast anisotropic etching of sincon in an inductively coupled plasma reactor. *Appl. Phys. Lett.* 55: 148–150.

Qin, X.V., Ting, Y.-H., and Wendt, A.E. (2010). Tailored ion energy distributions at an rf-biased plasma electrode. *Plasma Sources Sci. Technol.* 19: 065014 1–8.

Samukawa, S. (2006). Ultimate top-down etching processes for future nanoscale devices: advanced neutral-beam etching. *Jpn. J. Appl. Phys.* 45: 2395–2407.

Shannon, S., Hoffman, D., Yang, J.G. et al. (2005). The impact of frequency mixing on sheath properties: ion energy distribution and V_{dc}/V_{rf} interaction. *J. Appl. Phys.* 97: 103304 1–4.

Singh, H., Lill, T., Vahedi, V. et al. (2019). Internal plasma grid for semiconductor fabrication. US Patent 1,022,4221.

Subramonium, P. and Kushner, M. (2004). Pulsed plasmas as a method to improve uniformity during materials processing. *J. Appl. Phys.* 96: 82–93.

Surendra, M. and Graves, D.B. (1991). Capacitively coupled glow discharges at frequencies above 13.56 MHz. *Appl. Phys. Lett.* 59: 2091–2093.

Tokashiki, K., Cho, H., Banna, S. et al. (2009). Synchronous pulse plasma operation upon source and bias radio frequencys for inductively coupled plasma for highly reliable gate etching technology. *Jpn. J. Appl. Phys.* 48: 08HD01 1–11.

Walton, S.G., Boris, D.R., Hernandez, S.C. et al. (2015). Electron beam generated plasmas for ultra low T_e processing. *ECS J. Solid State Sci. Technol.* 4: N5033–N5040.

Wang, S.B. and Wendt, A.E. (2000). Control of ion energy distribution at substrates during plasma processing. *J. Appl. Phys.* 88: 643–646.

Xu, L., Economou, D.J., and Donnelly, V.M. (2005). Extraction of a nearly monoenergetic ion beam using a pulsed plasma. *Appl. Phys. Lett.* 87: 041502 1–3.

Zabeida, O. and Martinu, L. (1999). Ion energy distributions in pulsed large area microwave plasma. *J. Appl. Phys.* 85: 6366–6372.

Zhu, X.M., Chen, W.C., Zhang, S. et al. (2007). Electron density and ion energy dependence on driving frequency in capacitively coupled argon plasmas. *J. Phys. D: Appl. Phys.* 40: 7019–7023.

10

Emerging Etching Technologies

We will finally explore etching technologies that are not yet moved from research into semiconductor manufacturing. Using the framework of an etching process where the surface bonds are weakened by reactive species and broken by additional energy, it is understandable to also consider electrons and photons as the source of this energy.

The field of stimulating or enhancing chemical etching via electrons appears to be more developed than by using photons. A possible explanation is that bond energies are of the order of a few electron volts to tens of electron volts and hence the excitation must be achieved with radiation in the ultraviolet range of the spectrum (Chalker 2016). This requires ultraviolet light sources with sufficient intensities, which are more expensive than electron sources.

10.1 Electron-Assisted Chemical Etching

The classical paper by Coburn and Winters "Ion- and electron-assisted gas-surface chemistry: an important effect in plasma etching," which introduced the etching community to ion–neutral synergy, also discusses synergistic etching with electron and neutrals (Coburn and Winters 1979). Figure 10.1 shows experimental data for etching Si_3N_4 and SiO_2 with XeF_2 at a background pressure of 0.6 mTorr and 1500 eV electrons. In the case of Si_3N_4, only the electron beam is incident and etching is observed when XeF_2 is introduced into the chamber. For SiO_2, the gas is supplied first, and etching is induced with the electron beam. The authors propose a mechanism where the electron beam produces elemental silicon on the surface, which is etched by fluorine from XeF_2.

Electron-assisted chemical etching was also reported for silicon with SF_6 (Martinez et al. 1988) and with hydrogen gas (Gillis et al. 1995). Martinez et al. demonstrated anisotropic etching with a vertical-to-horizontal ratio of 2.5–3 for an electron energy of around 200 eV. The electrons were created in a hollow cathode discharge. The electron beam served two purposes: it created reactive radicals by electron impact dissociation of SF_6 and it provided directed energy to the wafer surface to assist surface reactions and desorption (Martinez et al. 1988). In the experiment by Gillis et al., the silicon sample was placed onto the anode of a dc

Atomic Layer Processing: Semiconductor Dry Etching Technology, First Edition. Thorsten Lill.

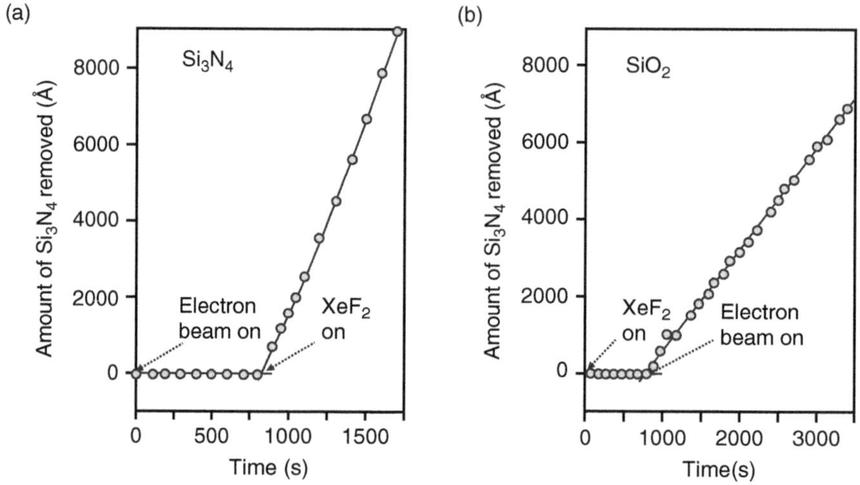

Figure 10.1 Electron-assisted etching of Si_3N_4 (a) and SiO_2 (b) using 1500 eV electrons and XeF_2. Source: Coburn and Winters (1979). © 1979 AIP Publishing.

plasma reactor, and thus received fluxes of low-energy electrons and hydrogen molecules and atoms (Gillis et al. 1995). Gillis et al. also demonstrated directional etching of GaN with chlorine, hydrogen, and a mixture of both gases (Gillis et al. 1999). In the experiments by Martinez and Gillis, radicals were created in the gas phase that chemisorb on the surface but can also etch silicon spontaneously, which is responsible for the isotropic etching component.

Electron-stimulated atomic layer etching (ALE) was demonstrated on GaAs (Meguro et al. 1990). The process used alternating exposure to chlorine gas and electrons. The etching per cycle (EPC) was one third of a monolayer per cycle and independent of electron current density and chlorine gas flow. The results indicate that surface modification with chlorine gas and electron stimulated desorption were self-limited. Figure 10.2 shows etching depth as a function of the number of cycles. The electron energy was 100 eV, the current density 13 mA/cm^2, the chlorine flow 2 sccm, and the base pressure 1 mTorr. The authors propose an etching mechanism where chlorine molecules chemisorb on the GaAs surface and $GaCl_3$ and $AsCl_3$ are desorbed under electron bombardment.

Veprek and Sarott conducted studies to differentiate the effects of electron-stimulated surface reactions vs desorption for silicon and hydrogen (Veprek and Sarott 1982). They showed that electron impact can initiate highly selective chemical reactions between a gas and a solid surface. The data suggests that the electron impact induces dissociation of adsorbed H_2 molecules, resulting in the formation of chemisorbed atomic hydrogen (Veprek and Sarott 1982). Thus, electron beams can be used to increase the process window of thermal etching processes (see Figure 2.18).

Electron beams can be focused, which allows spatially resolved etching, a technique that has been reviewed by Utke et al. (2008). This effect was demonstrated for etching of silicon and germanium with chlorine (Roediger et al. 2010; Shawrav et al.

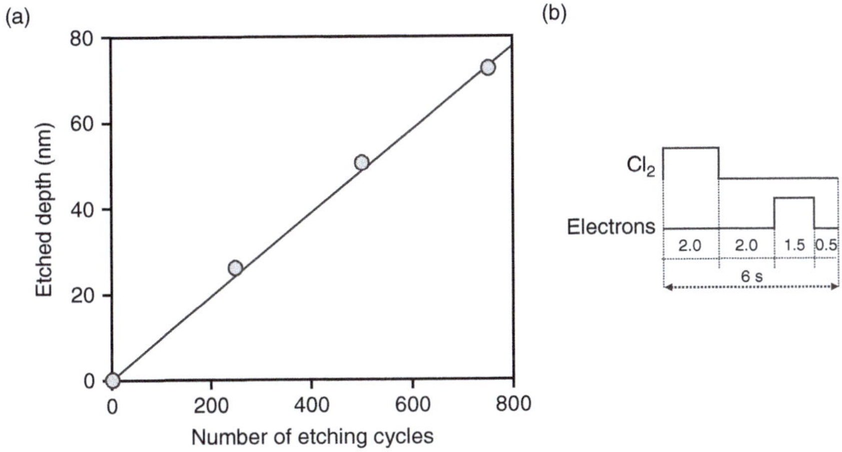

Figure 10.2 Etched depth of GaAs vs number of cycles for ALE with surface modification with chlorine and electron-stimulated removal. Source: Meguro et al. (1990). © 1990 AIP Publishing.

2016) and SiO_2 with XeF_2 (Randolph et al. 2005). This technique could enable direct etching without mask.

10.2 Photon-Assisted Chemical Etching

The effect of photon-assisted etching was demonstrated by Okano for etching of silicon with chlorine gas and a Hg–Xe lamp (Okano et al. 1985). They found that n-doped silicon etches without direct impact of light onto the surface with undercut profiles. Therefore, the etching is caused by chlorine radicals originating from the gas-phase dissociation of chlorine. In the case of undoped silicon, direct light exposure of the surface was required to stimulate etching. To explain their findings, Okano et al. evoke the effect of electrons in the conduction band of silicon to attract chlorine radicals and to stimulate diffusion into the surface via a Cabrera-Mott (CM) diffusion mechanism. We discussed the effect of electrons on n-doped silicon on etching with chlorine and fluorine in Section 7.2. Okano et al. argue that light generates electrons in undoped silicon, which have the same effect as n doping (Okano et al. 1985). Thus, this work illustrates two possible etching mechanisms for photon-induced etching: gas-phase generation of radicals and photon-induced surface reactions.

Iimori et al. confirmed the effect of photon-generated electrons in photon-assisted etching of silicon with chlorine gas in ultrahigh vacuum (UHV) experiments (Iimori et al. 1998). They found that surface adatom polychlorides are desorbed primarily as $SiCl_2$ species, and that adatom monochlorides are stable against irradiation. The effect of photogenerated charge carriers was also found to be an important mechanism in photon-assisted etching of silicon with XeF_2 (Houle 1989).

A third mechanism was proposed by Samukawa et al. based on silicon photon-assisted neutral beam etching experiments on silicon with chlorine and a xenon flash lamp (Samukawa et al. 2007). The wavelengths of the light was selected via a filter to be 220 nm and higher and 380 nm and higher. The study finds a monotonic increase of the etching rate with ion energy and a roughly fourfold increase of the etching rate when the surface was irradiated by 200 nm photons at 38 mW/cm^2. With a ultraviolet (UV) filter, this effect was absent. The authors proposed that UV irradiation generates defects on the surface of the silicon, which enhances the etching process.

A fourth mechanism is photon-induced thermal desorption. The transition between photochemical and photothermal reactions is dependent on the fluence of continuous radiation. For pulsed radiation, pulse frequency and duty cycle must also be considered (Kullmer and Baeuerle 1987; Mogyorosi et al. 1988; Ishii et al. 1993).

Ishii et al. conducted elegant experiments for etching of GaAs with chlorine and a 248 nm KrF excimer laser (Ishii et al. 1993). They alternated chlorine gas exposure and laser radiation. Both steps were saturated and hence the resulting process was an ALE process. The threshold fluence for etching was 13 mJ/cm^2 , above which the EPC abruptly increases. Figure 10.3 shows EPC as a function of laser repetition rate and laser fluence. It saturates at a value of 2 Å per cycle. Higher fluences and repetition rates result in larger EPC, which suggests a thermal desorption mechanism.

Thermal ALE using temperature cycling was also demonstrated using O_2 to oxidize germanium at room temperature followed by desorption with a rapid thermal pulse photon source (Paeng et al. 2019).

Shin et al. studied silicon etching with chlorine-added argon plasma with narrow ion energy distributions near the etching threshold of 16 eV (Shin et al. 2012). They found considerable etching even for ions with energies below the 16 eV threshold, which they could attribute to the effect of photons. This mechanism

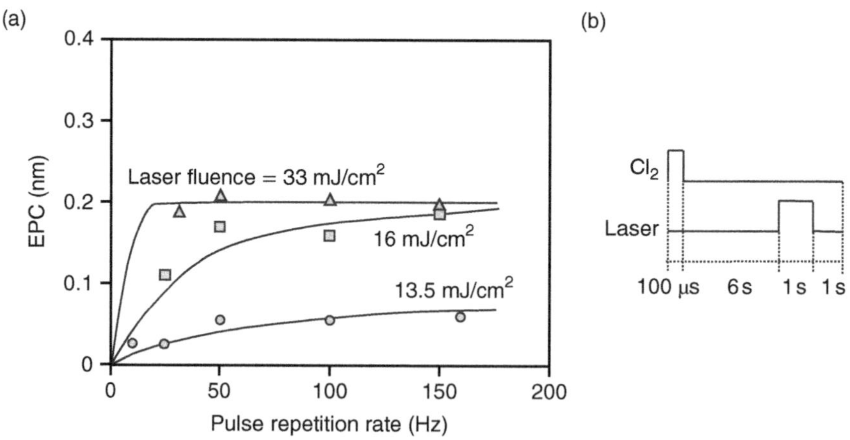

Figure 10.3 EPC of GaAs for ALE with chlorine gas and 248 nm excimer laser radiation as a function of pulse repetition rate and laser fluence. Source: Ishii et al. (1993).

can contribute to the loss of synergy in directional ALE of silicon with Cl_2/Ar (see Section 6.1.2).

In summary, electron- and photon-assisted etching are promising techniques to achieve low damage-free etching. The process can be implemented as continuous or ALE processes. In the case of electron stimulation, near vertical profiles have been demonstrated. Photons should be suited for isotropic etching if the surface is transparent for the chosen wavelength of the light.

Problems

P10.1 What are the mechanisms of how electrons can assist etching?

P10.2 What are the mechanisms of how photons can assist etching?

References

Chalker, P.R. (2016). Photochemical atomic layer deposition and etching. *Surf. Coat. Technol.* 291: 258–263.

Coburn, J.W. and Winters, H.F. (1979). Ion- and electron-assisted gas-surface chemistry – an important effect in plasma etching. *J. Appl. Phys.* 50: 3189–3196.

Gillis, H.P., Choutov, D.A., Steiner, P.A. et al. (1995). Low energy electron-enhanced etching of Si(100) in hydrogen/helium direct-current plasma. *Appl. Phys. Lett.* 66: 2475–2477.

Gillis, H.P., Christopher, M.B., Martin, K.P., and Choutov, D.A. (1999). Patterning III-V semiconductors by low energy electron enhanced etching (LE4). *MRS Internet J. Nitride Semicond. Res.* 4S1: G8.2 1–9.

Houle, F.A. (1989). Photochemical etching of silicon: the influence of photogenerated charge carriers. *Phys. Rev. B* 39: 10120–10132.

Iimori, T., Hattori, K., Shudo, K. et al. (1998). Laser-induced mono-atomic-layer etching on Cl-adsorbed Si(111) surfaces. *Appl. Surf. Sci.* 130–132: 90–95.

Ishii, M., Meguro, T., Gamo, K. et al. (1993). Digital etching using KrF excimer laser: approach to atomic-order-controlled etching by photo induced reaction. *Jpn. J. Appl. Phys.* 32: 6178–6181.

Kullmer, R. and Baeuerle, D. (1987). Laser-induced chemical etching of silicon in chlorine atmosphere. I. Pulsed irradiation. *Appl. Phys. A* 43: 227–232.

Martinez, R.O., Verhey, T.R., Boyer, P.K., and Rocca, J.J. (1988). Broad area electron-beam-assisted etching of silicon in sulfur hexafluoride. *J. Vac. Sci. Technol., B* 6: 1581–1583.

Meguro, T., Hamagaki, M., Modaressi, S. et al. (1990). Digital etching of GaAs: new approach of dry etching to atomic ordered processing. *Appl. Phys. Lett.* 56: 1552–1554.

Mogyorosi, P., Piglmayer, K., Kullmer, R., and Baeuerle, D. (1988). Laser-induced chemical etching of silicon in chlorine atmosphere. II. Continuous irradiation. *Appl. Phys. A* 45: 293–299.

Okano, H., Horiike, Y., and Sekine, M. (1985). *Photo-Excited Etching of Poly-Crystalline and Single-Crystalline Silicon in Cl2 Atmosphere. Jpn. J. Appl. Phys.* 24: 68–74.

Paeng, D., Zhang, H., and Kim, Y.S. (2019). Dynamic temperature control enabled atomic layer etching of titanium nitride. *ALD/ALE 2019*, Bellevue, VA, USA.

Randolph, S.J., Fowlkes, J.D., and Rack, P.D. (2005). Focused electron-beam-induced etching of silicon dioxide. *J. Appl. Phys.* 98: 034902 1–6.

Roediger, P., Hochleitner, G., Bertagnolli, E. et al. (2010). Focused electron beam induced etching of silicon using chlorine. *Nanotechnology* 21: 285306 1–10.

Samukawa, S., Jinnai, B., Oda, F., and Morimoto, Y. (2007). Surface reaction enhancement by UV irradiation during Si etching process with chlorine atom beam. *Jpn. J. Appl. Phys.* 46: L64–L66.

Shawrav, M.M., Gökdeniz, Z.G., Wanzenboeck, H.D. et al. (2016). Chlorine based focused electron beam induced etching: a novel way to pattern germanium. *Mater. Sci. Semicond. Process.* 42: 170–173.

Shin, H., Zhu, W., Donnelly, V.M., and Economou, D.J. (2012). Surprising importance of photo-assisted etching of silicon in chlorine-containing plasmas. *J. Vac. Sci. Technol., A* 30: 021306 1–10.

Utke, I., Hoffmann, P., and Melngailis, J. (2008). Gas-assisted focused electron beam and ion beam processing and fabrication. *J. Vac. Sci. Technol., B* 26: 1197–1276.

Veprek, S. and Sarott, F.A. (1982). Electron-impact-induced anisotropic etching ot silicon by hydrogen. *Plasma Chem. Plasma Process.* 2: 233–246.

Index

Atomic Layer Processing: Semiconductor Dry Etching Technology, First Edition. Thorsten Lill.